# ANALYTIC
# FUNCTION  THEORY

# ANALYTIC FUNCTION THEORY

BY

EINAR HILLE

VOLUME I

CHELSEA PUBLISHING COMPANY
NEW YORK, N.Y.

SECOND EDITION

The present, second edition of Volume I is a
fifth (corrected) printing of the first edition,
originally published in 1959. It is published at
New York, N.Y. in 1982 and it is printed on
recently developed 'long-life' acid-free paper

Library of Congress Cataloging in Publication Data

Hille, Einar, 1894-
     Analytic function theory.

     Includes bibliography.
     1.   Analytic functions.     I.   Title.
QA331.H54   1973      515      73-647
ISBN  0-8284-0269-8   (v. 1)

Printed in the United States of America

# Foreword

A new book, even a textbook, is supposed to have a message. Perhaps mine does have a message or two.

It is my hope that students of this book may come to respect the historical continuity of the subject. Analytic function theory is the result of a long development involving the contributions of many workers. Many theorems have names attached to them, names of men who were famous in their days. The student should know something about the historical development of the subject and about the men who took a major part in this development. He should understand mathematics as a living organism, a growing body of learning, fed by the efforts of many workers. If possible, he should acquire a sense of veneration for those who built up the structure that he is studying. This is the reason for the many historical footnotes.

There is also conceptual continuity: analytic function theory is just a part of the structure of mathematics. As the latter grows and changes, so does function theory. An abstract and postulational approach to mathematics is gradually penetrating into all fields, and it affects the instruction in mathematics at all levels. A modern treatise on function theory has to take this fact into account, both in building up the subject and in fitting it into the larger frame of mathematics. This book represents an effort to integrate the theory of analytic functions with modern analysis as a whole and, in particular, to present it as a branch of functional analysis, to which it gives concrete illustrations, problems, and motivation. Hence the emphasis on structural aspects such as linearity of the sets and of the operations under consideration. The algebraic aspects of the theory have been stressed whenever possible. Certain topological concepts, such as the notions of neighborhood, distance, length, and metric space, have also been stressed. On the other hand, this is not a textbook in topology: if intuition helps, an appeal is made to intuition.

The author believes that complex integration is the proper basis of function theory. The Cauchy integral is a much more pliable and versatile tool than the power series when it comes to doing things in function theory. But before the student can really grasp integrals of analytic functions, he should have at his disposal a large number of such functions, and here the power series is invaluable as a source. The power series also leads to important connections with real analysis, and it is indispensable for the problem of analytic continuation. The emphasis has to lie on diversity rather than on purity of method: the more methods the student can learn, the better he will be equipped. For this reason some of the important theorems, such as the maximum principle and the inverse function theorem, have been treated by several different methods.

v

These general considerations have led to the following arrangement of the subject matter of Volume I: After a preliminary study of number systems, the geometry of the complex plane is developed, and simple functions such as linear fractions, powers, and roots are studied. The main theory begins in Chapter 4 with the definition of holomorphic functions, the Cauchy–Riemann equations, inverse functions, and the elements of conformal mapping. This is followed by a chapter on power series and one on the elementary transcendental functions. The systematic study of holomorphic functions occupies the last three chapters, devoted to complex integration, representation theorems, and the calculus of residues. Supplementary material on point sets, polygons, and Riemann and Riemann–Stieltjes integration is to be found in three Appendixes. There is a brief Bibliography at the end of the volume, and suggestions for collateral reading are appended to the various chapters. An explanation of the symbols used precedes Chapter 1.

A word about the numbering is in order. Section 7.3 is the third section of Chapter 7; the theorems in this section are numbered 7.3.1 through 7.3.4. Lemmas and definitions are numbered in the same way.

A student who intends to use this book should have had a good course in advanced calculus. Familiarity with abstract mathematical reasoning and some skill in manipulating identities, integrals, and series are the main prerequisites. The book is to a large extent autonomous, and the student will find most of the factual information that he needs incorporated in the text. This means that there are considerable parts of the book which a well-prepared student can omit or use for reference only. This applies in particular to Chapters 1 and 6. It should be realized that Chapters 4, 5, 7, 8, and 9 form the core of the book, and that the rest is ancillary material. In this connection it should also be remembered that the present volume is preparatory for a second volume, which, it is hoped, will follow fairly soon.

With suitable omissions, the present volume can be used as a text for a one-term introductory course; in fact, the author has used a preliminary draft for this purpose. It should be possible to cover everything in one year at a leisurely tempo. The second volume will provide additional material for a course on a somewhat higher level.

It is my pleasant duty to bring thanks to the friends who have helped with this undertaking. In the first place I am deeply indebted to Dr. Ernest C. Schlesinger, who read the whole manuscript in detail and suggested numerous corrections and improvements. Thanks are due also to Professors Garrett Birkhoff, C. T. Ionescu-Tulcea, Shizuo Kakutani, and Angus E. Taylor for constructive criticism. Finally, I wish to thank Ginn and Company for the honor they have shown me by letting my book inaugurate their new series "Introductions to Higher Mathematics" as well as for sympathetic consideration of an author's whims and wishes.

*New Haven, Connecticut*                                    EINAR HILLE

# Contents

# 6 · SOME ELEMENTARY FUNCTIONS

# 7 · COMPLEX INTEGRATION

# 8 · REPRESENTATION THEOREMS

# 9 · THE CALCULUS OF RESIDUES

# Symbols

## 1. Set theory:

| | |
|---|---|
| $a \in A$ | $a$ is an element of the set $A$; $a$ belongs to $A$ |
| $B \subset A$ | $B$ is a subset of $A$ |
| $C \cap D$ | Intersection of sets $C$ and $D$ |
| $C \cup D$ | Union of sets $C$ and $D$ |
| $\bar{S}$ | Closure of the set $S$ |
| $S'$ | The derived set of a given set $S$ |
| $\mathbf{C}(S)$ | The complement of $S$ |
| $\partial S$ | The boundary of $S$ |
| Int $(S)$ | The interior of $S$ |

## 2. Complex variables:

| | |
|---|---|
| $a = \Re(a + bi)$ | $a$ is the real part of $a + bi$ |
| $b = \Im(a + bi)$ | $b$ is the imaginary part of $a + bi$ |
| $z = x + iy$ | $x = \Re(z)$; $y = \Im(z)$ |
| $\bar{z} = x - iy$ | The conjugate of $z$ |
| $\|z\| = (x^2 + y^2)^{\frac{1}{2}}$ | The absolute value of $z$ |
| arg $z$ | The argument of $z$ (differs from arc tan $\frac{y}{x}$ by a multiple of $\pi$) |
| $C$ | The field of complex numbers |

## 3. Curves, domains, regions:

| | |
|---|---|
| $C$ is a "scroc" | $C$ is a simple closed rectifiable oriented curve |
| $C_i$ | Interior of $C$ |
| $C_e$ | Exterior of $C$ |
| $C^*$ | $C \cup C_i$ |
| $D$ | Domain: open, arcwise connected set |
| $R$ | Region: a domain plus a subset of its boundary |
| $[z_1, z_2]$ | Closed interval or closed line segment with endpoints $z_1$ and $z_2$ |
| $[z_1, z_2, \cdots, z_n]$ | Polygonal line joining $z_1, z_2, \cdots, z_n$ in this order |
| $\Pi: [z_1, z_2, \cdots, z_n, z_1]$ | Oriented simple closed polygon with vertices at $z_1, z_2, \cdots, z_n$ |
| $d(z_1, z_2)$ | Euclidean distance of $z_1$ and $z_2$ |
| $\chi(z_1, z_2)$ | Chordal distance of $z_1$ and $z_2$ |
| $d(S_1, S_2)$ | Distance between the sets $S_1$ and $S_2$ |

## 4. Function spaces:

| | |
|---|---|
| $C[a, b]$ | Set of functions continuous in the interval $[a, b]$ |
| $BV[a, b]$ | Set of functions of bounded variation in $[a, b]$ |
| $C[D]$ | Set of functions continuous in the domain $D$ |
| $CB[D]$ | The subset of bounded functions of $C[D]$ |
| $H[D]$ | Set of functions holomorphic in $D$ |
| $HB[D]$ | The subset of bounded functions of $H[D]$ |
| $\|f\|$ | Norm of $f$ |

# 1

# NUMBER SYSTEMS

**1.1. The real number system.** In all branches of mathematics we are concerned with sets made up of elements of one type or another. If $a$ is an element of the set $A$, we write $a \in A$. We call $B$ a *subset* of $A$ if all elements of $B$ belong to $A$; and we denote this relationship by $B \subset A$. The set of elements common to two sets $C$ and $D$ is called their *intersection*; it is denoted by $C \cap D$. If $C$ and $D$ have no elements in common, we write $C \cap D = \emptyset$ and say that the intersection is *void*. The *empty*, or *void*, set has no elements and is denoted by $\emptyset$.

In every set of mathematical significance we have a notion of *equality* which enables us to distinguish between the elements of the set. This notion is subjected to three conditions, or *postulates*.

$E_1$. $a = a$ *(reflexive property)*.

$E_2$. $a = b$ *implies* $b = a$ *(symmetric property)*.

$E_3$. $a = b$ *and* $b = c$ *implies* $a = c$ *(transitive property)*.

If $a$ and $b$ are distinct we write $a \neq b$.

The sets with which we shall be concerned in the following are not mere amorphous collections of elements; they possess both algebraic and geometric structure. Algebraic operations will be defined in our sets, which are closed under these operations; that is, combining any two elements of the set according to the admissible laws of composition is always possible and leads to elements of the set. Further, we shall normally have a notion of distance defined in the set so that the vague notion of nearness can be given a precise meaning. In the case of the real and the complex number systems, the algebraic operations are well known to the student as *addition* and *multiplication*, and the distance between two elements will ultimately be defined as the absolute value of their difference.

DEFINITION 1.1.1. *The real number system $R$ is an ordered field in which bounded sets have least upper bounds.*

This definition contains a number of undefined terms. These will be delimited by further definitions in terms of restrictive conditions (i.e., postulates) which the notions in question have to satisfy. We start with the notion of a field.

DEFINITION 1.1.2. *A field $F$ is a set of elements, at least two in number, together with two binary laws of composition known as addition and multiplication. These laws are subject to the postulates $A_1$–$A_5$, $M_1$–$M_5$, and D listed below.*

1

The postulates for addition read as follows:

$A_1$. *To every ordered pair $(a, b)$ of elements of $F$ there corresponds a uniquely defined element $a + b$ of $F$, known as the sum of $a$ and $b$.*

$A_2$. $a + b = b + a$ *(commutative law).*

$A_3$. $(a + b) + c = a + (b + c)$ *(associative law).*

$A_4$. $a + c = b + c$ *implies $a = b$ (law of cancellation).*

$A_5$. *For every pair $(a, b)$ of elements in $F$ there is an element $x$ in $F$ such that $a + x = b$.*

According to $A_5$ every linear equation

$$a + x = b$$

has at least one root. In particular, if we take $b = a$ it is seen that there exists at least one $z$ such that

$$a + z = a.$$

There cannot be more than one such element $z$, however, for if

$$a + z = a \quad \text{and} \quad a + \theta = a,$$

then by $E_3$ and $A_4$ we have $z = \theta$. Secondly, $z$ cannot depend upon the particular choice of $a$, for if

$$a + z = a \quad \text{and} \quad b + w = b,$$

then

$$a + b + z = a + b = a + b + w$$

so that $z = w$. Thus there exists a unique element 0, called *zero*, such that

(1.1.1) $$a + 0 = a$$

for every $a \in F$. In particular, $0 + 0 = 0$, and $z = 0$ is the only solution of the equation

(1.1.2) $$z + z = z.$$

Using $A_5$ again we see that for every $a \in F$ there is (at least) one element $-a$, called the *negative* of $a$, such that

(1.1.3) $$a + (-a) = (-a) + a = 0.$$

## EXERCISE 1.1.1

**1.** Verify that $z = 0$ is the only solution of (1.1.2).

**2.** Prove that there is only one element $-a$ satisfying (1.1.3).

**3.** Prove that $-(-a) = a$.

**4.** Prove that the element $x$ of $A_5$ is unique.

The postulates for multiplication read as follows:

$M_1$. *To every ordered pair $(a, b)$ of elements of $F$ there corresponds a uniquely defined element $ab$ of $F$, known as the product of $a$ and $b$.*

$M_2$. $ab = ba$ *(commutative law).*

$M_3$. $(ab)c = a(bc)$ *(associative law).*

$M_4$. $ac = bc$ *with $c \neq 0$ implies $a = b$ (law of cancellation).*

$M_5$. *For every pair $(a, b)$ of elements in $F$ with $a \neq 0$, $b \neq 0$, there is an element $x$ in $F$, $x \neq 0$, such that $ax = b$.*

Just as $A_5$ implies the existence of a *neutral element with respect to addition*, $M_5$ implies the existence of a unique *neutral element with respect to multiplication*, called *unity* and denoted by 1. We have

$$(1.1.4) \qquad a \cdot 1 = a$$

for every $a$ of $F$, with the possible exception of $a = 0$. We have $1 \neq 0$ by $M_5$. Further, $x = 1$ is a solution of the equation

$$(1.1.5) \qquad x \cdot x = x.$$

We shall see below that $x = 0$ is also a solution of this equation and that these are the only solutions.

Using $M_5$ again we see that for every $a \in F$ with $a \neq 0$ there is a unique element $a^{-1}$, called the *reciprocal* of $a$, such that

$$(1.1.6) \qquad a \cdot a^{-1} = a^{-1} \cdot a = 1.$$

## EXERCISE 1.1.2

**1.** Verify that the equation $a \cdot x = a$, $a \neq 0$, has a unique solution and that this solution is independent of $a$, so that unity is really uniquely defined.

**2.** Verify that there is only one element $a^{-1}$ satisfying (1.1.6).

**3.** Prove that the solution $x$ of the equation $a \cdot x = b$ of $M_5$ is unique and is given by $x = b \cdot a^{-1}$.

The final field postulate is the *distributive law*.

D. *For any three elements $(a, b, c)$ we have*

$$(a + b)c = ac + bc.$$

Among the many consequences of the adjunction of this postulate we note the following:

$$(1.1.7) \qquad 0 \cdot a = 0 \quad \text{for every } a.$$

From $0 + 0 = 0$ it follows that $0 \cdot a + 0 \cdot a = 0 \cdot a$, that is, $z = 0 \cdot a$ is a solution of (1.1.2), the only solution of which is zero. In particular, (1.1.4) holds also for $a = 0$.

(1.1.8)     $a \cdot b = 0$   implies either   $a = 0$   or   $b = 0$.

For if $b \neq 0$ then $a \cdot b \cdot b^{-1} = 0 \cdot b^{-1} = 0$ or $a \cdot 1 = 0$ or $a = 0$. In particular $x \cdot x = x$ implies either $x = 0$ or $x = 1$ as asserted above. Finally we note the following relations, the proofs of which are left to the reader:

(1.1.9)     $(-a) \cdot b = -(a \cdot b) = a \cdot (-b), \quad (-a) \cdot (-b) = a \cdot b.$

These eleven postulates characterize a field. There is nothing to prevent a field from being a finite set. We have postulated the existence of two (distinct) elements at least, and actually there exists a field with only two elements. These must then be zero and unity, and the postulates require that

$$0 + 0 = 0, \quad 0 + 1 = 1 + 0 = 1, \quad 1 + 1 = 0,$$
$$0 \cdot 0 = 0, \quad 0 \cdot 1 = 1 \cdot 0 = 0, \quad 1 \cdot 1 = 1.$$

Actually, this table is not so strange as it looks. It gives the rules for adding and multiplying integers modulo two, that is, for the form of arithmetic where the actual value of the integer does not count; it matters only whether it is even or odd. Here 0 stands for "even" and 1 stands for "odd."

We pass now to the *postulates of order*. These establish an order of precedence between the elements of our set. This relation will be denoted by $a < b$ which, for the time being, will be read as "$a$ precedes $b$." This relation will also be denoted by $b > a$. Instead of ordering the elements directly, we shall first order them with respect to the zero element by designating certain elements of the set as *positive*, written $0 < a$, or equivalently $a > 0$. The labeling is largely arbitrary, except for the requirement that the following postulates should hold:

$P_1$. *If $a \neq 0$, then either $a$ or $-a$ is positive (law of trichotomy).*

$P_2$. *If $a$ and $b$ are positive, so is $a + b$.*

$P_3$. *If $a$ and $b$ are positive, so is $a \cdot b$.*

As a first consequence of these postulates we note that

(1.1.10)                $0 < 1.$

The contrary assumption, $0 < -1$, leads to a contradiction with $P_3$ since $(-1)(-1) = 1$ by (1.1.9), and 1 and $-1$ cannot both be positive by $P_1$.

Next we note that a field in which the elements have been labeled in this manner must *have infinitely many elements*. To see this, we consider the multiples of unity, which we denote by the classical symbols 2, 3, 4, $\cdots$. Here

$$2 = 1 + 1, \quad 3 = 2 + 1, \quad \cdots.$$

All these elements of the field are positive by (1.1.10) combined with $P_2$. Moreover, they are distinct, since, if $j + k = j$, we would have $k = 0$ instead of $k > 0$. We refer to these elements as the *positive integers of F*.

We note that

(1.1.11) $\qquad\qquad\qquad 0 < a$　implies that　$0 < a^{-1}$.

For the contrary assumption, by (1.1.9), would lead to $0 < -1$, contradicting (1.1.10).

DEFINITION 1.1.3.　*A field F in which positive elements are defined subject to* $P_1$–$P_3$ *is ordered by the convention:* $a < b$ *if and only if* $0 < b - a$.

We note that the order established in $F$ by this convention satisfies the following conditions:

$O_1$. *If $a \neq b$, then either $a < b$ or $b < a$.*

$O_2$. *If $a < b$, then $a + c < b + c$.*

$O_3$. *If $a < b$ and $0 < c$, then $a \cdot c < b \cdot c$.*

$O_4$. *If $a < b$ and $b < c$, then $a < c$.*

Actually we could have started out with these four postulates as defining order, and then we would verify that the positive elements, for which $0 < a$, satisfy $P_1$–$P_3$.

DEFINITION 1.1.4.　*If $a < b < c$, $b$ is said to lie between $a$ and $c$.*

The elements of an ordered field are *dense* in the sense that *between any two distinct elements of F there are infinitely many elements of the set.* It is sufficient to prove that if $a < b$, then we can find a $c$ such that $a < c < b$. From $a < b$ we get

$$a + a < a + b < b + b, \quad \text{or} \quad 2a < a + b < 2b.$$

Multiplying the last relation by the positive element $2^{-1}$ and using $O_3$, we see that

$$a < 2^{-1}(a + b) < b.$$

We saw above that an ordered field $F$ contains a set of integers. Much more can be asserted: If $n$ is a positive integer in $F$, then $n^{-1}$ is an element of $F$ as well as all its multiples $mn^{-1}$ and their negatives. Here $mn^{-1} = pq^{-1}$ if and only if $mq = np$. The subset of all these elements $\{\pm mn^{-1}\}$, with proper identifications, forms the *rational subfield of F*, which we shall denote by $Q$. Normally $Q$ is a proper subset of $F$, but it is conceivable that $F = Q$.

At this juncture a note of warning is in order. Suppose two ordered fields $F$ and $F^*$ are being considered simultaneously. $F$ has a zero element 0, unity 1, and an order relation denoted by $<$. For $F^*$ the corresponding entities are $0^*, 1^*$, and $<^*$. $F$ has a rational subfield $Q$, $F^*$ a rational subfield $Q^*$. Normally $Q$ and $Q^*$ are distinct, but they are *isomorphic with respect to addition, multi-*

*plication, and order.* By this is meant that we can establish a one-to-one correspondence between the elements of $Q$ and those of $Q^*$ such that

$$a \leftrightarrow a^*, \quad b \leftrightarrow b^*, \quad \text{and} \quad a < b$$

implies

$$a + b \leftrightarrow a^* + b^*, \quad a \cdot b \leftrightarrow a^* \cdot b^*, \quad \text{and} \quad a^* <^* b^*.$$

In fact, we can establish such a correspondence by starting with

$$0 \leftrightarrow 0^*, \quad 1 \leftrightarrow 1^*.$$

To the integer $n = n \cdot 1$ of $F$ we assign the integer $n^* = n \cdot 1^*$ of $F^*$, to the reciprocal of $n$ we assign the reciprocal of $n^*$, to multiples of $n^{-1}$, multiples of $(n^*)^{-1}$. It is easily seen that this is an isomorphism. Thus, the rational subfields of all ordered fields are isomorphic.

The ordered fields form a large class, and we need some distinguishing property in order to single out the real number system from this class. This is the object of the

DEDEKIND POSTULATE.[1]    *If $S$ is a non-void subset of the ordered field $F$ and if there exists a $c \in F$ such that $a \in S$ implies $a \leq c$, then there exists a $b \in F$ such that (1) $a \in S$ implies $a \leq b$, and (2) $d \in F$, $d < b$ implies that there exists an $a \in S$ such that $d < a$.*

DEFINITION 1.1.5.    *We call $b$ the least upper bound or the supremum of $S$. Notation:*

$$b = \text{l.u.b.} [a \mid a \in S], \quad b = \sup [a \mid a \in S]$$

*or*

$$b = \text{l.u.b.} S, \quad b = \sup S \text{ for short.}$$

As indicated by the terminology, the least upper bound of $S$ is uniquely defined. It should be observed that there are ordered fields in which bounded sets do not necessarily have least upper bounds. The rational field is a case in point. Here the set $S = [a \mid a^2 < 2]$ is bounded but does not have a least upper bound.

An ordered field in which the Dedekind postulate holds is called "a" *real number system.* We shall see shortly in what sense the indefinite article "a" may be replaced by the definite article "the." First we shall formulate the so-called *axiom of Archimedes.*

---

[1] Richard Dedekind (1831–1916) published his *Stetigkeit und irrationale Zahlen* in 1872, followed by *Was sind und was sollen die Zahlen?* in 1888. "Dedekind cuts" are the class divisions of the rational numbers which he used to introduce the real number system under the assumption that the rationals are given. Dedekind was a native of Braunschweig, Germany, and professor at the Technological Institute there. His main work was in number theory.

DEFINITION 1.1.6. *An ordered field $F$ is said to be Archimedean if, given any two elements $a$, $b$ of $F$ such that $0 < a < b$, there exists a multiple $na$ of $a$ such that $b < na$.*

An example of a non-Archimedean ordered field will be found in the exercise at the end of this section.

*A real number system is Archimedean.*

Suppose, contrariwise, that for some choice of $a$ and $b$ with $0 < a < b$, we have $na \leq b$ for every $n$. Then the set $S = \{na\}$ is bounded, and by the Dedekind postulate it has a least upper bound, $c$ say. We have then $na \leq c$ for every $n$, and since $c - a < c$ there exists an $ma$ such that $c - a < ma$ or $c < ma + a = (m + 1)a$. This is a contradiction.

On the basis of the Dedekind postulate we can obtain the representation of the elements of a real number system by means of decimal fractions. It is enough to give such a representation for the elements between zero and unity. Let $a$ be a given element, $0 < a < 1$, and form the element $e_1 = 10a < 10$. Consider the integers $0, 1, 2, \cdots, 9$ and let $a_1$ be the largest of these integers which precedes or equals $e_1$. Such an $a_1$ is uniquely determined by the Dedekind postulate since the set of integers $n$ with $n \leq e_1$ is non-void and hence has a least upper bound which is $a_1$. Next, take the element $e_2 = 10(e_1 - a_1)$, and let $a_2$ be the largest integer which precedes or equals $e_2$. Here $0 \leq a_2 \leq 9$. We proceed in this manner. If $a_1, a_2, \cdots, a_{n-1}$ have been determined, we let $a_n$ be the largest integer which precedes or equals

$$e_n = 10(e_{n-1} - a_{n-1}).$$

If at any stage of the process we arrive at an $e_n = 0$ we stop, otherwise the procedure continues indefinitely. In this manner we obtain a finite or an infinite set of integers

$$a_1, a_2, \cdots, a_n, \cdots, \quad 0 \leq a_n \leq 9.$$

If the procedure stops, we have

$$a = a_1 10^{-1} + a_2 10^{-2} + \cdots + a_n 10^{-n},$$

where, of course, $10^{-k} = (10^{-1})^k$. If the sequence is infinite, then any finite section gives an approximation of $a$ in the sense that

$$0 < a - a_1 10^{-1} - a_2 10^{-2} - \cdots - a_n 10^{-n} < 10^{-n}.$$

Since $10^{-n-1} < 10^{-n}$, these approximations become better the larger $n$ gets. More precisely, if $\varepsilon$ is any given positive element of the field, then by the axiom of Archimedes we can find an element of the form $10^n$ such that

$$\varepsilon^{-1} < 10^n \quad \text{or} \quad 10^{-n} < \varepsilon.$$

It follows that we can find a section of the decimal fraction of $a$ which differs from $a$ by less than any preassigned quantity $\varepsilon$ in the sense that the difference is positive but precedes $\varepsilon$.

We can now take the last step and speak of *the real number system* rather than *a real number system*. Suppose that $S$ and $S^*$ are two fields which are ordered and satisfy the Dedekind postulate. Each element $a$ of $S$ has an associated decimal fraction as shown above. Similarly, each element $a^*$ of $S^*$ has an associated decimal fraction. Now we can establish a one-to-one correspondence between $S$ and $S^*$ by requiring that $a$ and $a^*$ shall correspond to each other if and only if they have the same decimal fraction. This correspondence is obviously an isomorphism with respect to addition, multiplication, and order. Thus *the real number system is unique up to an isomorphism*. It will be denoted by $R$ in the following.

## EXERCISE 1.1.3

**1.** Verify that the least upper bound is unique when it exists.

The following example of a non-Archimedean ordered field goes back to D. Hilbert's *Grundlagen der Geometrie*, page 30:

**2.** Let $F$ be the field of rational functions of a variable $x$ with rational coefficients. Order is defined by saying that $f(x)$ is "positive" if $f(x)$ is ultimately positive in the usual sense for large positive values of $x$. Verify that this is an ordered non-Archimedean field.

**3.** The field $F$ of the preceding problem can be broken up into exclusive rank classes $F_n$ in the following manner: Let $n$ be an arbitrary fixed integer, and let $F_n$ be the class of all functions of the form $x^n g(x)$ where $g(x)$ is any rational function tending to a finite limit different from zero as $x \to \infty$. Then $F_m \cap F_n = \emptyset$, and every $f(x)$ in $F$, except the zero element, belongs to one and only one $F_n$. Show that the Archimedean property holds for positive elements of the same class $F_n$, but never for elements belonging to different classes.

**1.2. Further properties of real numbers.** In the preceding section the emphasis was on the algebraic structure and the order relations holding in $R$; only toward the end did we consider properties of a metric or topological nature. In the present section the latter will be stressed.

First, we note that a set $S$ of real numbers which is bounded below has a *greatest lower bound*, also known as the *infimum*, and that this is uniquely determined. (Notation: g.l.b. $S$ or $\inf S$.) We have

(1.2.1) $$\inf S = -\sup(-S)$$

where $-S = [-a \mid a \in S]$.

In dealing with real numbers we shall read the relation $a < b$ as "$a$ is less than $b$," and $b > a$ as "$b$ is greater than $a$." We use $a \leq b$ to indicate that either $a = b$ or $a < b$, and we refer to relations such as $a < b$ and $b \leq a$ as *inequalities*. We write

$$a^+ = \max(a, 0), \quad a^- = \max(-a, 0), \quad |a| = a^+ + a^-,$$

and note that $a = a^+ - a^-$. The *absolute value* $|a|$ satisfies:

(i) $|a| \geq 0$, and $|a| = 0$ if and only if $a = 0$.

(ii) $|a + b| \leq |a| + |b|$.

A *metric* is introduced in $R$ by defining the *distance* $d(a, b)$ from $a$ to $b$ as $|a - b|$. This distance has the properties:

$D_1$. $d(a, b) \geq 0$ *and* $= 0$ *if and only if* $a = b$.

$D_2$. $d(a, b) = d(b, a)$.

$D_3$. $d(a, b) \leq d(a, c) + d(c, b)$   (*triangular inequality*).

A set $X$ is called a *metric space* if for each pair of points $a$, $b$ there is defined a distance $d(a, b)$ satisfying $D_1$–$D_3$. If the elements of $X$ are those of $R$, and $d(a, b) = |a - b|$, we call the corresponding metric space *the real line* or *the Euclidean space of one dimension* and denote it by $E_1$. The elements of $E_1$ are interchangeably called *points* or *numbers*.

We say that $x$ *lies in an $\varepsilon$-neighborhood of $a$* if $|x - a| < \varepsilon$.

DEFINITION 1.2.1.    *If $\{a_n\}$ is an infinite sequence of real numbers, $a_n$ is said to converge to $a$ (or $a_n \to a$, or $\lim\limits_{n \to \infty} a_n = a$), provided that for every $\varepsilon > 0$ there is an integer $N = N(\varepsilon)$ such that*

$$|a_n - a| < \varepsilon \quad for \quad n > N(\varepsilon).$$

In other words, $a_n$ converges to $a$ if every $\varepsilon$-neighborhood of $a$ contains all but a finite number of elements of the sequence.

A sequence $\{a_n\}$ of real numbers is *monotone increasing (decreasing)* if for every $k$, $a_k \leq a_{k+1}$ ($a_k \geq a_{k+1}$).

THEOREM 1.2.1.    *A monotone increasing (decreasing) sequence bounded above (below) converges to its supremum (infimum).*

*Proof.*    It is enough to treat the increasing case. Set $a = \sup a_n$. If $\varepsilon > 0$ is given, the definition of the supremum shows the existence of an integer $N$ such that

$$a - \varepsilon < a_N \leq a,$$

and, since $\{a_n\}$ is monotone increasing, this inequality holds for any $n > N$. Thus every $a_n$ with $n > N$ belongs to an $\varepsilon$-neighborhood of $a$, and $a_n \to a$.

The set of all numbers $x$ such that $a < x < b$ defines the *open interval* $(a, b)$. The corresponding *closed interval* is written $[a, b]$ and contains all $x$ with

$a \leq x \leq b$. The notations $(a, b]$ and $[a, b)$ are then obvious. Next, we formulate and prove the *theorem on nested intervals*.

**THEOREM 1.2.2.** *If $\{I_n\}$ is a sequence of non-empty closed intervals such that* (1) $I_n \supset I_{n+1}$ *for each n and* (2) *the length of $I_n$ tends to zero as $n \to \infty$, then there is one and only one real number x which belongs to every $I_n$.*

*Proof.* Suppose $I_n = [a_n, b_n]$. Then the sequences $\{a_n\}$ and $\{b_n\}$ are bounded and

$$a_n \leq a_{n+1}, \quad b_{n+1} \leq b_n$$

for every $n$. We set

$$a = \sup a_n, \quad b = \inf b_n.$$

Since $a_n \leq a \leq b \leq b_n$ and $b_n - a_n \to 0$, it follows that $b = a$. From the construction it is clear that $a \in I_n$ for every $n$. If $c \in I_n$ for each $n$, it is seen that $|a - c| \leq b_n - a_n$ for each $n$, so that $c = a$. This completes the proof.

The following theorem is basic in the theory of convergence. It is known as Cauchy's *convergence principle*.[1]

**THEOREM 1.2.3.** *The sequence $\{a_n\}$ converges to a limit if and only if for every given $\varepsilon > 0$ there exists an integer $M = M(\varepsilon)$ such that*

$$(1.2.2) \qquad |a_m - a_n| < \varepsilon \quad \text{for } m \text{ and } n > M(\varepsilon).$$

*Proof.* The necessity is obvious, for if $a_n \to a$, then for $k > N(\varepsilon/2)$ we have $|a - a_k| < \varepsilon/2$, so that if $m$ and $n > N(\varepsilon/2)$, $|a_m - a_n| = |a_m - a + a - a_n| \leq |a_m - a| + |a - a_n| < \varepsilon/2 + \varepsilon/2 = \varepsilon$, and (1.2.2) holds. Conversely, suppose that (1.2.2) holds for every $\varepsilon > 0$. Here we take successively $\varepsilon = 2^{-k-1}$, $k = 0, 1, 2, \cdots$, and see that we can find integers $n_k$ such that every $a_n$ with $n > n_k$ lies in a $2^{-k-1}$-neighborhood of $a_{n_k}$. This gives a sequence of intervals

$$I_k = [a_{n_k} - 2^{-k}, a_{n_k} + 2^{-k}]$$

such that at most $n_k$ elements of the sequence $\{a_n\}$ lie outside of $I_k$; further, $I_k \supset I_{k+1}$, and the length of $I_k$ tends to zero. There is, consequently, one and only one point $a$ belonging to each of these intervals. We assert that $a_n \to a$. If an $\varepsilon$-neighborhood of $a$ is given, we can choose $k$ so large that $I_k$ is a subset of this neighborhood and every $a_n$ with $n > n_k$ belongs to the $\varepsilon$-neighborhood in question. Thus $\lim a_n = a$.

---

[1] Augustin Louis (baron de) Cauchy (1789–1857), the father of modern analysis, exercised a profound influence on the theory of series, real and complex function theory, differential equations, etc. Cauchy was a prolific but profound writer. As an ardent partisan of the Bourbons, Cauchy was excluded from public employment during the eighteen years of the July monarchy, part of which he spent in exile. He returned to a professorship at the École Polytechnique in Paris in 1848.

DEFINITION 1.2.2.    *If $S$ is an infinite set of real numbers, $a \in R$ is called a limit point of $S$ if every $\varepsilon$-neighborhood of $a$ contains at least two (and hence infinitely many) elements of the set.*

The following result is known as the *Bolzano-Weierstrass theorem*:[1]

THEOREM 1.2.4.    *An infinite bounded set of real numbers has at least one limit point.*

*Proof.* It is obvious that both conditions are necessary: a finite set cannot have a limit point, and the set of integers, though infinite, has no limit point because it is not bounded. Suppose then that $S$ is an infinite set of real numbers in the interval $I_0 = [a, b]$. Bisecting $[a, b]$ we obtain two subintervals at least one of which contains infinitely many elements from $S$. Let $I_1$ be the closed subinterval containing infinitely many points of $S$: if both subintervals have this property, $I_1$ shall be the left subinterval. We repeat this construction and obtain a sequence of nested intervals $\{I_n\}$ each of which contains infinitely many elements of $S$. The length of $I_n$ obviously tends to zero, so that Theorem 1.2.2 applies. Let $c$ be the point common to all intervals $I_n$. If $N_\varepsilon(c)$ is an $\varepsilon$-neighborhood of $c$, and if

$$2^{-n}(b - a) < \varepsilon,$$

then $I_n$ lies in $N_\varepsilon(c)$; that is, $N_\varepsilon(c)$ contains infinitely many points of $S$, and $c$ is a limit point of $S$. This completes the proof.

Two remarks should be added. The point $c$ constructed above is an element of $R$ but need not belong to $S$. Secondly, an infinite set may very well have several limit points, even infinitely many. The set of rationals between 0 and 1 is an example. Every point of this set is a limit point, but this does not exhaust the possibilities. In fact we see that the limit points fill the whole interval $[0, 1]$.

The set of limit points of a given set $S$ is called the *derived set* and is denoted by $S'$. $S$ is said to be *closed* if $S' \subset S$, *dense in itself* if $S \subset S'$, and *perfect* if $S = S'$. The *complement* of $S$, written $C(S)$, is the set of all real numbers not in $S$. A set $S$ (of real numbers) is *open* if $C(S)$ is closed, and vice versa. This convention requires that a finite or void set be regarded as closed. In this sense, the derived set $S'$ is always closed. If $S'$ is bounded and non-void, then it has a

---

[1] Bernard Bolzano (1781–1848) was professor of the philosophy of religion at Prague. His posthumous treatise *Paradoxien des Unendlichen* appeared in 1850, but its merits were not recognized for another twenty years.

Karl Weierstrass (1815–1897) after a long apprenticeship in Prussian secondary schools became professor at the University of Berlin in 1864. There he attracted many students. He is one of the three founders of the theory of analytic functions, Cauchy and Riemann being the others. An exponent of arithmetization and rigor, he contributed to all fields of analysis.

least element as well as a largest one and these coincide if $S$ has only one limit point. We set

$$\limsup S = \sup S', \quad \liminf S = \inf S',$$

and speak of the *superior* and the *inferior limits* of $S$.

A related concept is the following. A sequence $\{a_n\}$ of real numbers obviously is a function on positive integers to real numbers, that is, to each positive integer $n$ there corresponds a unique real number $a_n$. This function may of course assume the same value for several values of $n$, even for infinitely many. Suppose that the function is bounded, that is, that there exists an $M$ such that $|a_n| < M$ for all $n$. We set

(1.2.3)          $$\alpha = \liminf a_n, \quad \beta = \limsup a_n$$

if for any given $\varepsilon > 0$ the inequalities

(1.2.4)          $$\alpha - \varepsilon < a_n < \beta + \varepsilon$$

hold with at most a finite number of exceptions, while

(1.2.5)          $$a_m < \alpha + \varepsilon, \quad \beta - \varepsilon < a_p$$

hold for infinitely many values of $m$ and $p$ respectively.

## EXERCISE 1.2

**1.** Given the set $S$ of fractions of the form $\dfrac{1}{m} + \dfrac{1}{n}$, $\quad m, n = 1, 2, 3, \cdots$. Find $\inf S$, $\sup S$, $\liminf S$, $\limsup S$, and determine $S'$. Show that $S'$ is closed.

**2.** Same questions for $S = \left\{ \dfrac{1}{m} + \dfrac{\sqrt{2}}{n} \ \middle| \ m, n = 1, 2, 3, \cdots \right\}$.

**3.** A sequence is defined by

$$a_{2m} = \frac{1}{m}, \quad a_{2m-1} = \frac{m-1}{m}, \quad m = 1, 2, 3, \cdots.$$

Determine $\liminf a_n$ and $\limsup a_n$.

**4.** Let $m$ be a positive integer $> 1$ and set

$$a_n = \sin \frac{n\pi}{m}, \quad n = 1, 2, 3, \cdots.$$

Find $\liminf a_n$ and $\limsup a_n$.

**5.** Given $a_n = \log (n + 1) - \log n$. Find $\liminf a_n$ and $\limsup a_n$.

**6.** Same question for $a_n = n[\log (n + 1) - \log n]$.

**\*7.** Consider the set of numbers

$$R_{m, n} = (m^2 + n)^{\frac{1}{2}} - m, \quad m = 1, 2, 3, \cdots, \quad n = 0, 1, \cdots, 2m,$$

that is, the set of fractional parts of the square roots of the positive integers. Show that the derived set is the interval $[0, 1]$. If $\alpha$, $0 < \alpha < 1$, find a subset of $R_{m, n}$ having $\alpha$ as its only limit point. (*Hint:* For each $m$ take $n = [2\alpha m]$ where $[q]$ is the largest integer $\leq q$.)

**8.** Verify that $S'$ is closed for any set $S$.

The set of ordered pairs $(a, b)$ of real numbers with the distance function

$$d[(a_1, b_1), (a_2, b_2)] = [(a_1 - a_2)^2 + (b_1 - b_2)^2]^{\frac{1}{2}}$$

is a metric space known as the *Euclidean plane* and denoted by $E_2$.

**9.** Formulate analogues of Definitions 1.2.1 and 1.2.2 for convergence of sequences $\{(a_n, b_n)\}$.

**10.** Formulate and prove the analogue of Theorems 1.2.2 and 1.2.3 for sequences $\{(a_n, b_n)\}$.

**11.** Prove the Bolzano-Weierstrass theorem for sets in $E_2$.

**1.3. The complex number system.** We now turn to the complex field which is basic for analytic function theory.

DEFINITION 1.3.1.    *The complex number system $C$ is a field having as its elements the ordered pairs $(a, b)$ of real numbers with equality, addition, and multiplication defined by*

(1.3.1)        $(a_1, b_1) = (a_2, b_2)$   *if and only if*   $a_1 = a_2, b_1 = b_2$.

(1.3.2)        $(a_1, b_1) + (a_2, b_2) = (a_1 + a_2, b_1 + b_2)$.

(1.3.3)        $(a_1, b_1)(a_2, b_2) = (a_1a_2 - b_1b_2, a_1b_2 + a_2b_1)$

*respectively. There is also a notion of scalar multiplication*

(1.3.4)                        $c(a, b) = (ca, cb)$,

*where $c$ is any real number.*

We have to verify that these conventions actually define a field. The postulates for addition cause little difficulty; the addition defined by (1.3.2) is obviously commutative and associative and admits a law of cancellation, since the components $a$, $b$ of the complex $(a, b)$ are real numbers satisfying $A_2$, $A_3$, and $A_4$. Further we see that the equation

$$(a_1, b_1) + (x, y) = (a_2, b_2)$$

is satisfied by

$$(x, y) = (a_2 - a_1, b_2 - b_1).$$

In particular, $(0, 0)$ plays the role of zero element, and the negative of $(a, b)$ is $(-a, -b)$ as indicated by (1.3.4). Thus the postulates for addition are satisfied by the complex numbers $(a, b)$, by virtue of the fact that they are satisfied by the components $a$, $b$ and because of the definition of equality.

Next we consider the implications of (1.3.4). This gives

$$(a, 0) = a(1, 0), \quad (0, b) = b(0, 1),$$

and since

$$(a, b) = (a, 0) + (0, b),$$

we have finally the representation

(1.3.5)                    $$(a, b) = a(1, 0) + b(0, 1).$$

This formula should be considered against a broader frame of reference. We say that a set $X$ with elements $u$, $v$, $\cdots$ called *vectors* is a *linear vector space over the reals* if

$$u \in X, \quad v \in X, \quad \alpha \in R, \quad \beta \in R$$

implies that $\alpha u + \beta v$ is defined as an element of $X$. Here it is assumed that addition satisfies $A_1$–$A_5$ and that scalar multiplication satisfies

$$(\alpha + \beta)u = \alpha u + \beta u, \quad \alpha(u + v) = \alpha u + \alpha v, \quad \alpha(\beta u) = (\alpha\beta)u, \quad 1 \cdot u = u.$$

These assumptions imply that there exists a zero element with $0 \cdot u = 0$ for all $u$ and that $-u = (-1)u$. A set of $n$ vectors in $X$, say $u_1, u_2, \cdots, u_n$, is said to be *linearly independent* over the reals if

$$\beta_1 u_1 + \beta_2 u_2 + \cdots + \beta_n u_n = 0, \quad \beta_k \in R,$$

implies that

$$\beta_1 = \beta_2 = \cdots = \beta_n = 0.$$

$X$ is said to be *of dimension n over R if there exists a set* $u_1, u_2, \cdots, u_n$ *of n linearly independent vectors while every set of* $(n + 1)$ *vectors is linearly dependent.* Thus for any given $u \in X$ we can find real numbers $\alpha_1, \alpha_2, \cdots, \alpha_n$ such that

$$u = \alpha_1 u_1 + \alpha_2 u_2 + \cdots + \alpha_n u_n.$$

The vectors $u_1, u_2, \cdots, u_n$ are then said to form a *basis* of $X$.

In this frame of ideas, the set $C$ becomes a linear vector space of dimension two over the real field having

(1.3.6)                    $$u_1 = (1, 0), \quad u_2 = (0, 1)$$

as a basis.

The law of multiplication in $C$, defined by (1.3.3), is so chosen that the basis vectors have the following multiplication table:

(1.3.7) $\qquad (u_1)^2 = u_1, \quad u_1 u_2 = u_2 u_1 = u_2, \quad (u_2)^2 = -u_1.$

Further, (1.3.3) shows that

$$(\alpha_1 u_1 + \alpha_2 u_2)(\beta_1 u_1 + \beta_2 u_2) = \alpha_1 \beta_1 (u_1)^2 + \alpha_1 \beta_2 u_1 u_2 + \alpha_2 \beta_1 u_2 u_1 + \alpha_2 \beta_2 (u_2)^2$$
$$= (\alpha_1 \beta_1 - \alpha_2 \beta_2) u_1 + (\alpha_1 \beta_2 + \alpha_2 \beta_1) u_2.$$

Thus vector multiplication is distributive with respect to the basis vectors, and the latter commute with the real scalars.

It still remains to prove that the postulates for multiplication are satisfied. Multiplication is obviously commutative, and we can prove by elementary calculations that it is also associative and that the law of cancellation holds. Formula (1.3.7) shows that $u_1$ acts as unity in $C$ so that

(1.3.8) $\qquad (a, b)u_1 = u_1(a, b) = (a, b)$

for every $(a, b) \in C$. Next we have to show that every $(a, b) \neq (0, 0)$ has an inverse. But

$$(a, b)(x, y) = (1, 0)$$

gives

$$ax - by = 1, \quad bx + ay = 0,$$

whence

(1.3.9) $\qquad (x, y) = (a, b)^{-1} = \left( \dfrac{a}{a^2 + b^2}, -\dfrac{b}{a^2 + b^2} \right),$

and this is well defined since $a^2 + b^2 \neq 0$. Thus postulates $M_1$–$M_5$ hold. Since one can also verify the distributive law, it follows that $C$ is a field.

$C$ differs from $R$ in one very essential respect: $C$ *cannot be ordered with an order relation satisfying* $O_1$–$O_4$. Suppose contrariwise. Then we must have $0 < u_1$, and either $u_2$ or $-u_2$ must be positive. In either case, their squares $(u_2)^2 = (-u_2)^2 = -u_1$ would be positive, and this is impossible. There exists a subfield of $C$ which can be ordered, namely, $\{\alpha u_1, \alpha \in R\}$. First we note that

$$\alpha u_1 + \beta u_1 = (\alpha + \beta)u_1, \quad (\alpha u_1)(\beta u_1) = (\alpha \beta)u_1,$$

so we are actually dealing with a subfield. We define

$$\alpha u_1 < \beta u_1 \quad \text{if and only if} \quad \alpha < \beta.$$

This choice makes $\{\alpha u_1\}$ isomorphic with $R$ with respect to addition, multiplication, and order.

At the end of Section 1.1 we observed that the real number system is unique up to an isomorphism. In this sense we may identify the set $\{\alpha u_1\}$ with $R$. This is often expressed by saying that *the real number system may be embedded in the complex field*:

$$R \leftrightarrow Ru_1 \subset C.$$

We shall do so, and as a consequence we shall not distinguish between $\alpha$ and $\alpha u_1$. In particular we write 1 instead of $u_1$. We shall also use the customary notation $i$ for $u_2$. Thus

(1.3.10)
$$(1, 0) = 1, \quad (0, 1) = i$$

and

(1.3.11)
$$(a, b) = a + bi.$$

We can endow the complex field with a Euclidean metric. In fact, there is a one-to-one correspondence between the complex numbers $(a, b)$ and the points $(a, b)$ of the Euclidean plane $E_2$. The latter has its natural metric defined above in Exercise 1.2. We simply impose the same metric on $C$. This implies that we assign a length to the complex number

$$z = x + iy,$$

namely

(1.3.12)
$$|z| = (x^2 + y^2)^{\frac{1}{2}},$$

and define the distance between two complex numbers to be

(1.3.13)
$$d(z_1, z_2) = |z_1 - z_2|.$$

The quantity $|z|$ is also known as the *absolute value* of $z$ and has the properties

(1.3.14)
$$|z| > 0 \quad \text{unless} \quad z = 0,$$

(1.3.15)
$$|z_1 + z_2| \leq |z_1| + |z_2|,$$

and $d(z_1, z_2)$ satisfies $D_1$–$D_3$.

Once $C$ has been assigned the metric of $E_2$, we can define the notions of convergence and limits in $C$ with the aid of the corresponding notions in $E_2$. The reader is supposed to be familiar with the latter, having worked Problems 9, 10, and 11 of Exercise 1.2. We shall consequently restrict ourselves to formulating the appropriate concepts and theorems in the terminology of $C$.

First a note of warning. The student should always keep in mind that complex numbers cannot be ordered, so that inequalities between such numbers are meaningless. Inequalities may hold between the absolute values of complex numbers, however, and such inequalities play an important role. The following definitions serve to illustrate this statement:

*A set $S$ of complex numbers is said to be bounded if the set $S_0 = [|z| \mid z \in S]$ is bounded.*

An $\varepsilon$-neighborhood of $z_0$ is the set of complex numbers $z$ such that $|z - z_0| < \varepsilon$.

**DEFINITION 1.3.2.** *The sequence $\{z_n\}$, $z_n \in C$, converges to the limit $z_0$ (or $z_n \to z_0$ or $\lim z_n = z_0$) if for every $\varepsilon > 0$ there exists an integer $N = N(\varepsilon)$ such that*

$$|z_0 - z_n| < \varepsilon \quad \text{for} \quad n > N(\varepsilon).$$

DEFINITION 1.3.3.　*If $S$ is an infinite set of complex numbers, $z_0 \in C$ is called a limit point of $S$ if every $\varepsilon$-neighborhood of $z_0$ contains at least two elements of $S$.*

THEOREM 1.3.1.　[CAUCHY'S CONVERGENCE PRINCIPLE.] *The sequence $\{z_n\}$, $z_n \in C$, converges to a limit if and only if for every $\varepsilon > 0$ there is an integer $M = M(\varepsilon)$ such that*

$$|z_m - z_n| < \varepsilon \quad for \quad m, n > M(\varepsilon).$$

Such a sequence $\{z_n\}$ is said to be *fundamental*, or a *Cauchy sequence*.

THEOREM 1.3.2.　[BOLZANO-WEIERSTRASS THEOREM.] *A bounded infinite set of complex numbers has at least one limit point.*

For an extension of the theorem on nested intervals to arbitrary closed sets, see Appendix A.

### EXERCISE 1.3

**1.** Prove $|\cos \theta + i \sin \theta| = 1$, $\theta$ real.

**2.** What is the distance of $2 - i$ from its negative?

**3.** Find $S'$ if $S = \left\{ \left. \dfrac{1}{m} + \dfrac{i}{n} \right| m, n = 1, 2, 3, \cdots \right\}$.

**4.** Find $\lim\limits_{n \to \infty} n \left[ \cos \dfrac{\theta}{n} + i \sin \dfrac{\theta}{n} - 1 \right]$.

**5.** Verify the associative law for multiplication and the law of cancellation.

**6.** Verify the distributive law.

**7.** Prove Theorem 1.3.1.

**8.** Prove Theorem 1.3.2.

### COLLATERAL READING

BEGLE, E. G. *Introductory Calculus with Analytic Geometry*, Chap. 1. Henry Holt and Company, Inc., New York, 1954.

BIRKHOFF, G., and MAC LANE, S. *A Survey of Modern Algebra*, Revised Edition, Chaps. 2, 4, and 5. The Macmillan Company, New York, 1953.

GRAVES, L. M. *The Theory of Functions of Real Variables*, Second Edition, Chaps. 2 and 3. McGraw-Hill Book Company, Inc., New York, 1956.

OSTROWSKI, A. *Vorlesungen über Differential- und Integralrechnung*, Vol. I, Chaps. 1 and 2. Verlag Birkhäuser, Basel, 1950.

# 2

# THE COMPLEX PLANE

**2.1. Geometry of complex numbers.** The present chapter will revolve around the idea of representing complex numbers by points or vectors in a Euclidean plane.

It is a very simple and natural idea, but it took a long time to break through. When it finally came, it occurred at nearly the same time to three different persons. A Norwegian surveyor and cartographist, Caspar Wessel (1745–1818), presented a paper on the subject to the Danish Academy of Sciences and Letters in 1797. It was published but remained unnoticed until the Academy issued a French translation in 1897. An essay on the geometrical interpretation of imaginary quantities by a French Swiss mathematician, Jean Robert Argand (1768–1822), appeared in 1806; it was rescued from near-oblivion in 1813 and provoked much discussion and unfavorable comment. The third discoverer, Karl Friedrich Gauss[1] (1777–1855), a genius of the first order, carried enough authority to establish the idea. In his dissertation of 1799, he employed the idea without explicit mention; in a letter of 1811 to Bessel it is clearly outlined; and finally in discussing the integers in the complex field, later known as Gaussian integers, in 1831, he gave the geometric interpretation explicitly. As a consequence we find French mathematicians referring to *le diagramme d'Argand*, while Germans speak of *die Gaussische Ebene*. The Norwegians with becoming modesty avoid claiming *det Wesselske planet*. To us it will be the *complex plane*.

After this historical digression let us return to the substance of the matter. We take a Euclidean plane, choose an origin $O$ and a pair of orthogonal axes through $O$, and introduce Cartesian coordinates in the usual manner. Then with every point $P$ in the plane there is associated an ordered number pair $(a, b)$. This point $P$ is regarded as the geometrical representative or carrier of the complex number $a + bi$. An equivalent representative is furnished by the vector $OP$. See Figure 1.

---

[1] Gauss was professor of astronomy and director of the observatory for astronomy and terrestrial magnetism in Göttingen from 1807 until his death. His contributions to celestial mechanics, geodesy, electricity, and magnetism (gauss, degaussing!) were outstanding. As a mathematician he enriched algebra (four proofs of the fundamental theorem), number theory (law of quadratic reciprocity, biquadratic residues, cyclotomy), analysis (hypergeometric series), differential geometry, mathematical physics, etc. His pupils were mostly astronomers, some quite famous; though his writings, especially in number theory, had a profound influence on German mathematicians, he had few direct personal contacts with them, and Riemann was perhaps the only one who was directly affected by him. Gauss was the forerunner of the later so famous mathematical school of Göttingen.

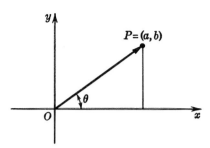

**Figure 1**

We call $a$ the *real* and $b$ the *imaginary part*[1] of $a + bi$ and write

(2.1.1)                    $a = \Re(a + bi), \quad b = \Im(a + bi)$.

American students have to be warned that the second German letter is an $I$ and not a $T$. In this representation, the $x$-axis is called the *real axis*, the $y$-axis the *imaginary axis*. As remarked earlier,

(2.1.2)                    $r = \mid a + bi \mid = (a^2 + b^2)^{\frac{1}{2}}$

is known as the *absolute value* or, by an older name, as the *modulus* of $a + bi$. It is simply the length of the vector representing the complex number. The angle or arc $\theta$ from the positive real axis to the vector $OP$ is called the *argument*, or the *amplitude*, of the number. We shall write

(2.1.3)                    $\theta = \arg (a + bi)$.

This is determined only up to multiples of $2\pi$. To avoid ambiguities it is at times convenient to select a "principal branch" or a "principal determination" of the argument; there are two conventions in use, depending on the purpose. Either one takes $0 \leq \theta < 2\pi$ or $-\pi < \theta \leq \pi$ for the principal determination. Of course, both these schemes have the drawback that they are not continuous functions of position in the plane. We have

(2.1.4)
$$a = r \cos \theta, \quad b = r \sin \theta,$$
$$a + bi = r(\cos \theta + i \sin \theta).$$

We have of course also

$$\theta = \arc \tan \frac{b}{a},$$

but in using this formula it must be remembered that the arc tangent is determined up to multiples of $\pi$, not of $2\pi$ as for the argument. Thus, a value of $\theta$

---

[1] There is no other branch of mathematics where the terminology shows such a marked distrust of the objects named as the field of numbers. The terms "radical," "surd," "negative," "irrational," and "imaginary" all have uncomplimentary connotations, and in most cases these terms indicate the opposition which once upon a time met these revolutionary innovations.

computed from this formula must be tested, since it could just as well be an argument of $-(a + bi)$.

The vectors $z$ and $-z$ are opposite in direction, but have the same absolute value. As just observed, arg $z$ and arg $(-z)$ differ by an odd multiple of $\pi$.

The quantity

$$(2.1.5) \qquad \bar{z} = a - bi = r(\cos\theta - i\sin\theta)$$

is called the *conjugate* of $z = a + bi$. For typographical reasons the symbol $z*$ instead of $\bar{z}$ has come into use, especially if $z$ is a complicated expression. Geometrically speaking, *the points $z$ and $\bar{z}$ are images of each other under a reflection of the complex plane in the real axis.* We have

$$(2.1.6) \qquad |\bar{z}| = |z|, \quad \arg\bar{z} = -\arg z,$$

$$(2.1.7) \qquad z + \bar{z} = 2a = 2\Re(z), \quad z - \bar{z} = 2bi = 2i\Im(z).$$

Further we have

$$(2.1.8) \qquad z\bar{z} = |z|^2.$$

It should also be observed that

$$(2.1.9) \qquad |\cos\theta + i\sin\theta| = 1.$$

We now proceed to the geometry of the arithmetical operations. Since the sum of

$$z_1 = x_1 + iy_1 \quad \text{and} \quad z_2 = x_2 + iy_2$$

is

$$(x_1 + x_2) + i(y_1 + y_2),$$

addition of complex numbers corresponds to vector addition. We plot the two vectors representing $z_1$ and $z_2$ and construct the diagonal from the origin in

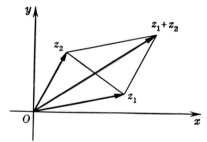

**Figure 2**

the parallellogram having these vectors as sides. In other words, we construct the resultant of two forces applied at the origin. The parallelogram construction breaks down if the vectors have either the same or opposite direction, but the composition of forces is obvious. As asserted in (1.3.15) we have

$$(2.1.10) \qquad |z_1 + z_2| \leq |z_1| + |z_2|.$$

We now see that equality can hold here if and only if arg $z_1$ = arg $z_2$. It should be noted that the other diagonal in the parallelogram is parallel to and equal in length to the vector representing $z_2 - z_1$. Thus, we verify again that $d(z_1, z_2) = |z_2 - z_1|$.

Using a device from graphical statics, we can construct the sum of several complex numbers. We start with the vector $z_1$. At its endpoint we apply a vector parallel to and of the same length as $z_2$. This takes us to the point $z_1 + z_2$. At this point we apply a vector parallel to and of the same length as $z_3$. This leads to the point $z_1 + z_2 + z_3$. In this manner we add consecutively one vector at a time, and after $n$ operations we are at $z_1 + z_2 + \cdots + z_n$.

Formula (1.3.3) defines the product by

$$(2.1.11) \qquad (a_1, b_1) \cdot (a_2, b_2) = (a_1 a_2 - b_1 b_2, \ a_1 b_2 + a_2 b_1).$$

This definition has nothing to do with either the dot product or the cross product of vector analysis. Our product of two vectors is a vector, just as the cross product is. There, however, the analogy ends. The cross product is neither commutative nor associative, and it vanishes when the two factors are equal, while multiplication of complex vectors according to (2.1.11) will be found to be commutative and associative; moreover the product vanishes if and only if one of the factors does. The last property follows from the identity

$$(a_1 a_2 - b_1 b_2)^2 + (a_1 b_2 + a_2 b_1)^2 = (a_1{}^2 + b_1{}^2)(a_2{}^2 + b_2{}^2),$$

so that

$$(2.1.12) \qquad |z_1 z_2| = |z_1| \, |z_2|,$$

or:

*The absolute value of a product is the product of the absolute values of the factors.*

To this we shall now add:

*The argument of a product is the sum of the arguments of the factors.*

Since the argument of $z$ is not a single-valued function of $z$, this statement has to be understood in the following mannner. If $\theta_1$ and $\theta_2$ are admissible values of arg $z_1$ and arg $z_2$ respectively, then $\theta_1 + \theta_2$ is an admissible value of arg $(z_1 z_2)$. This is elementary trigonometry: from

$$z_1 = r_1(\cos \theta_1 + i \sin \theta_1), \quad z_2 = r_2(\cos \theta_2 + i \sin \theta_2)$$

follows

$$z_1 z_2 = r_1 r_2 \{[\cos \theta_1 \cos \theta_2 - \sin \theta_1 \sin \theta_2] + i[\cos \theta_1 \sin \theta_2 + \cos \theta_2 \sin \theta_1]\}$$
$$= r_1 r_2[\cos (\theta_1 + \theta_2) + i \sin (\theta_1 + \theta_2)],$$

so that

$$(2.1.13) \qquad \arg (z_1 z_2) = \arg z_1 + \arg z_2$$

in the sense asserted above.

We can now give a geometrical construction of the product. While the construction of sums of vectors does not presuppose any knowledge of the scale, we do have to know the length of the unit vector in order to construct products. In the adjacent Figure 3 we have marked $0$, $1$, $z_1$, $z_2$. We can label the latter

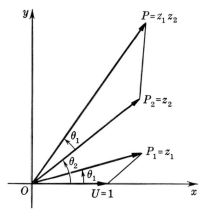

**Figure 3**

two points in such a way that $\arg z_1 \leqq \arg z_2$. On the vector $OP_2$ as basis, we construct a triangle $OP_2P$ similar to the triangle $OUP_1$ and oriented similarly. The construction breaks down if $z_1$ is real, but then the problem reduces to the construction of a line segment $OP$ such that $OP : OP_1 = OP_2 : OU$. The point $P$ is located on $OP_2$, produced to the same side of the origin as $P_2$ if $z_1 > 0$, otherwise to the opposite side.

Since the quotient $z_1/z_2$ is the product of $z_1$ and the reciprocal of $z_2$, it is enough to construct the reciprocal. From formula (1.3.9) we get

$$\frac{1}{z} = \frac{\bar{z}}{|z|^2} = \frac{1}{r}(\cos \theta - i \sin \theta) \quad \text{if} \quad z = r(\cos \theta + i \sin \theta).$$

Thus,

(2.1.14) $$\left|\frac{1}{z}\right| = \frac{1}{|z|}, \quad \arg \frac{1}{z} = -\arg z.$$

These two relations suggest various ways and means of carrying out the geometrical construction of the reciprocal of $z$.

One way is to construct a triangle similar to $OUP$ on $OU$ in the opposite half-plane. This is the analogue of the construction for multiplication. The following observation puts the problem under a larger heading. In passing from $z$ to $1/z$ there are really two operations involved: *a reflection in the unit circle and a reflection in the real axis*. The order in which these operations are carried out is immaterial: they commute. The first operation is known in the theory of geometric transformations as an *inversion* or a *reciprocation*. A circle is given, in this case the so-called *unit circle*. Every point $P \neq O$ has a unique inverse with respect to the circle, that is, a point $Q$, collinear with $O$ and $P$ such that

$$OP \cdot OQ = (OR)^2.$$

Such a transformation is an *involution*, that is, *the inverse of the inverse is the original point*. The inversion leaves the circle invariant point for point. Every point outside the circle has its inverse inside and vice versa. It is then enough to give a construction of the inverse of an exterior point. This is given in Figure 4. We draw a tangent from $P$ to the circle; from the point of contact $T$ we drop a perpendicular to $OP$. The foot point $Q$ is the desired inverse.

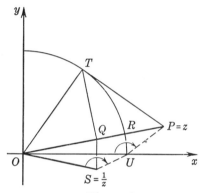

**Figure 4**

If this operation is applied to the point $z$, we obtain $1/\bar{z}$. We get to $1/z$ by the second operation of *conjugation*, or reflection in the real axis. The transformation

$$w = 1/z$$

is of basic importance in function theory, and we shall study further aspects of it later.

We can, of course, use the method given above in order to construct the product of three or more complex numbers and to construct powers. The definition of multiplication gives

(2.1.15)          $|z^n| = |z|^n, \quad \arg(z^n) = n \arg z,$

where $n$ is an integer. The last relation has to be taken with a grain of salt: every determination of $\arg z$ will upon multiplication by $n$ give a determination of $\arg(z^n)$, but not every determination of the latter is obtainable in this manner. The correct relation is

(2.1.16)          $\arg(z^n) \equiv n \arg z \pmod{2\pi},$

which says that the two sides are congruent modulo $2\pi$. In other words, any admissible values of the two sides will differ by an integral multiple of $2\pi$.

We note for $|z| = 1$ that

$$(2.1.17) \qquad (\cos \theta + i \sin \theta)^n = \cos n\theta + i \sin n\theta,$$

where $n$ is any positive or negative integer. This is known as the theorem of de Moivre.[1] Supposing $n$ to be a positive integer, and separating reals and imaginaries, we obtain the two trigonometric identities

$$\cos n\theta = \cos^n \theta - \binom{n}{2} \cos^{n-2} \theta \sin^2 \theta + \binom{n}{4} \cos^{n-4} \theta \sin^4 \theta + \cdots,$$

$(2.1.18)$

$$\sin n\theta = \binom{n}{1} \cos^{n-1} \theta \sin \theta - \binom{n}{3} \cos^{n-3} \theta \sin^3 \theta + \cdots.$$

Powers will be discussed in greater detail in Chapter 3. At this juncture we recall that a polynomial in $z$ of degree $n$ is an expression of the form

$$P(z) = a_0 + a_1 z + a_2 z^2 + \cdots + a_n z^n,$$

where $a_0, a_1, \cdots, a_n$ are given real or complex numbers and $a_n \neq 0$. The so-called fundamental theorem of algebra is the assertion that there exists at least one complex number, $z_0$ say, such that $P(z_0) = 0$. The "so-called" refers to the fact that modern algebraists are inclined to deny both its algebraic and its fundamental character. It is a theorem of function theory, however, and we shall give several proofs as illustrations of function theoretical principles in later chapters. At this stage a couple of elementary remarks are in order. We note that:

*The conjugate of a sum (product) is the sum (product) of the conjugates.*
From this one gets

$$\overline{P(z)} = \bar{a}_0 + \bar{a}_1 \bar{z} + \bar{a}_2 \bar{z}^2 + \cdots + \bar{a}_n \bar{z}^n.$$

If the coefficients $a_k$ are real, $\bar{a}_k = a_k$, so that

$$\overline{P(z)} = a_0 + a_1 \bar{z} + a_2 \bar{z}^2 + \cdots + a_n \bar{z}^n = P(\bar{z}).$$

In particular, if $P(z_0) = 0$ so is $\overline{P(z_0)} = P(\bar{z}_0)$. Hence:

*If a polynomial has real coefficients, then the values of $z$ for which it is zero either are real or form pairs of conjugate complex numbers.*
Secondly we observe that separating reals and imaginaries gives

$$P(x + iy) = U(x, y) + iV(x, y),$$

---

[1] Abraham de Moivre (1667–1754), of a French Huguenot family, settled in London in 1685 after the revocation of the Edict of Nantes. He lived on private lessons in mathematics and games of chance. He was one of the founders of the theory of probability.

where $U(x, y)$ and $V(x, y)$ are polynomials in $x$ and $y$ of degree $n$ and of very special structure. The trigonometric identities (2.1.18) show, in particular, that

(2.1.19)
$$\Re(z^n) = \sum (-1)^k \binom{n}{2k} x^{n-2k} y^{2k},$$

$$\Im(z^n) = \sum (-1)^k \binom{n}{2k+1} x^{n-2k-1} y^{2k+1},$$

with the aid of which $U(x, y)$ and $V(x, y)$ can easily be formed. The curves

$$U(x, y) = 0 \quad \text{and} \quad V(x, y) = 0$$

divide the complex plane into a number of regions, in each of which $U(x, y)$ and $V(x, y)$ keep constant signs; the roots of the equation

$$P(x + iy) = 0$$

are precisely the roots of the simultaneous equations

(2.1.20)
$$\begin{cases} U(x, y) = 0, \\ V(x, y) = 0. \end{cases}$$

For equations of low degree, plotting these curves and finding their intersections is a convenient method of finding approximate values of the complex roots.

### EXERCISE 2.1

**1.** Prove that $|\,z_1 + z_2\,| \geq |\,|\,z_1\,| - |\,z_2\,|\,|$. When does equality hold?

**2.** Find necessary and sufficient conditions in order that

$$V(x, -y) = -V(x, y).$$

Show that $y$ is a factor of $V(x, y)$ in this case.

**3.** Prove that the equation

$$z^3 + 2z + 4 = 0$$

has its roots outside the unit circle. (*Hint:* What is the maximum value of the modulus of the sum of the first two terms if $|\,z\,| \leq 1$?)

**4.** With $P(z)$ defined as in Problem 3, discuss the curves $U = 0$, $V = 0$ in sufficient detail to locate the roots with an error not exceeding 0.2.

**5.** If $P$ and $Q$ are inverse points with respect to the unit circle, prove that the circle of diameter $PQ$ is orthogonal to the unit circle.

**6.** $Q$ is an arbitrary point on the circle with center at $z = \frac{1}{2}$ and radius $\frac{1}{2}$. $P$ is the inverse of $Q$ with respect to the unit circle. Prove that $P$ lies on the straight line tangent to the unit circle at $z = 1$.

**7.** Prove that

$$| z_1 - z_2 |^2 + | z_1 + z_2 |^2 = 2 \, ( \, | z_1 |^2 + | z_2 |^2).$$

**8.** Interpret $\Re(i\bar{z}) = 2$ geometrically.

**9.** Find $\sup \Re(i\bar{z})$ if $| z | < 1$.

**10.** If $z_1$, $z_2$, $z_3$ are three noncollinear points, construct $\frac{1}{3}(z_1 + z_2 + z_3)$ and interpret your result.

**2.2. Curves and regions in the complex plane.** A curve in the complex plane is normally defined by an equation, but this may very well take on an unfamiliar form, since we have the possibility of expressing the coordinates in terms of the complex variable. Thus, instead of

$$F(x, y) = 0$$

we may write

$$(2.2.1) \qquad\qquad F\left[\frac{1}{2}\,(z + \bar{z}),\, \frac{1}{2i}\,(z - \bar{z})\right] = 0.$$

In this section we shall explore this possibility.

We start with straight lines. The equation

$$(2.2.2) \qquad\qquad \arg\,(z - z_0) = \theta$$

defines a *ray* from the point $z_0$ making the angle $\theta$ with the positive real axis. The same set of points may also be represented parametrically by

$$(2.2.3) \qquad\qquad z = z_0 + \alpha(\cos\theta + i\sin\theta), \quad 0 \leqq \alpha < \infty.$$

If we allow $\alpha$ to take on all real values, we get the whole straight line. If $z_1 \neq z_0$ is a point on the straight line, we can also write the equation of the line

$$(2.2.4) \qquad\qquad z = z_0 + \beta(z_1 - z_0), \quad -\infty < \beta < \infty.$$

If $0 \leq \beta \leq 1$, we get the line segment from $z_0$ to $z_1$, values of $\beta > 1$ give the infinite segment beyond $z_1$, and values of $\beta < 0$ give the other infinite segment. Since $\beta$ is real, the equation may also be written

$$(2.2.5) \qquad\qquad \Im\left(\frac{z - z_0}{z_1 - z_0}\right) = 0.$$

This is the complex form of the familiar two-point formula for the equation of the straight line.

The normal form of the equation of the straight line involves the length and direction of the perpendicular from the origin to the line. If the foot point of the perpendicular is $z_0 \neq 0$, the equation may be written

$$(2.2.6) \qquad\qquad \Re\left(\frac{z}{z_0}\right) = 1.$$

This straight line divides the complex plane into the two half-planes

$$(2.2.7) \qquad \Re\left(\frac{z}{z_0}\right) > 1 \quad \text{and} \quad \Re\left(\frac{z}{z_0}\right) < 1,$$

the second of which contains the origin. Similarly for (2.2.5) the two half-planes are

$$(2.2.8) \qquad \Im\left(\frac{z - z_0}{z_1 - z_0}\right) > 0 \quad \text{and} \quad \Im\left(\frac{z - z_0}{z_1 - z_0}\right) < 0.$$

If we travel along the line from $z_0$ to $z_1$, we have the first half-plane on the left, the second on the right. These are open half-planes; if the boundary is to be included, $<$ is replaced by $\leq$.

The equation

$$(2.2.9) \qquad |z - z_0| = r$$

obviously defines a circle with center at $z_0$ and radius $r$. The interior of the circle is defined by

$$|z - z_0| < r,$$

the exterior plus boundary by

$$|z - z_0| \geq r.$$

The same circle is given parametrically by

$$(2.2.10) \qquad z = z_0 + r(\cos \alpha + i \sin \alpha), \quad 0 \leq \alpha < 2\pi.$$

Various locus problems in elementary geometry lead to circles. Thus we see that the equation

$$(2.2.11) \qquad |z - z_1| = C |z - z_2|, \quad C > 0$$

defines a circle, unless $C = 1$ when it represents the perpendicular bisector of the line segment from $z_1$ to $z_2$. If here we replace $=$ by $<$ we get the domain bounded by the circle which contains the point $z_1$. This is the interior of the circle when $C < 1$, otherwise the exterior. Representing the familiar general equation of a circle in the form (2.2.1), we see that

$$(2.2.12) \qquad |z|^2 - 2\Re(\bar{z}_0 z) + C = 0, \quad |z_0|^2 > C$$

represents a circle with center at $z_0$ and radius $(|z_0|^2 - C)^{\frac{1}{2}}$.

From the definitions of the conic sections we are able to identify

$$(2.2.13) \qquad |z - z_1| + |z - z_2| = 2a$$

as an ellipse and

$$|z - z_1| - |z - z_2| = 2a$$

as a hyperbola.

Curves given by an equation of the form

(2.2.14) $$|z - z_1| |z - z_2| \cdots |z - z_n| = C, \quad C > 0$$

are known as *general lemniscates*. On such a curve the absolute value of the polynomial

$$P(z) = (z - z_1)(z - z_2) \cdots (z - z_n)$$

is constant. For small values of $C$ the curve consists of $n$ separate small ovals, one around each root if they are distinct; for large values of $C$ there is a single large oval surrounding all the roots. As $C$ increases, the small ovals expand; they coalesce two or more at a time until a single oval is formed. The latter keeps on expanding, inward bends are straightened out, and for large values of $C$ the oval looks very much like a circle.

EXERCISE 2.2

**1.** Plot the straight lines

    **a.** $\Re(z) = 1.$

    **b.** $\Im\left(\dfrac{z - 1}{2i}\right) = 0.$

    **c.** $|z + i| = |z - i|.$

    **d.** $z = 1 + i\alpha, \quad -\infty < \alpha < \infty.$

**2.** Plot the circles

    **a.** $|z| = 1.$

    **b.** $|z - 1| = 1.$

    **c.** $|z - 1| = 2|z - 2|.$

    **d.** $|z|^2 = 2\Re(z).$

**3.** Describe the regions defined by

    **a.** $|z| < 1.$

    **b.** $|z + i| \geq |z - i|.$

    **c.** $|z - 1| < 2|z - 2|.$

    **d.** $|z - 2| \geq 2|z - 1|.$

**4.** Give in complex form the equations of the circles through $z = i$ and $z = -i$.

**5.** Find the equations of the orthogonal trajectories of the circles in Problem 4.

**6.** If $|z_0| < 1$, describe the region $\left|\dfrac{z - z_0}{1 - \bar{z}_0 z}\right| \leq 1.$

**7.** What conditions must $z_1$, $z_2$, and $a$ satisfy in equation (2.2.13) in order that a real ellipse result?

**8.** Find the curves **(a)** $\Re(z^2) = C$; **(b)** $\Im(z^2) = C.$

**9.** Discuss the curves **(a)** $|z^2 - 1| = C$; **(b)** $|z^3 - 1| = C.$ For what value or values of $C$ does coalescence occur? Plot carefully the corresponding curves and determine the tangents at the nodes. What property of the polynomial determines the position of the nodes?

**2.3. Regions and convexity.** We start with a number of definitions. All the sets mentioned are supposed to be subsets of the complex plane, though similar conventions hold in any Euclidean space.

DEFINITION 2.3.1.     (1) *A point $z_0$ of the set $S$ is said to be an interior point of $S$ if some $\varepsilon$-neighborhood of $z_0$ also belongs to $S$. The subset of interior points of $S$ is called the interior of $S$ and is denoted by* Int $(S)$.

(2) *A set $G$ is open if $G =$ Int $(G)$.*

(3) *A set $S$ is arcwise connected if each pair of points $z_1$ and $z_2$ of $S$ may be joined by a polygonal line, all the points of which are in $S$.*

(4) *A domain is an open, arcwise connected set.*

(5) *The complement of $S$, $\mathbf{C}(S)$, is the set of all points in the plane not in $S$.*

(6) *The closure of $S$ is the union of $S$ and its derived set $S'$:* $\quad \bar{S} = S \cup S'$.

(7) *A set $S$ is said to be compact if every infinite subset of $S$ has at least one limit point which belongs to $S$.*

(8) *The frontier, or boundary, of $S$ is the set*

$$\partial(S) = \bar{S} \cap \overline{\mathbf{C}(S)}.$$

(9) *A region is an open set plus a subset (void, proper, or improper) of its boundary.*

The reader is warned that many authors use the term "region" for what we call a domain; others make no distinction between the two terms. It seems to be common practice to speak of the region of convergence of a power series. This is a region in our terminology, but it need not be a domain.

Condition (3) above is to be understood as follows: For the given pair $z_1$, $z_2$ it is possible to find a finite number of points $a_1$, $a_2$, $\cdots$, $a_n$ such that the line segments $[z_1, a_1]$, $[a_1, a_2]$, $\cdots$, $[a_n, z_2]$ are made up of points belonging to $S$. The union of these line segments is the polygonal line. We shall often use the symbol $[z_1, a_1, a_2, \cdots, a_n, z_2]$ for such a line.

Again the reader is warned that there are less restrictive definitions of connectedness than the arcwise variety. A set $S$ is said to be *connected* if it does not allow any decomposition $S = S_1 \cup S_2$, with $(\bar{S}_1 \cap S_2) \cup (S_1 \cap \bar{S}_2) = \emptyset$. In particular, a closed set which is not the union of two disjoint non-empty closed sets is called a *continuum*. The degenerate case of a single point is usually excluded.

As a drill in terminology let us consider the set

(2.3.1) $$S: \quad [z \mid |z^2 - 1| \le 1, z \neq 0].$$

This is the region of convergence of the series

$$\sum_{n=1}^{\infty} \frac{1}{n} (1 - z^2)^n,$$

which is a power series in $1 - z^2$. $S$ is a region; it is not connected, much less arcwise connected. Its interior, the set $| z^2 - 1 | < 1$, is open but not arcwise connected, so it is not a domain. Each of the two components of Int $(S)$, the one in the left half-plane and the one in the right, is a domain. The boundary of $S$ is the lemniscate $| z^2 - 1 | = 1$, all the points of which belong to $S$ except $z = 0$. $S$ is not closed; its closure is obtained by adding the one missing point $z = 0$.

DEFINITION 2.3.2.    *A set $R$ is convex if whenever $z_1$ and $z_2$ are two points of $R$, then the points*

$$[z \mid z = z_1 + \alpha(z_2 - z_1),   0 \leqq \alpha \leqq 1]$$

*also belong to $R$.*

It is obvious that a convex set must be arcwise connected, but it need not be a domain. The interior[1] of a circle or an ellipse is a convex domain. The subset of the set $S$ of (2.3.1) which lies in the left (right) half-plane is a convex region.

A more elementary example of a convex domain is the interior of a triangle. If the vertices are $z_1$, $z_2$, $z_3$, the interior $\triangle_i$ is the point set

$$(2.3.2) \quad \triangle_i = [z \mid z = \alpha_1 z_1 + \alpha_2 z_2 + \alpha_3 z_3; \\ \alpha_1 > 0, \alpha_2 > 0, \alpha_3 > 0;   \alpha_1 + \alpha_2 + \alpha_3 = 1].$$

Intuitively this is obvious: *z is the center of mass of the system of three particles of masses $\alpha_1$, $\alpha_2$, $\alpha_3$ at $z_1$, $z_2$, $z_3$ respectively.* In particular, the *center of gravity* of the triangle corresponds to the case when all the $\alpha$'s are equal to 1/3.

A triangle is the simplest example of a closed polygon. If the $n$ vertices are properly numbered, the symbol $\Pi = [z_1, z_2, \cdots, z_n, z_1]$, in our previous notation for a polygonal line, defines a closed polygon. The same polygon is of course also defined by $[z_2, z_3, \cdots, z_n, z_1, z_2]$, etc. The same set of points is defined by $[z_1, z_n, z_{n-1}, \cdots, z_2, z_1]$, but this symbol indicates that the polygon is described in the opposite sense, and we shall normally distinguish between the two possible orientations.

We speak of a *simple closed polygon* if $\Pi$ does not intersect itself. More precisely expressed, if $\Pi$ is given by the equation

$$(2.3.3) \qquad z = z(t),   0 \leq t \leq 1,   z(0) = z(1),$$

and $t_1 < t_2$, then $z(t_1) = z(t_2)$ shall imply $t_1 = 0$, $t_2 = 1$.

A simple closed polygon may be *convex*. This property can be defined in two

---

[1] This is the notion of "interior" in the sense of the Jordan curve theorem (Theorem 2.4.1 and following) and not in the sense of Definition 2.3.1 (1). The same remark applies to the interpretation of the term "interior of a triangle" or "interior of a closed polygon," used below.

different ways. We may call $\Pi$ *a convex polygon if* $\Pi$ *is the boundary of a convex domain*. A more constructive definition is the following:

DEFINITION 2.3.3.    $\Pi = [z_1, z_2, \cdots, z_n, z_1]$ *is a convex polygon if for each* $k$, $k = 1, 2, \cdots, n$, *the straight line* $l_k$ *passing through the vertices* $z_k$ *and* $z_{k+1}$ $(z_{n+1} = z_1)$ *separates the plane into two open half-planes one of which contains no point of* $\Pi$.

Suppose that $H_k$ is the half-plane determined by $l_k$ which contains a part of $\Pi$, and form the intersection $\Pi_i$ of these open half-planes. This is a convex domain, for if $Z_1$ and $Z_2$ lie in $H_k$ so does $\gamma Z_1 + (1 - \gamma)Z_2$ for every $\gamma$ with $0 < \gamma < 1$. Hence, if $Z_1$ and $Z_2$ belong to the intersection, $\gamma Z_1 + (1 - \gamma)Z_2$ does too. Each line segment $[z_k, z_{k+1}]$ clearly belongs to the boundary of $\Pi_i$, and $\partial(\Pi_i)$ is the union of these line segments, that is, $\partial(\Pi_i) = \Pi$, in agreement with our first-proposed definition. We refer to $\Pi_i$ as the *interior* of the convex polygon $\Pi$. We have

$$(2.3.4) \qquad \begin{aligned} \Pi_i &\equiv [z \mid z = \alpha_1 z_1 + \alpha_2 z_2 + \cdots + \alpha_n z_n; \\ &\quad \alpha_1 > 0, \cdots, \alpha_n > 0; \quad \alpha_1 + \cdots + \alpha_n = 1]. \end{aligned}$$

Again the physical interpretation is obvious: *z is the center of mass of n particles of masses* $\alpha_1, \alpha_2, \cdots, \alpha_n$ *at* $z_1, z_2, \cdots, z_n$ *respectively*. In particular, the *center of gravity of the vertices of the polygon* is obtained if all the $\alpha$'s are equal to $1/n$.[1]

The idea of using the intersection of half-planes can evidently be exploited in more general situations. Such considerations lead to the notion of the *function of support of a closed convex set*. This concept was first introduced by Hermann Minkowski[2] (1864–1909). To him we owe the recognition of the importance of convex geometry as well as the discovery and development of the basic facts of this discipline.

Let $C$ be a closed, bounded convex region in the complex plane and define the following real-valued function of the complex variable $w$:

$$(2.3.5) \qquad F(w) = \sup [\Re(\bar{z}w) \mid z \in C].$$

---

[1] The rest of this section may be omitted in a first reading. Knowledge of it is not required until Volume II.

[2] At the age of 18 Minkowski won a prize from the Academy of Sciences in Paris for his work on quadratic forms. His interest in this field led him in the early 1890's to create a *geometry of numbers* based on the properties of convex bodies in $n$ dimensions. He introduced the abstract notion of distance and analyzed the concepts of surface and volume. In this connection he also established a number of important inequalities which carry his name. His later work threw much light on the geometry of special relativity. Minkowski was a product of the brilliant Königsberg school; he was professor at Göttingen until his untimely death.

For the given value of $w$ we plot the straight line

(2.3.6) $$L_w: \quad \Re(\bar{z}w) = F(w),$$

or in real form

$$ux + vy = F(w), \quad w = u + iv, \quad z = x + iy.$$

By the definition of $F(w)$, we have

$$ux_1 + vy_1 \leqq F(w)$$

for every point $z_1 = x_1 + iy_1$ of $C$, that is, $C$ lies to one side of the line $L_w$. Since $ux_1 + vy_1$ is a bounded continuous function of $(x_1, y_1)$ in the closed set $C$, this function attains its maximum value $F(w)$, that is, there exists at least one point $z_0$ of $C$ which lies on the line $L_w$, and this point evidently lies on the boundary $\partial(C)$ of $C$. For this reason, $L_w$ is known as a *line of support* of $C$ at $z_0$. There may be infinitely many lines of support at $z_0$. If $\partial(C)$ is a convex polygon, for instance, then there are infinitely many lines of support at each of the vertices.

The function $F(w)$ is known as the *function of support* of $C$ though often this name is reserved for the restriction of $F(w)$ to the unit circle of the $w$-plane. We shall write

(2.3.7) $$\sigma(\varphi) = F(w), \quad l_\varphi = L_w \quad \text{for} \quad w = \cos \varphi + i \sin \varphi,$$

so that $\sigma(\varphi)$ is the distance of $l_\varphi$ from the origin. To $l_\varphi$ corresponds a closed half-plane $H_\varphi$ bounded by $l_\varphi$ and containing $C$. We have then that $C$ is the intersection of all these half-planes $H_\varphi$ with $0 \leqq \varphi < 2\pi$, or

(2.3.8) $$C = \bigcap_\varphi H_\varphi.$$

The function $\sigma(\varphi)$ is continuous and periodic with period $2\pi$. Its difference quotient has right- and left-hand limits everywhere, but $\sigma(\varphi)$ normally does not have a continuous derivative or even an everywhere-existing derivative.

## EXERCISE 2.3

**1.** Suppose $z_2 - z_1$ and $z_3 - z_1$ are complex numbers linearly independent over the real field. If $z$ is any complex number, show that

$$z - z_1 = \beta(z_2 - z_1) + \gamma(z_3 - z_1),$$

where $\beta$ and $\gamma$ are uniquely determined real numbers which are continuous functions of $z$. Hence show that there exist three real numbers $\alpha_1, \alpha_2, \alpha_3$ of sum unity such that

$$z = \alpha_1 z_1 + \alpha_2 z_2 + \alpha_3 z_3,$$

where the $\alpha$'s are uniquely determined and continuous in $z$. The $\alpha$'s are known as *barycentric coordinates* of $z$.[1]

**2.** With $z_1$, $z_2$, $z_3$ as in Problem 1 and $\triangle_i$ defined by (2.3.2), prove analytically that $\triangle_i$ is a convex domain.

**3.** If $\triangle_e$ is defined as the interior (in the sense of Definition 2.3.1 (1)) of the complement of $\triangle_i$, $\triangle_e = \text{Int } [\mathbf{C}(\triangle_i)]$, represent $\triangle_e$ as the union of (overlapping) half-planes, and verify that any two points $Z_1$ and $Z_2$ in $\triangle_e$ may be joined in $\triangle_e$ by a polygonal line $[Z_1, Z_3, Z_2]$ for appropriate choice of $Z_3$.

**\*4.** Using continuity of the coordinates, prove that any polygonal line joining a point $Z_1$ in $\triangle_i$ with a point $Z_2$ in $\triangle_e$ must have at least one point in common with $\triangle = \partial\triangle_i = \partial\triangle_e$.

**5.** Verify that the function $F(w)$ of (2.3.5) is *positive-homogeneous*, that is, $F(\alpha w) = \alpha F(w)$ if $\alpha > 0$.

**6.** Verify that $F(w)$ is *subadditive*, that is, $F(w_1 + w_2) \leq F(w_1) + F(w_2)$.

**7.** What relation holds between the functions of support $\sigma(\varphi)$ of two convex regions $C_1$ and $C_2$ if $C_2 = [z + a \mid z \in C_1]$, $a$ being a fixed complex number?

**8.** Find the function $\sigma(\varphi)$ for the regions (**a**) $\mid z \mid \leq 1$; (**b**) $\mid x \mid + \mid y \mid \leq 1$.

**2.4. Paths.** To the analyst the complex plane is not merely a vast expanse, it is also the site of many obstacles among which he has to navigate with circumspection. He is always on the move from one point to another and sometimes back to the starting point. The path that he follows normally has to satisfy restrictions on location, direction, intersections, and length imposed by the problem under consideration. When mathematicians started to become aware of such questions during the last quarter of the nineteenth century, they soon found that the underlying basic concepts had to be re-examined, redefined, or, in most cases, created.

The concepts of a *continuous arc* and a *continuous closed curve* were among the troublemakers. A line segment and a circle are the prototypes of these notions, and any continuous image of these sets would seem to deserve the name of continuous curve. In other words, given any complex-valued function $z(t)$, defined and continuous for $0 \leq t \leq 1$, then the set of points

$$(2.4.1) \qquad\qquad z = z(t), \quad 0 \leq t \leq 1,$$

should be called a continuous arc if $z(0) \neq z(1)$, otherwise a continuous closed

---

[1] Introduced in 1827 by August Ferdinand Möbius (1790–1868), a pupil of Gauss and astronomer in Leipzig from 1815 on. He contributed to celestial mechanics, but his main work was in pure mathematics: number theory, combinatorics, inversion formulas, the barycentric calculus, geometric transformations, non-orientable manifolds (the Möbius strip!).

curve. This definition turns out to be far too wide. It does include all "respectable" bounded curves which one would want to call "curves." In addition, however, many geometrical objects that do not correspond to our intuitive idea of a curve satisfy the conditions of the definition. Thus, in 1890 the Italian mathematician and logician Giuseppe Peano (1858–1932) produced a space-filling curve, and many constructions of such "Peano curves" have appeared since. Thus, there exists a continuous function $z(t)$ such that every complex number $z_0$ with $0 \leq x_0 \leq 1$, $0 \leq y_0 \leq 1$, is taken on by $z(t)$ at least once in the interval $[0, 1]$. This is a continuous mapping; it is not one-to-one, but $z(t)$ may be chosen so that no value corresponds to more than three values of $t$. In the opposite direction Georg Cantor[1] (1845–1918) proved that a line segment can be mapped in a one-to-one but highly discontinuous manner onto a square. No one-to-one and continuous mapping of a line segment onto a square can exist; this is a consequence of the *invariance of dimensionality* proved by L. E. J. Brouwer (1881—).

As far back as 1887, Camille Jordan[2] (1838–1922) had indicated a way out of the difficulties by excluding curves having multiple points.

DEFINITION 2.4.1.    *Equation* (2.4.1) *defines a simple arc, if $z(t)$ is a continuous function such that $t_1 \neq t_2$ implies $z(t_1) \neq z(t_2)$. It is a simple closed curve if $z(0) = z(1)$ and $t_1 < t_2$, $z(t_1) = z(t_2)$ implies $t_1 = 0$, $t_2 = 1$.*

Simple arcs and simple closed curves are often known as *Jordan arcs and Jordan curves* respectively. *A simple closed curve is the homeomorphic image of a circle* in the terminology of the topologists, that is, the correspondence is one-to-one and continuous in both directions. A circle separates the plane, and this property carries over to Jordan curves.

THEOREM 2.4.1.    *If $J$ is a simple closed curve* (in the complex plane), *the complement of $J$ is the union of exactly two mutually exclusive domains, and every point of $J$ is a boundary point of each domain.*

---

[1] Born in what was then called St. Petersburg, Cantor came to Germany as a boy, studied in Berlin under Weierstrass, and was attached to the University of Halle after 1869. He proposed in 1872 a theory of irrational numbers based on nested intervals. Incidentally, in the same year there also appeared two accounts of the theory of Weierstrass as well as the basic paper of Dedekind and a publication by Charles Méray on real numbers. Cantor's early work on the convergence of trigonometric series led him to the epoch-making creation of the theory of sets: first, point sets; later, abstract sets, transfinite cardinals and ordinals. After much resistance, Cantor's daring ideas broke through in the 1880's and led to the foundation crisis in mathematics and logic which is still with us.

[2] Professor at the École Polytechnique in Paris, during the nineteenth century the training center of all French mathematicians. His work on the theory of substitutions is basic (the Jordan normal form of matrices, the Jordan–Hölder theorem in group theory). His discussions of simple curves, length, and functions of bounded variation were included in his *Cours d'Analyse*, a treatise widely read by the mathematicians of his time.

This is known as the Jordan curve theorem. The argument given by Jordan was insufficient, and the first valid proof, published in 1905, is due to the American topologist Oswald Veblen (1880–1960). We shall not attempt to give a proof here. The reader will find a discussion of the special case of a closed simple polygon in Appendix B.

One of the two domains determined by $J$ is bounded; it is known as the *interior of J*. The other, the unbounded domain, is called the *exterior*. Every point of $J$ is accessible both from the interior and from the exterior in the sense that the point is limit point of the vertices of a polygonal line $[z_1, z_2, \cdots, z_n, \cdots]$ located in the domain in question.

Most of the geometric properties of a circle are lost under a homeomorphic mapping. Thus a Jordan curve need not have a tangent at any point. This perhaps is not so surprising: a continuous function $z(t)$ need not have a derivative for any value of $t$, as first shown by Weierstrass in 1880, and a simple closed polygon lacks a unique tangent at each of its $n$ vertices. As $n$ increases, the number of such points becomes infinite. This observation has been used by a number of writers for the effective construction of a Jordan curve without tangent. The following procedure, due to the German analyst Konrad Knopp (1882–1957), is particularly elegant. We start with an isosceles triangle with two angles of 36° and one of 108°. We trisect the obtuse angle and discard the middle third of the original triangle. What is left is two congruent triangles similar to the original one and having one vertex in common. Each of these triangles is treated in the same manner: after each trisection we discard the middle third. This process is continued indefinitely. The vertices of the resulting triangles and their limit points form a Jordan arc $K$. Each point $P$ of $K$ lies in a nested sequence of triangles $(A_n, B_n, C_n)$ all similar to one another and to the original triangle. Let us draw secants from $P$ to $A_n$, $B_n$, and $C_n$. A simple argument shows that these secants make angles with each other of which at least one lies between 36° and 108° for each $n$. Thus the secants cannot tend to any limit, and $K$ has no tangent at $P$, that is, no tangent at any point.

This is also an example of a Jordan arc such that the length between any two points is infinite. Intuitively this is fairly obvious, for if $P$ and $Q$ are two adjacent vertices at the $n$th stage of the construction, then a new vertex $R$ is inserted between them at the next stage and $l[P, R, Q] = l[P, Q]$ sec 36°.

But before we can talk about the length of a curve, we really have to define this concept. Our starting point naturally is the simplest of all arcs, namely, the line segment $[z_1, z_2]$ of length $|z_1 - z_2|$. We can join line segments, obtaining a polygonal line, and the length of $[z_1, z_2, \cdots, z_n]$ has the value

$$|z_1 - z_2| + |z_2 - z_3| + \cdots + |z_n - z_{n-1}|.$$

Next, we take a simple arc given parametrically by

$$\Gamma: \quad z = z(t),$$

where $z(t)$ is a continuous function in the interval $[0, 1]$ and $t_1 \neq t_2$ implies $z(t_1) \neq z(t_2)$. We use the parametrization to establish an orientation of $\Gamma$ by saying that the point $z(t_1)$ precedes the point $z(t_2)$ if and only if $t_1 < t_2$. Suppose now that $z_0, z_1, \cdots, z_n$ are points on $\Gamma$ where the enumeration is such that $z_{k-1}$ precedes $z_k$ in the prescribed orientation. Thus the polygonal line $\Pi: \ [z_0, z_1, \cdots, z_n]$ is "inscribed" in $\Gamma$ and its orientation agrees with that of $\Gamma$ insofar as their common points are concerned. Each such inscribed polygonal line $\Pi$ has a length $l(\Pi)$. We shall say that $\Gamma$ *has finite length* or $\Gamma$ *is a rectifiable curve* provided the set $\{l(\Pi)\}$ is bounded and the least upper bound of this set is defined as the length of $\Gamma$. In analytical formulation:

DEFINITION 2.4.2. *Let* $0 = t_0 < t_1 < \cdots < t_n = 1$ *be any partition of the interval* $[0, 1]$. *To this partition* $\pi$ *corresponds the sum*

$$(2.4.2) \qquad\qquad S_\pi = \sum_{k=1}^{n} |\, z(t_k) - z(t_{k-1})\,|.$$

*The curve* $\Gamma: \ z = z(t)$ *is said to have a finite length, or to be rectifiable, if the set of all such sums* $\{S_\pi\}$ *is bounded, and the length of* $\Gamma$ *is by definition*

$$(2.4.3) \qquad\qquad l(\Gamma) = \sup S_\pi.$$

This formulation leads directly to the analytic characterization of rectifiable curves. The sums $S_\pi$ are bounded if and only if $z(t)$ is a *function of bounded variation* in the sense of Jordan, and the supremum of $S_\pi$ is the *total variation* of $z(t)$ *in the interval* $[0, 1]$. For these notions the reader is referred to Appendix C. We have thus

THEOREM 2.4.2. *The simple arc* $\Gamma$ *defined by* $z = z(t)$ *is rectifiable if and only if* $z(t)$ *is a function of bounded variation. In this case the length of* $\Gamma$ *is the total variation of* $z(t)$, *or*

$$(2.4.4) \qquad\qquad l(\Gamma) = \mathrm{Var}\,[z(t) \mid 0 \leq t \leq 1] = \int_0^1 |\, dz(t)\,|.$$

In particular cases this formula takes on the more familiar form

$$(2.4.5) \qquad\qquad l(\Gamma) = \int_0^1 |\, z'(t)\,|\, dt = \int_0^1 \{[x'(t)]^2 + [y'(t)]^2\}^{\frac{1}{2}}\, dt$$

given in elementary calculus. The most important case is that in which $z'(t)$ is piecewise continuous; that is, the interval $[0, 1]$ breaks up into a finite number of subintervals $[\alpha_0, \alpha_1], [\alpha_1, \alpha_2], \cdots, [\alpha_{m-1}, \alpha_m]$, and $z'(t)$ is continuous in each of them. This implies that at the common endpoints $z'(t)$ has right- and left-hand limits; these may be distinct, however.

THEOREM 2.4.3. *If* $z'(t)$ *is piecewise continuous,*

$$(2.4.6) \qquad\qquad l(\Gamma) = \sum_{k=1}^{m} \int_{\alpha_{k-1}}^{\alpha_k} |\, z'(t)\,|\, dt,$$

*where each integral is a proper Riemann integral.*

*Proof.* Since the argument is the same for each subinterval, we may just as well assume that there is only one so that $z'(t)$ is continuous in $[0, 1]$. The reader will be familiar with the fact that a real-valued function is uniformly continuous in any bounded closed interval where it is pointwise continuous. Thus, $\Re[z'(t)]$ and $\Im[z'(t)]$ have this property, and consequently also $z'(t)$ itself. Hence, if $\varepsilon > 0$ is given, there exists a sufficiently fine partition $\pi$ of $[0, 1]$ into subintervals $[t_{k-1}, t_k]$ such that for every $k$

$$| z'(t) - z'(t_k) | < \varepsilon, \quad t_{k-1} \leqq t \leqq t_k,$$

or

$$z'(t) = z'(t_k) + \eta_k(t), \quad | \eta_k(t) | < \varepsilon.$$

Hence

$$z(t_k) - z(t_{k-1}) = \int_{t_{k-1}}^{t_k} z'(t)\, dt = z'(t_k)(t_k - t_{k-1}) + \int_{t_{k-1}}^{t_k} \eta_k(t)\, dt,$$

and

$$\left| \sum | z(t_k) - z(t_{k-1}) | - \sum | z'(t_k) | (t_k - t_{k-1}) \right|$$

$$\leqq \sum \int_{t_{k-1}}^{t_k} | \eta_k(t) |\, dt \leqq \varepsilon \sum (t_k - t_{k-1}) = \varepsilon.$$

But for a sufficiently fine partition $\pi$ the two sums on the left differ from sup $S_\pi$ and $\int_0^1 | z'(t) |\, dt$ respectively by amounts less than $\varepsilon$, so that

$$\left| l(\Gamma) - \int_0^1 | z'(t) |\, dt \right| < 3\varepsilon.$$

Since $\varepsilon$ is arbitrary, formula (2.4.6) follows.[1]

We have defined the length of the curve $\Gamma$ in terms of a particular representation $z = z(t)$. Now it is clear that the same curve $\Gamma$ admits of infinitely many parametric representations. Thus if $\alpha$ is any positive number, all the functions

(2.4.7) $\qquad\qquad z(t^\alpha), \quad 0 \leqq t \leqq 1, \quad 0 < \alpha < \infty,$

represent the same curve $\Gamma$ with the same orientation. In this case it is an easy matter to prove that if one of these functions is of bounded variation so are the rest, and that they all have the same total variation. Incidentally, the functions $z(1 - t^\alpha)$ also represent $\Gamma$, but now with opposite orientation. It is clear that all these representations lead to the same notion of rectifiability and to the same numerical value for the length. It may be shown, but we shall not

---

[1] It was proved by Henri Lebesgue (1875–1941), the father of the modern theory of integration, that a function of bounded variation has a derivative almost everywhere. The integral in (2.4.5) exists always in the sense of Lebesgue; its value is normally less than $l(\Gamma)$, and it equals $l(\Gamma)$ if and only if $z(t)$ is absolutely continuous.

take time to do it, that all representations of a rectifiable curve $\Gamma$, given in the first instance by $z = z(t)$, are of the form

(2.4.8) $$z = z[f(t)] \quad \text{or} \quad z = z[1 - f(t)],$$

where $f(t)$ is a continuous strictly increasing function of $t$ with $f(0) = 0$, $f(1) = 1$. A particularly advantageous representation of a rectifiable curve is in terms of the arc length as parameter so that

$$z = Z(s), \quad 0 \leq s \leq l(\Gamma).$$

Then $Z(s)$ is an absolutely continuous function of $s$, and

$$| Z'(s) | = 1$$

for almost all $s$.

## EXERCISE 2.4

**1.** A curve is defined by $z = a \cos \theta + ib \sin \theta$, $-\pi < \theta \leq \pi$, where $a$ and $b$ are fixed positive numbers. Show that the curve is rectifiable. What is the curve?

**2.** Same questions for $z = \dfrac{1 - t^2 + 2it}{1 + t^2}$, $-\infty < t < \infty$.

**3.** Show that
$$z = t[\cos (1/t) + i \sin (1/t)], \quad 0 < t \leq 1,$$

with $z(0) = 0$, defines a Jordan arc. Does the arc have finite length?

**4.** Verify that equation (2.4.7) represents the same curve for all $\alpha > 0$ and that the functions $z(t^\alpha)$ have the same total variation.

**5.** Same question for (2.4.8) and $z[f(t)]$.

**\*6.** The equations $z = z(t), z = Z(s), 0 \leq s, t \leq 1$, are supposed to represent the same rectifiable curve with the same orientation. A function $s = f(t)$ is defined implicitly by $z(t) = Z(s)$. Show that $f(t)$ is continuous and strictly increasing, and that $f(0) = 0, f(1) = 1$.

## 2.5. The extended plane; stereographic projection.
Mathematicians like to make general statements and do not believe that exceptions prove the rule. The assertion that a quadratic equation has two roots is not true if we restrict ourselves to real numbers. This was a powerful reason for introducing the complex numbers; in the enlarged number field the statement is true. "Two straight lines intersect in a point" does not hold for parallel lines; by introducing as ideal element a line at infinity in projective geometry, the exceptional role of the parallel lines is eliminated. Similar ideal elements are introduced in algebraic number theory in order to restore unique factorization.

That a similar situation holds in complex functions theory is shown by the elementary transformation

$$(2.5.1) \qquad w = \frac{1}{z}.$$

This defines a one-to-one mapping of the $z$-plane onto the $w$-plane with two notable exceptions: the point $z = 0$ has no image and the point $w = 0$ has no pre-image. In order to remove this anomaly we form the *extended complex plane* by adding to the ordinary plane an ideal element, called the *point at infinity* and denoted by $\infty$. The transformation (2.5.1) maps the extended plane onto itself without exceptions: $z = 0$ goes into $w = \infty$, and $w = 0$ is the image of $z = \infty$. A neighborhood of $z = \infty$ is defined by $|z| > R$. It is mapped onto a neighborhood $|w| < 1/R$ of the origin by (2.5.1).

We introduced the complex plane as the geometric representation of complex numbers. For the extended set, complex numbers and infinity, we can find a geometric representation which is easier to visualize than the plane with an adjoined ideal point. We can use a sphere for this purpose or, for that matter, any surface topologically equivalent to the sphere. It is customary to set up the correspondence with the aid of *stereographic projection*. There are two commonly used alternatives, according as the complex plane is tangent to the sphere or passes through its center. We shall use the first alternative.

In a three-dimensional Euclidean space with coordinates $(x, y, u)$ we identify the $(x, y)$-plane with the complex plane and consider the sphere

$$(2.5.2) \qquad x^2 + y^2 + u^2 = u.$$

The $(x, y)$-plane is tangent to the sphere at the origin. It is common practice to refer to the points $(0, 0, 0)$ and $(0, 0, 1)$ as the *south pole* and the *north pole* of the sphere respectively; the great circle in the plane $u = \frac{1}{2}$ is called the *equator*.

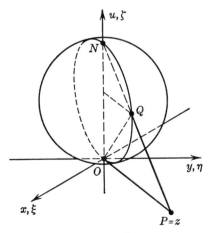

**Figure 5**

The north pole is used as the center of projection. Through the points $N = (0, 0, 1)$ and $P = (x, y, 0)$ we draw a straight line and note its second intersection with the sphere. Let this be the point $Q = (\xi, \eta, \zeta)$. Then $(\xi, \eta, \zeta)$ is called the stereographic projection, or image, of $(x, y, 0)$ on the sphere and is taken as the spherical representation of the complex number $z = x + iy$. This procedure assigns a unique point $Z = Q$ on the sphere to every given complex number $z$. Conversely, every point on the sphere with the exception of the north pole corresponds to a unique complex number. We complete the mapping by assigning the north pole and the point at infinity to each other. There is then a one-to-one correspondence between the sphere on one hand and the extended complex plane on the other.

Consideration of similar triangles in Figure 5 shows that

$$(2.5.3) \qquad \xi = \frac{x}{1 + r^2}, \quad \eta = \frac{y}{1 + r^2}, \quad \zeta = \frac{r^2}{1 + r^2}, \quad r^2 = x^2 + y^2,$$

and conversely

$$(2.5.4) \qquad x = \frac{\xi}{1 - \zeta}, \quad y = \frac{\eta}{1 - \zeta}, \quad r^2 = \frac{\zeta}{1 - \zeta}.$$

Stereographic projection maps the unit circle onto the equator of the sphere, the interior of the unit circle onto the southern hemisphere, and the exterior onto the northern hemisphere. Reflection in the unit circle corresponds to reflection in the equator. We list some further properties of this mapping as theorems.

THEOREM 2.5.1.    *Under stereographic projection the circles of the sphere go into the circles and straight lines of the plane and vice versa.*

*Proof.*    A circle on the sphere is the intersection of the sphere with a plane:

$$(2.5.5) \qquad Ax + By + Cu = D, \quad \text{with} \quad A^2 + B^2 > 4D(D - C)$$

to ensure actual intersection. Thus, if the point $(\xi, \eta, \zeta)$ describes the circle, its coordinates must satisfy the equation of the plane. Using formulas (2.5.3) we see that the corresponding points $(x, y, 0)$ must satisfy the equation

$$(2.5.6) \qquad (C - D)(x^2 + y^2) + Ax + By = D, \quad u = 0.$$

This is a real circle in the $(x, y)$-plane unless $C = D$, when it is a straight line. We note that in the latter case the plane (2.5.5) and hence also the circle on the sphere pass through the north pole. To prove the converse proposition, we start with equation (2.5.6), with $A^2 + B^2 > 4D(D - C)$, and retrace our steps. Equation (2.5.6) represents all lines and circles in the $(x, y)$-plane. Using formulas (2.5.4) to express $x$ and $y$ in terms of $(\xi, \eta, \zeta)$, we see that the latter point on the sphere must lie in the plane (2.5.5). Thus, the stereographic projection on the sphere is always a circle, and this circle goes through the north pole if the pre-image is a straight line.

THEOREM 2.5.2. *The stereographic projection is an isogonal transformation, that is, it preserves angles.*

*Proof.* The statement implies that if two curves in the $(x, y)$-plane intersect and their tangents at the point of intersection form an angle $\alpha$ with each other, then the tangents of the stereographic images at their point of intersection form the same angle $\alpha$. Strictly speaking we should first prove that a curve's property of having a tangent is preserved under stereographic projection. This we shall assume, however, and simplifying still further, we shall prove the theorem only for the case of straight lines. Actually the last step involves no restriction of the generality. Now two straight lines in the $(x, y)$-plane passing through the point $z_0$ map onto two circles on the sphere through the points $(0, 0, 1)$ and $(\xi_0, \eta_0, \zeta_0)$, and these circles make the same angle with each other at their two intersections. If the two lines are

$$a_1 x + a_2 y + a_3 = 0, \quad u = 0,$$

(2.5.7)

$$b_1 x + b_2 y + b_3 = 0, \quad u = 0,$$

then their stereographic images lie in the planes

$$a_1 x + a_2 y + a_3(1 - u) = 0,$$
$$b_1 x + b_2 y + b_3(1 - u) = 0,$$

respectively. The tangents to the corresponding circles at the north pole are the intersections of the planes with the plane $u = 1$, that is, their equations are

$$a_1 x + a_2 y = 0, \quad u = 1,$$

(2.5.8)

$$b_1 x + b_2 y = 0, \quad u = 1,$$

respectively. It is obvious that the angle between the two lines (2.5.7) is the same as the angle between the lines (2.5.8), and this proves the assertion.

THEOREM 2.5.3. *The stereographic projection is a pure magnification, that is, the ratio between corresponding line elements in the plane and on the sphere is a function of position only.*

*Proof.* Given a line segment $[z_1, z_2]$ in the plane and the corresponding circular arc $\{Z_1, Z_2\}$ on the sphere, we plan to prove that

(2.5.9)
$$\lim_{z_2 \to z_1} \frac{a(Z_1, Z_2)}{|z_1 - z_2|} = \frac{1}{1 + |z_1|^2},$$

where $a(Z_1, Z_2)$ is the length of the arc. The theorem obviously is implied by this relation.

Before proceeding to the proof of (2.5.9), let us note the following consequence of this relation. Let $C$ be a rectifiable curve in the $z$-plane given in terms of arc length by

$$C: \ z = z(s), \quad 0 \leq s \leq L.$$

Let $\Gamma$ be the stereographic projection of $C$. Then $\Gamma$ is also rectifiable and its length is

$$(2.5.10) \qquad l(\Gamma) = \int_0^L \frac{|\,dz(s)\,|}{1 + |\,z(s)\,|^2}.$$

In order to prove (2.5.9), we observe first that $a(Z_1, Z_2)$ may be replaced by $d(Z_1, Z_2)$, for the ratio between the arc and its chord tends to 1 as $z_2 \to z_1$. We set

$$(2.5.11) \qquad d(Z_1, Z_2) = \chi(z_1, z_2).$$

This expression is called the *chordal distance* of $z_1$ and $z_2$. We have

$$(2.5.12) \qquad \chi(z_1, z_2) = \frac{|\,z_1 - z_2\,|}{(1 + |\,z_1\,|^2)^{\frac{1}{2}}(1 + |\,z_2\,|^2)^{\frac{1}{2}}}.$$

This expression obviously implies (2.5.9) in the limit.

To derive (2.5.12), we lay a plane through the points $(0, 0, 1)$, $(x_1, y_1, 0)$, and $(x_2, y_2, 0)$. This plane also contains the points $Z_1$, $Z_2$, the line segment $[Z_1, Z_2]$, and the arc $\{Z_1, Z_2\}$. See Figure 6, where the circle is the intersection of the plane with the sphere. We note first that if the angle $Z_1 N Z_2$ equals $\alpha$, then

$$\frac{d(Z_1, Z_2)}{a(Z_1, Z_2)} = \frac{\sin \alpha}{\alpha},$$

that is, this ratio is arbitrarily near to 1 if $Z_2$ is near to $Z_1$. This justifies our assertion that $a(Z_1, Z_2)$ may be replaced by $d(Z_1, Z_2)$ in (2.5.9).

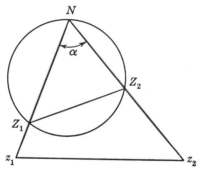

**Figure 6**

Figures 5 and 6 show that

$$d(N, z_1) = (1 + |\,z_1\,|^2)^{\frac{1}{2}}, \quad d(N, z_2) = (1 + |\,z_2\,|^2)^{\frac{1}{2}},$$
$$d(N, Z_1) = (1 + |\,z_1\,|^2)^{-\frac{1}{2}}, \quad d(N, Z_2) = (1 + |\,z_2\,|^2)^{-\frac{1}{2}},$$

where the last two relations follow from the first two since

$$d(N, Z) : d(N, z) = (1 - \zeta) : 1 = 1 : (1 + |\,z\,|^2).$$

From the relation

$$d(N, Z_1)d(N, z_1) = d(N, Z_2)d(N, z_2) = 1,$$

we conclude that the triangles $Nz_1z_2$ and $NZ_2Z_1$ are similar. It follows that

$$d(Z_1, Z_2) : d(z_1, z_2) = d(N, Z_2) : d(N, z_1).$$

Solving for $d(Z_1, Z_2)$ we get (2.5.12). In the derivation we have assumed $z_1 \neq \infty$, $z_2 \neq \infty$, but Figure 5 gives

(2.5.13) $$\chi(z_1, \infty) = (1 + | z_1 |^2)^{-\frac{1}{2}} = \lim_{z_2 \to \infty} \chi(z_1, z_2).$$

It follows from (2.5.11) that $\chi(z_1, z_2)$ is a distance function, that is,

(i) $\chi(z_1, z_2) \geqq 0$ with equality only for $z_1 = z_2$.

(ii) $\chi(z_2, z_1) = \chi(z_1, z_2)$.

(iii) $\chi(z_1, z_2) \leqq \chi(z_1, z_3) + \chi(z_3, z_2)$.

The first two are evident from (2.5.12), but the third property is a less obvious consequence. The following simple argument, due to Shizuo Kakutani, uses the identity

$$(a - b)(1 + \bar{c}c) = (a - c)(1 + \bar{c}b) + (c - b)(1 + \bar{c}a)$$

so that

$$| a - b | [1 + | c |^2] \leqq | a - c | | 1 + \bar{c}b | + | c - b | | 1 + \bar{c}a |$$

$$\leqq | a - c | [1 + | c |^2]^{\frac{1}{2}}[1 + | b |^2]^{\frac{1}{2}} + | c - b | [1 + | c |^2]^{\frac{1}{2}}[1 + | a |^2]^{\frac{1}{2}},$$

where we have used the elementary inequality

(2.5.14) $$(1 + zw)(1 + \overline{zw}) \leqq [1 + | z |^2][1 + | w |^2].$$

It follows that

$$| a - b | [1 + | c |^2]^{\frac{1}{2}} \leqq | a - c | [1 + | b |^2]^{\frac{1}{2}} + | c - b | [1 + | a |^2]^{\frac{1}{2}}$$

and hence

$$\chi(a, b) \leqq \chi(a, c) + \chi(c, b).$$

*The extended complex plane is a complete metric space with respect to the chordal metric* (see Problem 7 below). Among other things this implies the validity of Cauchy's convergence principle and the Bolzano-Weierstrass theorem. It should be observed that in this metric the extended plane is bounded, since

(2.5.15) $$\max \chi(z_1, z_2) = 1.$$

It has the even stronger property of being *totally bounded*: for every $\varepsilon > 0$

there exists a finite point set the $\varepsilon$-neighborhoods of which cover the space. Finally a historical remark: The complex sphere (2.5.2) is often called the *Riemann sphere*.[1]

## EXERCISE 2.5

**1.** If $z_1 = (1 + i)\sqrt{2}/2$, $z_2 = \bar{z}_1$, find the coordinates of the corresponding points $Z_1$ and $Z_2$ on the sphere. Compute $d(Z_1, Z_2)$ and check your result with formula (2.5.12).

**2.** Same questions for $z_1 = 2i$, $z_2 = i/2$.

**3.** Give proofs of formulas (2.5.3) and (2.5.4).

**4.** What is the length of the stereographic projection of the unit circle? Check your result with formula (2.5.10).

**5.** Describe the relative positions of

$$z, \quad -z, \quad \bar{z}, \quad -\bar{z}, \quad 1/z, \quad -1/z, \quad 1/\bar{z}, \quad -1/\bar{z}$$

in the plane and on the Riemann sphere.

**6.** A sector $0 \leq |z| \leq R$, $\alpha \leq \arg z \leq \beta$, is projected onto the sphere. What is the area of the stereographic projection?

**7.** Given a sequence $\{z_n\}$ such that $\lim \chi(z_m, z_n) = 0$ for $m \to \infty$, $n \to \infty$, then there exists a unique $z_0$, which may be $\infty$, such that $\chi(z_n, z_0) \to 0$. This is what is meant by saying that the extended plane is a complete metric space in the chordal metric.

**8.** What is an $\varepsilon$-neighborhood of $z_0$ in the chordal metric?

**9.** Verify that $\chi(z_1, z_2) = \chi(\bar{z}_1, \bar{z}_2) = \chi(1/z_1, 1/z_2)$.

**10.** Verify that $\chi(z, -1/\bar{z}) = 1$.

**11.** If $a$ and $b$ are fixed complex numbers and $z$ ranges over the sphere, find the supremum and the infimum of $\chi(z + a, z + b)$.

---

[1] Bernhard Riemann (1826–1866) has his name attached to more concepts and theories than does any other mathematician. The student is familiar with Riemann integrals and sums. Later he will meet Riemann surfaces, the Cauchy-Riemann equations, the Riemann zeta function, the famous Riemann hypothesis connected with the latter, and the Riemann conformal mapping problem. There are other Riemann problems and functions, there is Riemannian geometry, Riemann's non-Euclidean geometry, etc., etc. Riemann was one of the three fathers of function theory and represented a physico-geometrical standpoint. Professor in Göttingen after 1859, he was shy, in poor health, and started many more things than he could ever finish. The robust perfectionist Weierstrass was in most respects his opposite. Weierstrass stands for logic, Riemann is intuition at its best.

**12.** Formula (2.5.14) is a special case of Cauchy's inequality

$$\left| \sum_{m=1}^{n} z_m w_m \right|^2 \leq \sum_{m=1}^{n} | z_m |^2 \sum_{m=1}^{n} | w_m |^2.$$

Prove the latter.

**13.** Find necessary and sufficient conditions for equality to hold in Problem 12.

## COLLATERAL READING

For a more formal treatment of the subject matter of Sections 2.3 and 2.4, see

THRON, W. J.  *Introduction to the Theory of Functions of a Complex Variable*, Sections 11, 13, and 14.  John Wiley & Sons, Inc., New York, 1953.

See also Appendixes B and C and

KNOPP, K.  "Einheitliche Erzeugung der Kurven von Peano, Osgood und v. Koch," *Archiv der Mathematik und Physik*, Series 3, Vol. 26 (1917), pp. 103–115.

NEWMAN, M. H. A.  *Elements of the Topology of Plane Sets of Points.*  Cambridge University Press, New York, 1951.

For the properties of general lemniscates and related questions, see

WALSH, J. L.  "Lemniscates and Equipotential Curves of Green's Function," *American Mathematical Monthly*, Vol. 42 (1935), pp. 1–17.

# 3

# FRACTIONS, POWERS, AND ROOTS

**3.1. Fractional linear transformations.** In this chapter we shall study some elementary mappings of the complex plane into itself. It is convenient to imagine that we have two copies of the complex plane or of the sphere, one for $z$ and one for its image, $w$. We start with the mapping defined by the linear fraction

(3.1.1)
$$w = \frac{az + b}{cz + d},$$

where $a$, $b$, $c$, $d$ are given complex numbers and $ad - bc \neq 0$. Here (3.1.1) is often called a Möbius transformation, for in 1853 Möbius inaugurated the study of an equivalent class of geometrical transformations which he called *Kreisverwandtschaften*. Other names are "homographic transformation" and "homography."

We shall start with some special cases. The simplest is perhaps given by $a = d = 1$, $c = 0$, and $b$ arbitrary so that

(3.1.2)
$$w = z + b.$$

This transformation is known as a *translation*, or a *shift*: we get $w$ by adding the fixed vector $b$ to the vector $z$. The effect is a rigid motion carrying the $z$-plane the distance $|b|$ in the direction arg $b$. We note that every point in the finite plane is shifted, but the point at infinity is, of course, left invariant.

There is one other type of rigid motion given by a simple Möbius transformation, namely the *rotations* about the origin. A rotation through the angle $\alpha$ is obtained by setting

(3.1.3)
$$w = az, \quad a = \cos \alpha + i \sin \alpha.$$

Here the origin and the point at infinity are left invariant; all other points are displaced.

The transformation

(3.1.4)
$$w = az, \quad a > 0, \quad a \neq 1,$$

is generally known as a *dilation*, or a magnification, though it is a *contraction* for $a < 1$ and a *stretching* of the plane only if $a > 1$. Again, the origin and the point at infinity are the only invariant points.

Combining the last two transformations, we see that

(3.1.5)
$$w = az, \quad a = r(\cos \alpha + i \sin \alpha),$$

46

is, in general, a rotation followed by a dilation. Actually the order of the two transformations is immaterial: they commute.

We encountered the transformation

$$(3.1.6) \qquad w = \frac{1}{z}$$

in Section 2.1; it is an *inversion*. Geometrically it is composed of two reflections, one in the real axis and one in the unit circle, neither of these reflections being a Möbius transformation. Here the points $z = \pm 1$ are left invariant; all other points are displaced.

We shall show that any Möbius transformation is obtainable by composing the three elementary transformations (3.1.2), (3.1.5), and (3.1.6). We have two essentially different cases according as $c = 0$ or $c \neq 0$. In the first case

$$w = \frac{a}{d} z + \frac{b}{d},$$

which is equivalent to

$$w_1 = \frac{a}{d} z, \quad w = w_1 + \frac{b}{d}.$$

This is a dilation and/or rotation followed by a translation.

If $c \neq 0$ we have

$$w = \frac{bc - ad}{c^2} \cdot \frac{1}{z + d/c} + \frac{a}{c},$$

and this is equivalent to

$$w_1 = z + \frac{d}{c}, \quad w_2 = \frac{1}{w_1}, \quad w_3 = \frac{bc - ad}{c^2} w_2, \quad w = w_3 + \frac{a}{c},$$

that is, a translation, an inversion, a dilation and/or rotation, and finally a translation. This proves the assertion.

The correspondence between $z$ and $w$ is one-to-one and everywhere defined in the extended planes. If $w$ is given by (3.1.1), then conversely,

$$(3.1.7) \qquad z = \frac{-dw + b}{cw - a},$$

so the inverse is also a Möbius transformation. The determinant $(-d)(-a) - bc$ obviously equals $ad - bc$ and thus is different from zero. The identical mapping

$$w = z$$

is trivially a Möbius transformation. Further, if we first map the $z$-plane onto the $w$-plane by (3.1.1) and then map the $w$-plane onto the $w_1$-plane by

$$w_1 = \frac{a_1 w + b_1}{c_1 w + d_1}, \quad a_1 d_1 - b_1 c_1 \neq 0,$$

then the result is a Möbius transformation of the $z$-plane onto the $w_1$-plane

$$w_1 = \frac{Az + B}{Cz + D}$$

with

$$AD - BC = (ad - bc)(a_1 d_1 - b_1 c_1) \neq 0.$$

We express these three properties by saying:

*The set of all Möbius transformations is a group $M$.*[1]

There are certain subgroups of $M$ which are of particular importance for the applications. One of these is the *group of rigid motions of the plane* to which (3.1.2) and (3.1.3) belong. Now, any rigid motion of the plane is of the form

(3.1.8)        $w = w_0 + \omega(z - z_0), \quad \omega = \cos \alpha + i \sin \alpha,$

that is, a combination of two translations and a rotation. It is clear that these transformations form a group $E$ which is a subgroup of $M$.

A more interesting subgroup is the *group of rotations of the sphere* about its center. Suppose that the axis of rotation passes through the two antipodal points $Z_0$ and $Z_1$ on the sphere, corresponding to the two points $z_0$ and $-1/\bar{z}_0$ in the plane. Let us consider the transformation defined by

(3.1.9)        $\dfrac{w - z_0}{1 + \bar{z}_0 w} = \omega \dfrac{z - z_0}{1 + \bar{z}_0 z}, \quad \omega = \cos \alpha + i \sin \alpha.$

This may be solved for $w$, and the result is obviously a Möbius transformation which leaves $z_0$ and $-1/\bar{z}_0$ invariant. Hence $Z_0$ and $Z_1$ also remain fixed. Further,

$$\left| \frac{w - z_0}{1 + \bar{z}_0 w} \right| = \left| \frac{z - z_0}{1 + \bar{z}_0 z} \right|, \quad \arg \frac{w - z_0}{1 + \bar{z}_0 w} = \arg \frac{z - z_0}{1 + \bar{z}_0 z} + \alpha.$$

The first of these equalities says that if $z$ lies on the circle

(3.1.10)        $|z - z_0| = \rho \, | 1 + \bar{z}_0 z |, \quad 0 < \rho,$

so does its image $w$. The second equality says that if $z$ lies on the circular arc

---

[1] We recall that an *abstract group* $G$ is a system of elements, together with a single-valued binary law of combination such that for any two elements $a$ and $b$ the "product" $a \circ b$ is a well-defined element of $G$, the operation is associative, $(a \circ b) \circ c = a \circ (b \circ c)$, there is a neutral or identity element $e$ such that $a \circ e = e \circ a = a$ for every $a \in G$, and every element $a$ has an inverse $a^{-1}$ such that $a \circ a^{-1} = a^{-1} \circ a = e$. In the present case we are dealing with a *transformation group*. The elements of the group are the various Möbius transformations $T$, and the product $T_1 \circ T_2$ is the result of first carrying out the transformation $T_2$ and then following this by $T_1$. Here the order is normally essential. Associativity is easily verified. A subgroup of $G$ is a subset of $G$ which itself is a group under the same operation and with the same identity element as $G$.

(3.1.11)                    $\arg \dfrac{z - z_0}{1 + \bar{z}_0 z} = \theta, \quad 0 \leq \theta \leq 2\pi,$

then its image lies on the arc

$$\arg \frac{w - z_0}{1 + \bar{z}_0 w} = \alpha + \theta.$$

If $\rho$ and $\theta$ take on all values consistent with the stated inequalities, we obtain two *pencils of circles* in the plane.[1] The second of these is the set of all circles through the fixed points $z_0$ and $-1/\bar{z}_0$, the first is the set of its orthogonal trajectories. By Theorems 2.5.1 and 2.5.2 these pencils of circles are mapped onto two families of circles on the sphere which are orthogonal trajectories of each other. One of these is the family of great circles passing through the antipodal points $Z_0$ and $Z_1$. Using the chordal distance, we can write the equation of the trajectories in the form

(3.1.12)          $\chi(z, z_0) = d[Z, Z_0] = \gamma = \dfrac{\rho}{\sqrt{\rho^2 + 1}}$

where $Z$ is the image on the sphere of the point $z$ in the plane. Denoting the image of $w$ by $W$, we see that the effect of the transformation (3.1.9) on the sphere is to carry $Z$ into $W$ in such a manner that

$$\chi(z, z_0) = d[Z, Z_0] = d[W, Z_0] = \chi(w, z_0),$$

while the angle between the two planes through $Z_0 Z Z_1$ and through $Z_0 W Z_1$ equals $\alpha$. In other words, (3.1.9) defines a rotation of the sphere about the axis $Z_0 Z_1$ through an angle of $\alpha$. It may be verified that the set of all transformations of the form (3.1.9) forms a group, $R$ say.

The third and last subgroup to be considered is the set of all *linear transformations leaving the unit circle invariant*. Such a transformation is defined by

(3.1.13)     $\dfrac{w - z_0}{1 - \bar{z}_0 w} = \omega \dfrac{z - z_0}{1 - \bar{z}_0 z}, \quad \omega = \cos \alpha + i \sin \alpha, \quad |z_0| < 1.$

This is evidently a Möbius transformation, and since the equation of the unit circle may also be written in the form

(3.1.14)                         $|z - z_0| = |1 - \bar{z}_0 z|,$

it follows that the transformation leaves this circle invariant. Proceeding as in the case of (3.1.9) we see that there are two invariant pencils of circles associated with (3.1.13). One is the set of all circles through the points $z_0$ and $1/\bar{z}_0$ which correspond to each other under a reciprocation in the unit circle. The other is the family of orthogonal trajectories. The unit circle belongs to the

---

[1] If $C_1 = 0$ and $C_2 = 0$ are equations of distinct circles, the family $\alpha C_1 + \beta C_2 = 0$, $-\infty < \alpha/\beta \leq \infty$, is a pencil of circles.

second family. We denote this subgroup by $U$. Its elements leave the northern and southern hemispheres invariant as well as the equator on the sphere.

## EXERCISE 3.1

**1.** Verify that (3.1.14) represents the unit circle for any choice of $z_0$ with $|z_0| \neq 1$.

**2.** If (3.1.1) has real coefficients, and $ad - bc > 0$, show that the upper half-plane is left invariant. (*Hint:* Show that the imaginary part of $w$ has the same sign as the imaginary part of $z$.)

**3.** If $a$, $b$, $c$, $d$ are integers, and $ad - bc = 1$, show that the corresponding transformations form a group. This is known as the *modular group*.

**4.** Show that the rotations of the sphere are given by

$$w = \frac{az + b}{-\bar{b}z + \bar{a}} \quad \text{with} \quad |a|^2 + |b|^2 = 1.$$

(*Hint:* Solve (3.1.9) for $w$ and divide numerator and denominator by a suitably chosen factor.)

**5.** Show similarly that the transformations which leave the unit circle invariant may be written in the form

$$w = \frac{az + b}{\bar{b}z + \bar{a}} \quad \text{with} \quad |a|^2 - |b|^2 = 1.$$

**3.2. Properties of Möbius transformations.** There are three essential constants in formula (3.1.1). As a consequence it is possible to impose three suitably chosen conditions to determine the transformation. Thus, given three pairs of points

$$(z_1, w_1), \quad (z_2, w_2), \quad (z_3, w_3)$$

with

$$(z_1 - z_2)(z_2 - z_3)(z_3 - z_1)(w_1 - w_2)(w_2 - w_3)(w_3 - w_1) \neq 0,$$

there exists one and only one Möbius transformation $T$ such that

(3.2.1) $$w_k = T[z_k], \quad k = 1, 2, 3.$$

This $T$ is given by

(3.2.2) $$\frac{w - w_1}{w - w_2} \frac{w_3 - w_2}{w_3 - w_1} = \frac{z - z_1}{z - z_2} \frac{z_3 - z_2}{z_3 - z_1}.$$

Solving for $w$ we obtain a Möbius transformation having the desired mapping properties. The following argument shows that $T$ is unique:

For the determination of the coefficients $a$, $b$, $c$, $d$ of $T$ we have the system of equations   $z_k a + b - z_k w_k c - w_k d = 0$,   $k = 1, 2, 3$.   The corresponding matrix

$$\begin{bmatrix} z_1 & 1 & -z_1 w_1 & -w_1 \\ z_2 & 1 & -z_2 w_2 & -w_2 \\ z_3 & 1 & -z_3 w_3 & -w_3 \end{bmatrix}$$

is of rank three. For, if we multiply the first and the second columns by $w_1$ and add to the third and fourth columns respectively, we obtain a matrix of the same rank, and now the determinant of the last three columns equals

$$(w_1 - w_2)(w_1 - w_3)(z_2 - z_3) \neq 0.$$

It follows that the ratio $a : b : c : d$ is uniquely determined, and hence that $T$ is unique.

Next, we prove

THEOREM 3.2.1.    *A Möbius transformation maps the family $F$ of straight lines and circles in the plane onto itself.*

*Proof.* It was shown in the preceding section that any Möbius transformation is obtainable as the result of performing consecutively a suitable sequence of elementary transformations: translations, rotations, dilations, and inversions. The first three will carry straight lines into straight lines and circles into circles, so it suffices to examine what an inversion will do to an element of $F$. As observed earlier (see (2.5.6)), such a locus is given by an equation of the form

$$(C - D)(x^2 + y^2) + Ax + By = D.$$

We now take the inversion

$$w = \frac{1}{z}, \quad w = u + iv,$$

and obtain as image the locus

$$C - D + Au - Bv = D(u^2 + v^2),$$

which belongs to the family $F$ in the $w$-plane. We note in particular that a straight line in the $z$-plane, not through the origin, is mapped onto a circle in the $w$-plane through the origin, while a straight line through the origin is simply reflected in the real axis. Since an inversion takes $F$ into itself, every Möbius transformation will also have this property.

COROLLARY.    *If $L_1$ and $L_2$ are any two elements of $F$, then there exists a Möbius transformation mapping $L_1$ onto $L_2$.*

For we can choose three points $z_1$, $z_2$, $z_3$ on $L_1$ and three points $w_1$, $w_2$, $w_3$ on $L_2$, and the transformation (3.2.2) will then map $L_1$ onto $L_2$.

We come finally to the classification of Möbius transformations based upon their fixed points. We say that $z_0$ *is a fixed point of the transformation* $T$ *if* $z = z_0$ *is a root of the equation*

$$(3.2.3) \qquad\qquad z = T[z];$$

that is,

$$(3.2.4) \qquad\qquad cz^2 + (d - a)z - b = 0$$

in the present case. In general this equation has two roots $z_1$ and $z_2$ which coincide if $(d - a)^2 + 4bc = 0$. There are two exceptional cases:

(i) $c = 0$, $d \neq a$. Then there is one finite root, and since $T$ maps infinity into itself, we take infinity as the second root.

(ii) $c = 0$, $d = a$. This is the case of a translation; infinity is the only invariant point, and we take $z = \infty$ as a double root of (3.2.4).

With these conventions we may say that the *transformation* $T$ *has always two fixed points* $z_1$ *and* $z_2$.

DEFINITION 3.2.1. *The Möbius transformation* $T$ *is said to be loxodromic or parabolic according as* $T$ *has distinct or coincident fixed points.*

We shall restrict ourselves to the case of finite fixed points and start with the loxodromic case. We have now $c \neq 0$, and equation (3.2.4) has two distinct roots $z_1$ and $z_2$. We can determine a complex number $\mu \neq 0$ such that $w = T[z]$ is given by

$$(3.2.5) \qquad\qquad \frac{w - z_1}{w - z_2} = \mu \, \frac{z - z_1}{z - z_2}.$$

In fact,

$$\mu = \frac{a - cz_1}{a - cz_2}$$

since $z = \infty$ is mapped into $w = a/c$. To bring out the essential parameters we denote this transformation by

$$(3.2.6) \qquad\qquad T(\mu; \; z_1, z_2).$$

Let $z_1$ and $z_2$ be fixed, and consider the set $GL(z_1, z_2)$ of all transformations $T(\mu; \; z_1, z_2)$ where $\mu$ ranges over the finite complex plane omitting the origin. We have

$$(3.2.7) \qquad\qquad T(1; \; z_1, z_2)[z] \equiv z,$$

so that $T(1; \; z_1, z_2)$ is the identical transformation. Further,

$$(3.2.8) \qquad\qquad T(\mu; \; z_1, z_2)T(\nu; \; z_1, z_2) = T(\mu\nu; \; z_1, z_2).$$

In particular, $T(\mu;\ z_1, z_2)$ and $T(\mu^{-1};\ z_1, z_2)$ are inverse transformations. Since the composition is associative, we conclude that $GL(z_1, z_2)$ is a group—more precisely, a one-parameter transformation group with respect to $\mu$ (note that $z_1$ and $z_2$ are fixed). This is, of course, a subgroup of $M$. There are various subgroups of $GL(z_1, z_2)$ which are worthy of attention.

Let $\alpha$, $\beta$, $\sigma$ be arbitrary real numbers, and suppose that

$$(3.2.9) \qquad \mu = \mu(\sigma) = e^{\alpha\sigma}[\cos{(\beta\sigma)} + i \sin{(\beta\sigma)}].$$

Here we keep $\alpha$, $\beta$ fixed but let $\sigma$ range over $(-\infty, \infty)$. The corresponding transformations $T(\mu;\ z_1, z_2)$ we denote by

$$(3.2.10) \qquad S(\sigma;\ \alpha, \beta;\ z_1, z_2).$$

For fixed $\alpha$, $\beta$, $z_1$, $z_2$ this is a one-parameter transformation group, the law of composition being

$$(3.2.11) \qquad S(\sigma;\ \alpha, \beta;\ z_1, z_2)S(\tau;\ \alpha, \beta;\ z_1, z_2) = S(\sigma + \tau;\ \alpha, \beta;\ z_1, z_2).$$

In order to study the mapping problems posed by these groups, we proceed as in the study of the groups $R$ and $U$ in the preceding section. We introduce two conjugate pencils of circles, $P_1(z_1, z_2)$ and $P_2(z_1, z_2)$, or $P_1$ and $P_2$ for short. Here $P_1$ is the set of all circles

$$C(\rho): \quad \left|\frac{z - z_1}{z - z_2}\right| = \rho, \quad 0 < \rho,$$

while $P_2$ is the set

$$\Gamma(\theta): \quad \arg\frac{z - z_1}{z - z_2} = \theta, \quad -\infty < \theta < +\infty.$$

Actually, $\Gamma(\theta)$ is a circular arc from $z_1$ to $z_2$, and the complementary arc is $\Gamma(\theta + \pi)$. We have, of course, $\Gamma(\theta + 2\pi) = \Gamma(\theta)$ for each $\theta$. The two sets $P_1$ and $P_2$ are orthogonal trajectories of each other. Define

$$w = w(\sigma) = S(\sigma;\ \alpha, \beta;\ z_1, z_2)[z], \quad -\infty < \sigma < \infty.$$

This locus is the path curve of $z$ under the transformation group $\{S(\sigma)\}$. We have $w(0) = z$. From the defining formulas we conclude that if $z$ lies at the intersection of $C(\rho)$ with $\Gamma(\theta)$, then $w(\sigma)$ lies at the intersection of $C(\rho e^{\alpha\sigma})$ with $\Gamma(\theta + \beta\sigma)$.

Let us settle first the two most important special cases. If $\beta = 0$, then $w(\sigma)$ also lies on $\Gamma(\theta)$. The path curve then coincides with $\Gamma(\theta)$. To fix the ideas, suppose that $\alpha > 0$. Then, as $\sigma$ goes from 0 to $\infty$, $w(\sigma)$ goes along $\Gamma(\theta)$ from $z$ to $z_2$, and if $\sigma$ goes from 0 to $-\infty$, $w(\sigma)$ goes from $z$ to $z_1$. In this case the fixed points can be thought of as *centers of attraction or repulsion*. For a fixed positive $\sigma$, $w(\sigma)$ is closer to $z_2$ and farther from $z_1$ than $z$ was, distances being taken along the arc $\Gamma(\theta)$. $S(\sigma;\ \alpha, 0;\ z_1, z_2)$ is said to be a *hyperbolic* transformation. See Figure 7 for this and the next case.

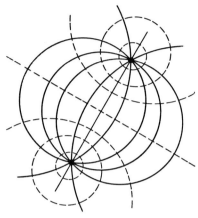

**Figure 7**

If $\alpha = 0$ instead, the situation is reversed. Now $C(\rho)$ is the path curve, and this curve is described infinitely often as $\sigma$ goes from 0 to $\pm\infty$. The fixed points are now *centers of rotation*. Such a transformation is said to be *elliptic*.

We return now to the general case in which $\alpha \neq 0$, $\beta \neq 0$. We shall show that the path curve $w = w(\sigma)$ cuts every circular arc $\Gamma(\theta)$ of the family $P_2$ under the fixed angle $\arg(\alpha + i\beta)$.[1] It is enough for this purpose to determine the tangent vector of the path curve $w = w(\sigma)$ at $\sigma = 0$ and to compare its inclination with that of the tangent vector of $\Gamma(\theta)$ at the same point $w(0) = z$. If the angle between the two vectors turns out to be independent of $z$, the assertion will be proved.

Now the tangent vector of $w(\sigma)$ at $\sigma = 0$ is, by definition,

$$w'(0) \equiv \lim_{\sigma \to 0} \frac{1}{\sigma}[w(\sigma) - w(0)].$$

Setting $\mu = \mu(\sigma)$ and $w = w(\sigma)$ in (3.2.5) and solving for $w(\sigma)$, we get

(3.2.12)
$$w(\sigma) = \frac{\mu(\sigma)z_2(z_1 - z) + z_1(z - z_2)}{\mu(\sigma)(z_1 - z) + (z - z_2)}.$$

This is a complex-valued function of the real variable $\sigma$, and a moment's

---

[1] This is the reason for the term "loxodromic." A *loxodrome* is the general name for a curve on a surface of revolution cutting the meridians under a fixed angle. In the case of a sphere we obtain what in navigation is known as a *rhumb line*.

reflection shows that we can apply the ordinary rules of the calculus to obtain

$$w'(\sigma) = \frac{(z_1 - z_2)(z - z_1)(z - z_2)}{[\mu(\sigma)(z_1 - z) + (z - z_2)]^2} \mu(\sigma)(\alpha + \beta i),$$

where the last two factors represent the derivative of $\mu(\sigma)$. Hence,

$$(3.2.13) \qquad w'(0) = \frac{(z - z_1)(z - z_2)}{z_1 - z_2} (\alpha + \beta i).$$

On the other hand, $\Gamma(\theta)$ is represented parametrically by

$$\frac{z - z_1}{z - z_2} = \rho(\cos \theta + i \sin \theta),$$

where $\rho$ goes from 0 to $+\infty$. Here $z$ is evidently a differentiable function of the real variable $\rho$, and again we can justify using the rules of the calculus to obtain

$$\frac{z_1 - z_2}{(z - z_2)^2} \frac{dz}{d\rho} = \cos \theta + i \sin \theta = \frac{1}{\rho} \frac{z - z_1}{z - z_2},$$

so that

$$(3.2.14) \qquad \frac{dz}{d\rho} = \frac{1}{\rho} \frac{(z - z_1)(z - z_2)}{z_1 - z_2}, \quad \rho > 0,$$

and this is the tangent vector of $\Gamma(\theta)$ at the point $z$. It follows that

$$\arg w'(0) - \arg \frac{dz}{d\rho} = \arg (\alpha + i\beta),$$

that is, independent of $z$. Thus, we see that the path curves intersect the circles of the family $P_2$ under the constant angle $\arg (\alpha + \beta i)$. Each path curve is a double spiral, one arm of which winds around and approaches $z_1$ while the other approaches $z_2$.

The parabolic case with a double fixed point $z_0$ remains. Such a transformation $T(\alpha; z_0)$ is given by

$$(3.2.15) \qquad \frac{1}{w - z_0} = \frac{1}{z - z_0} + \alpha.$$

The family $\{T(\alpha; z_0)\}$, where $\alpha$ ranges over the finite complex plane and $z_0$ is fixed, is a one-parameter transformation group $GP(z_0)$ with the law of composition

$$(3.2.16) \qquad T(\alpha; z_0)T(\beta; z_0) = T(\alpha + \beta; z_0).$$

Here $T(0; z_0) = I$, the identical transformation, and $T(-\alpha; z_0)$ is the inverse of $T(\alpha; z_0)$. Let $\alpha$ and $z_0$ be fixed, $\alpha \neq 0$, and set

$$(3.2.17) \qquad w(\sigma) = T(\sigma\alpha; z_0)[z], \quad -\infty < \sigma < +\infty.$$

The curve described by $w(\sigma)$ as $\sigma$ goes from $-\infty$ to $+\infty$ is a circle passing through $z$ and $z_0$. At the latter point its tangent has the direction $-\arg \alpha$. See Figure 8.

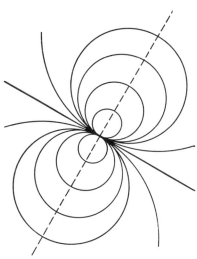

**Figure 8**

One final result:

THEOREM 3.2.2. *A Möbius transformation is isogonal everywhere.*

*Proof.* Suppose that $T[z]$ is given by (3.1.1). If $c = 0$ the result is immediate, since translations, rotations, and dilations obviously preserve angles. Suppose then that $c \neq 0$ and, to start with, that $z_0 \neq -d/c$ and $\infty$. If $h$ is a complex quantity tending to zero in such a manner that

$$\lim_{h \to 0} \arg h = \alpha$$

exists, then

$$T[z_0 + h] - T[z_0] = \frac{ad - bc}{[c(z_0 + h) + d](cz_0 + d)} h$$

and

(3.2.18)    $$\lim_{h \to 0} \arg \{T[z_0 + h] - T[z_0]\} = \arg Q(z_0) + \alpha$$

where

(3.2.19)    $$Q(z_0) = \frac{ad - bc}{(cz_0 + d)^2}.$$

It follows that if two curves $C_1$ and $C_2$ intersect at $z = z_0$, where they have tangents making an angle of $\gamma$ with each other, then their images $T[C_1]$ and $T[C_2]$ intersect at $w = T[z_0]$, where their tangents make the same angle $\gamma$ with each other. It should be noted that the sense as well as the magnitude of the angle is preserved.

In passing let us remark that the transformation $T[z]$ is a pure magnification at $z_0$ in the sense of Theorem 2.5.3, the magnification being given by $|Q(z_0)|$. The reader has probably observed that $Q(z_0)$ is the value of the derivative of $T[z]$ at $z_0$.

Consider now the excluded cases. If $z_0 = -d/c$, we have $T[z_0] = \infty$ and

$$T[z_0 + h] = \frac{1}{ch}(az_0 + ah + b)$$

so that

$$\lim_{h \to 0} \arg T[z_0 + h] = \arg \frac{bc - ad}{c^2} - \lim_{h \to 0} \arg h,$$

when the right side exists. To the tangents of $C_1$ and $C_2$ at $z_0$ now correspond asymptotes of $T[C_1]$ and $T[C_2]$, and the angle between the asymptotes is the negative of the angle between the tangents. Finally if $z \to \infty$ in such a way that $\lim \arg z$ exists, then a simple computation shows that

$$\lim_{z \to \infty} \arg \left\{ T[z] - \frac{a}{c} \right\} = \arg \frac{bc - ad}{c^2} - \lim \arg z,$$

and the angle between the asymptotes of $C_1$ and $C_2$ is the negative of the angle between the tangents of $T[C_1]$ and $T[C_2]$ at $w = a/c$.

## EXERCISE 3.2

**1.** When is a Möbius transformation its own inverse?

**2.** A mapping $T$ such that $T^2 = T$ is called a projection. Are any Möbius transformations projections?

**3.** Verify (3.2.8) by computation and prove (3.2.11).

**4.** Verify (3.2.16) by computation.

**5.** Find a parabolic transformation mapping the circle $|z| = 1$ onto the circle $|z - \frac{1}{2}| = \frac{1}{2}$.

**6.** Prove that two Möbius transformations commute if they have the same fixed points. This condition is also necessary in the parabolic case, but not in the loxodromic one. Find necessary and sufficient conditions in the latter case. Prove that a loxodromic transformation never commutes with a parabolic one.

**7.** Verify that the pencils of circles $P_1$ and $P_2$ are orthogonal to each other.

**8.** Prove that the path curves in the parabolic case have the properties stated in the text.

**9.** What reasons would you give in justification of formulas (3.2.13) and (3.2.14)?

**10.** A Möbius transformation is said to be real if it maps the real axis into itself. What conditions does this impose on the coefficients and on the fixed points? How does it affect the parameters $\mu$ in (3.2.5) and $\alpha$ in (3.2.15)?

**11.** Prove that a Möbius transformation leaves the cross ratio $R(z_1, z_2, z_3, z_4)$ of any four points invariant, where

$$R(z_1, z_2, z_3, z_4) = \frac{z_1 - z_2}{z_1 - z_4} \frac{z_3 - z_4}{z_3 - z_2}.$$

**3.3. Powers.** Let $n$ be a positive integer, and consider the mapping defined by

$$(3.3.1) \qquad\qquad w = z^n.$$

Using formula (2.1.17) we see that to $z = r(\cos\theta + i\sin\theta)$ corresponds

$$(3.3.2) \qquad\qquad w = r^n(\cos n\theta + i\sin n\theta).$$

This says that the ray

$$\arg z = \theta$$

is mapped onto the ray

$$\arg w = n\theta,$$

and the circle

$$|z| = r$$

goes into the circle

$$|w| = r^n$$

covered $n$ times. In fact, each of the $n$ circular arcs

$$(3.3.3) \qquad |z| = r, \quad k\frac{2\pi}{n} \leqq \arg z < (k+1)\frac{2\pi}{n}, \quad k = 0, 1, 2, \cdots, n-1,$$

is mapped onto the full circle $|w| = r^n$.

This situation shows that in order to obtain a geometric representation of the mapping we really need $n$ copies of the $w$-plane, which, however, have to be joined together in a suitable manner. Such representations were introduced by B. Riemann in his Göttingen dissertation of 1851, and this class of manifolds is consequently known as *Riemann surfaces*. This is not the right place for a formal definition of these somewhat abstruse concepts. At this juncture we shall merely describe in some detail the surface generated by the mapping (3.3.1).

We start with $n$ copies of the extended $w$-plane, denoted by $S_1, S_2, \cdots, S_n$, out of which we form a connected surface $R$ by the following conventions: $S_k$ is referred to as the $k$th sheet of $R$. All $n$ sheets shall have the points $w = 0$ and $w = \infty$ in common; these two points are known as the *branch points*, or *points of ramification*, of the surface and are said to be branch points of order $(n-1)$; $n$ sheets are joined there. In each sheet we draw a curve joining $0$ and $\infty$, the same curve in each sheet. The curve is known as a *branch line* of the surface since the various sheets will be connected with one another along this line. To simplify the discussion we take the positive real axis as the branch line. We proceed to specify how the sheets are to be joined along the branch line, and this convention will be amplified by a description of what constitutes an $\varepsilon$-neighborhood of a point of $R$. Figure 9 (a) gives the conventional pictorial representation of a portion of $R$ near the origin for the simplest case $n = 2$.

Consider two points $a + bi$ and $a - bi$ in $S_k$, where $a > 0$, $b > 0$. These points may be connected by going around three sides of a rectangle along the successive segments $[a + bi, -a + bi], [-a + bi, -a - bi], [-a - bi, a - bi]$; this path lies entirely in $S_k$. If, instead of following this path from $a + bi$ to $a - bi$, we follow the vertical down from $a + bi$, then we leave $S_k$ when we cross the real axis and enter $S_{k-1}$, where we set $S_0 = S_n$ by definition. In this manner we reach the point $a - bi$ of the sheet $S_{k-1}$. If we start from $a - bi$ in the sheet $S_k$ and proceed upward along the vertical, we leave $S_k$ when we cross the real axis and enter $S_{k+1}$, where $S_{n+1} = S_1$. In this manner we reach the point $a + bi$ of the sheet $S_{k+1}$.

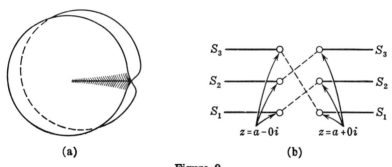

(a)                     (b)

**Figure 9**

Starting at $w = a + bi$ in $S_k$ and describing the circle $|w| = |a + bi|$ $n$ times in the positive sense, that is, so that $\arg w$ is steadily increasing, then we encounter the point $a + bi$ consecutively in the sheets

$$S_k, S_{k+1}, \cdots, S_n, S_1, S_2, \cdots, S_{k-1}, S_k.$$

Thus, we return to the starting point after $n$ positive circuits. If we describe the circle in the negative sense instead, the sheets follow each other in the order

$$S_k, S_{k-1}, \cdots, S_1, S_n, S_{n-1}, \cdots, S_{k+1}, S_k.$$

It is difficult to visualize the surface $R$ in the neighborhood of the branch line. To help the understanding it is customary to use a diagram such as Figure 9 (b) for the case $n = 3$. This is supposed to give a cross section of the surface in three dimensions by a plane $x = a$ ($> 0$) as seen from the end of the positive real axis. The three sheets have been separated, as have the edges of the branch line. The dashed lines indicate which edges are to be joined or, better, identified. Such a representation is suggestive, but it may also be misleading. In Euclidean space the dashed lines do intersect, but this reflects no property of the abstract surface. It is simply an imperfection of the model.

To complete our description of the surface, we assign neighborhoods to the points of $R$ in the following manner: If $a + bi$ lies in $S_k$ and its distance from the positive real axis exceeds $\varepsilon$, then the set of all points of $S_k$ whose distance from $a + bi$ is less than $\varepsilon$ constitutes an $\varepsilon$-neighborhood of $a + bi$. For points on the branch line, that is, on the positive real axis, we have to distinguish between the two edges of the line, because $a + 0i$ in $S_k$ is the same point as $a - 0i$ in $S_{k-1}$, and $a - 0i$ in $S_k$ is the same point as $a + 0i$ in $S_{k+1}$. In the first case, an $\varepsilon$-neighborhood is given by

$$[w \mid \mid w - a \mid < \varepsilon, \Im(w) \geqq 0, w \in S_k] \cup [w \mid \mid w - a \mid < \varepsilon, \Im(w) < 0, w \in S_{k-1}],$$

and in the second case by

$$[w \mid \mid w - a \mid < \varepsilon, \Im(w) < 0, w \in S_k] \cup [w \mid \mid w - a \mid < \varepsilon, \Im(w) \geqq 0, w \in S_{k+1}],$$

where $a \neq 0, \infty, a > \varepsilon$. An $\varepsilon$-neighborhood of $w = 0$ is the set of all points $w$ with $\mid w \mid < \varepsilon$ and $w$ ranging over all $n$ sheets $S_k$. Replacing $\mid w \mid < \varepsilon$ by $1/\mid w \mid < \varepsilon$, we get a neighborhood for $w = \infty$. This completes the description of the surface $R$.

We now see that there is a one-to-one correspondence between the extended $z$-plane and the Riemann surface $R$ under the mapping (3.3.1). Instead of constructing $R$ over the extended $w$-plane, we could just as well have spread the surface over the Riemann sphere.

We shall prove that the mapping (3.3.1) is isogonal, except at zero and infinity. At the latter points, formula (3.3.2) shows that angles are multiplied by the factor $n$. Suppose then that $z_0 \neq 0, \infty$, and consider the mapping in a small neighborhood of $z = z_0$. If

$$w = (z_0 + h)^n,$$

the binomial theorem shows that

(3.3.4)                     $$w - z_0{}^n = n z_0{}^{n-1} h + O(h^2)$$

as $h \to 0$, where the notation $O(h^2)$ means that for small values of $h$ the remainder does not exceed some fixed quantity, independent of $h$, times $\mid h \mid^2$.

We remark in passing that $o(h^2)$ would mean that the remainder divided by $h^2$ tends to zero with $h$.[1]

From (3.3.4) we see that

$$(3.3.5) \qquad \lim_{h \to 0} \arg [w - z_0{}^n] = \arg z_0{}^{n-1} + \lim_{h \to 0} \arg h,$$

if the last limit exists. From this it may be concluded that if two curves $C_1$ and $C_2$ intersect at $z = z_0 \ (\neq 0, \infty)$, forming an angle $\gamma$ with each other, then their images intersect at $w = z_0{}^n$, forming the same angle $\gamma$. Thus, the mapping is isogonal. By formula (3.3.4) the magnification at $z = z_0$ equals

$$n \mid z_0 \mid^{n-1}.$$

The reader naturally suspects that $nz^{n-1}$ must be the derivative of $z^n$ also for complex values of the variable. This is indeed so.

## EXERCISE 3.3

**1.** How is the half-plane $\Re(z) > 1$ mapped by $w = z^2$? Discuss also $\Re(z) > -1$.

**2.** How are the straight lines $x = a$, $y = b$ mapped by $w = z^3$?

**3.** Sketch the curves $\Re(z^3) = 1$ and $\Im(z^3) = 1$.

**4.** Show that $w = z + \frac{1}{2}z^2$ maps $\mid z \mid < 1$ in a one-to-one manner on the interior of a cardioid. Show that the mapping ceases to be one-to-one in any larger circle.

**5.** Discuss the mapping $w = z + \frac{1}{3}z^3$ for $\mid z \mid < 1$ and prove that the mapping is not one-to-one in any larger circle.

**6.** If $\mid z \mid < r$ show that $\left| 1 + \dfrac{z}{n} \right|^n < e^r$.

The following problems are intended as drill in the use of the notation "$O$" and "$o$":

**7.** The sum of the first $n$ positive integers is $O(n^2)$ as $n \to \infty$.

**8.** The sum of their squares is $\frac{1}{3}n^3 + O(n^2)$. Generalize!

**9.** $\log (n + 1) - \log n = o(1)$ as $n \to \infty$. Can you make a stronger statement?

**10.** $\log (1 + h) = h + o(h)$ as $h \to 0$. Can you make a stronger statement?

---

[1] More generally, $F(t) = O[g(t)]$ as $t \to a$ means that there exists a constant $M$ such that $\mid F(t) \mid \leq M \mid g(t) \mid$ in some neighborhood of $t = a$, while $f(t) = o[g(t)]$ means that $f(t)/g(t) \to 0$ as $t \to a$. Here the variables may be real or complex, and $a$ may be any finite number or $\infty$.

**3.4. Roots.** If $w = z^n$, where $n$ is a positive integer $> 1$, then we have conversely

(3.4.1)
$$z = \sqrt[n]{w} \equiv w^{1/n},$$

but this function of $w$ is not single-valued and must be properly defined. We shall accept a complex number $z$ as an $n$th root of a given complex number $w$ if and only if the $n$th power of $z$ equals $w$. Suppose that

(3.4.2)
$$w = R(\cos \Theta + i \sin \Theta), \quad 0 \leqq \Theta < 2\pi.$$

We then define

(3.4.3)
$$z_k = z_k(w) = R^{1/n}\left\{\cos\left[\frac{\Theta}{n} + (k-1)\frac{2\pi}{n}\right] + i \sin\left[\frac{\Theta}{n} + (k-1)\frac{2\pi}{n}\right]\right\}$$

for $k = 1, 2, \cdots, n$, and we find that

(3.4.4)
$$[z_k(w)]^n = w$$

for every $k$. Thus, formula (3.4.3) defines $n$ determinations of the $n$th root of $w$. This exhausts the possibilities, for if

$$[r(\cos \theta + i \sin \theta)]^n = R(\cos \Theta + i \sin \Theta)$$

we must have

$$r^n = R, \quad n\theta \equiv \Theta \pmod{2\pi},$$

the distinct solutions of which are given by (3.4.3).

We note that the points

$$z_1(w), z_2(w), \cdots, z_n(w)$$

form the vertices of a regular polygon of $n$ sides having its center at the origin. It is clear that this polygon varies continuously with $w$. In particular, if $w$ describes the circle $|w| = R$ in such a manner that arg $w$ increases by $2\pi$, then the root configuration rotates about its center through an angle of $2\pi/n$. This leaves the configuration as such unchanged, but it implies that the various determinations of the roots are permuted. Thus $z_k(w)$ becomes

$$R^{1/n}\left\{\cos\left[\frac{\Theta + 2\pi}{n} + (k-1)\frac{2\pi}{n}\right] + i \sin\left[\frac{\Theta + 2\pi}{n} + (k-1)\frac{2\pi}{n}\right]\right\} = z_{k+1}(w),$$

and we have a cyclical permutation of the roots

$$z_1 \to z_2, \; z_2 \to z_3, \; \cdots, \; z_{n-1} \to z_n, \; z_n \to z_1.$$

If we let $\Theta$ keep on increasing, we find that each root changes continuously; and after $j$ circuits, $z_k$ is carried into $z_{j+k}$, where the subscript $j + k$ is to be replaced by the least positive residue modulo $n$. After $n$ circuits, $z_k$ is carried into $z_{n+k} = z_k$, so all roots are back in their original positions.

In order to make $z_k(w)$ into a single-valued function of $w$, we have two alternatives at our disposal: either we restrict the domain of $w$ or we consider

$w$ on the Riemann surface $R$ of the preceding section rather than in the $w$-plane. According to the first alternative, let $D$ be a domain in the $w$-plane such that if $\Pi$ is an arbitrary simple closed polygon the points of which belong to $D$, then $w = 0$ is never a point of $\Pi$ or of its interior. If $w_0 \in D$ and $\Theta_0$ is a determination of arg $w_0$, the function arg $w \equiv \Theta$ is uniquely determined in $D$ by the requirement that it shall be a continuous function of $w$ and take on the value $\Theta_0$ for $w = w_0$. The function $z_k(w)$ defined by (3.4.3) using this value of $\Theta$, is then uniquely defined and is a continuous single-valued function of $w$ in $D$. In particular, we may take $D$ as the sector

$$0 < \arg w < 2\pi$$

and use (3.4.3). Each of the resulting $n$ functions $z_k(w)$ is single-valued and continuous in the sector. It is customary to call $z_1(w)$ the *principal determination* of the $n$th root of $w$. Since this choice of $D$ leaves out real positive values of $w$, we complete the definition by allowing $\Theta = 0$ in (3.4.3).

The second alternative is to let $w$ range over the Riemann surface $R$. Then to each point $w_0$ of the surface there corresponds a unique point $z_0$ in the $z$-plane which is the $n$th root of $w_0$. If $w_0$ lies in $S_k$, then we have $z_0 = z_k(w_0)$.

Since the mapping of the $z$-plane onto $R$ is one-to-one and isogonal except at zero and infinity, we conclude that the inverse map has the same properties. We shall later be able to verify the isogonality by direct computation.

In conclusion, we shall make some remarks concerning the *roots of unity*. If $w = 1$, the $n$ quantities

$$(3.4.5) \qquad \omega_n = \cos (k - 1) \frac{2\pi}{n} + i \sin (k - 1) \frac{2\pi}{n}, \quad k = 1, 2, \cdots, n,$$

are known as the $n$th roots of unity. They obviously form the vertices of a regular polygon of $n$ sides.

This brings up the question of when a regular polygon may be constructed by the methods of Euclidean geometry, that is, by the use of ruler and compass alone. Such a construction involves finding the intersections of straight lines and circles. Algebraically this amounts to solving linear equations or quadratic equations having real roots. Consequently the construction problem for the regular $n$-gon is equivalent to the algebraic problem of when the equation

$$z^n + z^{n-1} + \cdots + z + 1 = 0$$

is solvable by repeated extraction of square roots of positive quantities. This algebraic problem was solved by Gauss (*Disquisitiones Arithmeticae*, Leipzig, 1801), who showed that $n$ must be the product of a power of 2 and of distinct primes of the special form

$$p_k = 2^{2^k} + 1, \quad k = 0, 1, 2, \cdots.$$

These are the so-called Fermat numbers.[1] The first five numbers in this sequence are

$$3, \quad 5, \quad 17, \quad 257, \quad 65{,}537,$$

and all these numbers are primes. Fermat believed that $p_k$ is always a prime; actually no prime $p_k$ has been found beyond $p_4$, and about a dozen Fermat numbers are known to be composite. The construction of regular $n$-gons for the cases where $n$ has no other prime factors than 2, 3, and 5 was known to the Greeks; the case $n = 17$ was one of Gauss' earliest discoveries and formed the first entry in his famous "Tagebuch."

## EXERCISE 3.4

**1.** How does $w = \sqrt{z}$ map the half-plane $\Re(z) \geq 1$? $(\sqrt{1} = +1.)$

**2.** Express the real and the imaginary parts of the square root of $x + iy$ as functions of $x$ and $y$.

**3.** Discuss the curve

$$| \sqrt{z} - 1 | = 1$$

if $| \arg z | < \pi$ and $\sqrt{1} = +1$.

**4.** If the $n$th root is determined in the same manner as the square root in Problem 3, find

$$\lim_{n \to \infty} n[\sqrt[n]{z} - 1],$$

by separating real and imaginary parts or otherwise.

**5.** What is the sum of the $n$ $n$th roots of unity? Of their $k$th powers?

**6.** Given a regular $n$-gon inscribed in the unit circle. Find the product of the lengths of the line segments that can be drawn from a fixed vertex to all the other vertices.

**7.** The square root of $1 - z^2$ has been assigned the value $+1$ at $z = 0$. Find the values of the square root at $z = +2$ and at $z = -2$ for approach along a path in the upper half-plane or in the lower half-plane.

**8.** If $z \to \infty$ along the ray $\arg z = \theta$, $\theta \neq 0$, $\pi$, and the square root is defined as in Problem 7, find

$$\lim z^{-1} \sqrt{1 - z^2}.$$

---

[1] Pierre de Fermat (1601–1665), a lawyer and member of the parliament (regional superior court) of Toulouse, is famous for his work on number theory, in particular for his still unproved assertion that the equation

$$x^n + y^n = z^n$$

has no solutions in positive integers when $n > 2$. Fermat was also in possession of a method of finding maxima and minima.

**9.** Discuss the Riemann surface of $w = \sqrt{z^2 - 1}$. There are two sheets and two branch points. The branch line may be taken as the line segment $[-1, +1]$.

**10.** Discuss the Riemann surface of $w = \sqrt{z^3 - 1}$. There are two sheets and four branch points. Note that $z = \infty$ is a branch point. There are two branch lines, which may be taken as the line segments $\left[ -\dfrac{1}{2} - \dfrac{i}{2}\sqrt{3}, \ -\dfrac{1}{2} + \dfrac{i}{2}\sqrt{3} \right]$ and $[1, +\infty]$.

**3.5. The function $(z^2 + 1)/(2z)$.** The mapping defined by

$$(3.5.1) \qquad\qquad w = \frac{1}{2}\left( z + \frac{1}{z} \right)$$

is interesting in itself and is basic for the more complicated mapping problems presented by the elementary trigonometric functions. See Section 6.4.

We note that $w$ is unchanged if $z$ is replaced by $1/z$. It follows that these two values give the same value for $w$. Since $1/z$ lies inside the unit circle when $z$ lies outside, it is sufficient to study the mapping of $|z| > 1$. Let us set

$$(3.5.2) \qquad\qquad z = r(\cos\theta + i\sin\theta), \quad w = u + iv.$$

We have then

$$(3.5.3) \qquad u = \frac{1}{2}\left( r + \frac{1}{r} \right)\cos\theta, \quad v = \frac{1}{2}\left( r - \frac{1}{r} \right)\sin\theta.$$

Hence if $z$ describes the circle

$$|z| = r > 1,$$

$w$ describes the ellipse

$$(3.5.4) \qquad\qquad \left[ \frac{2u}{r + 1/r} \right]^2 + \left[ \frac{2v}{r - 1/r} \right]^2 = 1,$$

while the ray

$$\arg z = \theta, \quad r > 1,$$

is mapped onto an arc of the hyperbola

$$(3.5.5) \qquad\qquad \left[ \frac{u}{\cos\theta} \right]^2 - \left[ \frac{v}{\sin\theta} \right]^2 = 1,$$

namely, from the point $w = \cos\theta$ on the real axis to the point at infinity in the upper or the lower half-plane according as $\sin\theta > 0$ or $< 0$. The four arcs of the hyperbola correspond to the four rays

$$\arg z = \pm\theta, \quad \pi \pm \theta, \quad r > 1.$$

There are a couple of extreme cases. The unit circle $|z| = 1$ is mapped onto the interval $[-1, +1]$ covered twice. The intervals $[-\infty, -1]$ and $[+1, +\infty]$ are invariant under the mapping, but $z = \pm 1$ and $\infty$ are the only fixed points.

From equations (3.5.3) we obtain

$$\frac{\partial u}{\partial r} = \frac{r^2 - 1}{2r^2} \cos \theta, \qquad \frac{\partial v}{\partial r} = \frac{r^2 + 1}{2r^2} \sin \theta,$$

$$\frac{\partial u}{\partial \theta} = -\frac{r^2 + 1}{2r} \sin \theta, \qquad \frac{\partial v}{\partial \theta} = \frac{r^2 - 1}{2r} \cos \theta.$$

On the ellipse (3.5.4), $r$ is constant, and the slope equals

$$\frac{dv}{du} = \frac{\partial v}{\partial \theta} \bigg/ \frac{\partial u}{\partial \theta} = -\frac{r^2 - 1}{r^2 + 1} \cot \theta,$$

while on the hyperbola (3.5.5), $\theta$ is constant, and the slope is

$$\frac{dv}{du} = \frac{\partial v}{\partial r} \bigg/ \frac{\partial u}{\partial r} = \frac{r^2 + 1}{r^2 - 1} \tan \theta.$$

Thus the two curves intersect at right angles.

Since

$$\left(\frac{r^2 + 1}{2r}\right)^2 - \left(\frac{r^2 - 1}{2r}\right)^2 = 1, \quad (\cos \theta)^2 + (\sin \theta)^2 = 1,$$

it follows that the ellipses and hyperbolas have the same foci. We say that equations (3.5.4) and (3.5.5) represent a system of *confocal conics* with foci at $z = \pm 1$. In such a system, the ellipses and the hyperbolas are orthogonal trajectories of each other.

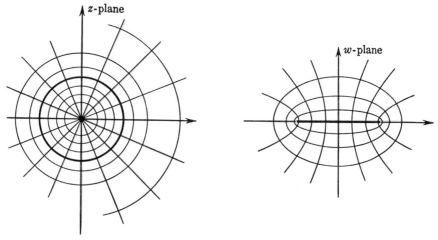

**Figure 10**                      **Figure 11**

The interior of the unit circle is mapped in the same manner. The complete map of the $z$-plane is a Riemann surface of two sheets $S_1$ and $S_2$ with branch points at $w = \pm 1$ and a branch line joining these two points, say the line segment $[-1, +1]$. $S_1$ is the map of $|z| > 1$, $S_2$ of $|z| < 1$. We pass from $S_1$ to $S_2$ by crossing the line segment $[-1, +1]$, the image of $|z| = 1$, in either direction, and we return in the same manner.

We refer to Figures 10 and 11 for the mapping of the $z$-plane onto the $w$-plane. We note that

(3.5.6) $$z = w + (w^2 - 1)^{\frac{1}{2}}.$$

This explicit formula for the inverse function confirms what has already been said about the ramification of this function. It should be observed that $w = \infty$ is not a branch point.

## COLLATERAL READING

For Sections 3.1–3.3 consult:

CARATHÉODORY, C. *Conformal Representation*, Chaps. 1 and 2. Cambridge University Press, New York, 1932.

———. *Theory of Functions of a Complex Variable*, Vol. I, Chaps. 1 and 2. Chelsea Publishing Co., New York, 1954.

FORD, L. R. *Automorphic Functions*, Chaps. 1 and 2. Chelsea Publishing Co., New York, 1951.

Further examples of elementary Riemann surfaces are given in

BEHNKE, H., and SOMMER, F. *Theorie der analytischen Funktionen einer komplexen Veränderlichen*, Chap. 5, Section 1. Springer-Verlag, Berlin, 1955.

# 4

# HOLOMORPHIC FUNCTIONS

**4.1. Complex-valued functions and continuity.** Let $S$ be an arbitrary point set in the complex plane or, equivalently, on the sphere. Suppose that to each point $z_0$ in $S$ there corresponds a uniquely determined complex number $f(z_0)$. We then say that a *function $f(z)$* of the complex variable $z$ is defined in $S$ and that $f(z)$ defines a *mapping* of $S$ into the complex plane. We shall not be concerned with general mappings—on the contrary, quite severe restrictions will be imposed on $f(z)$. As to $S$, it will normally be a domain $D$ or an arc of a simple curve.

DEFINITION 4.1.1. *The function $f(z)$ is said to be continuous at the point $z_0$ of $S$ if, given any $\varepsilon > 0$, there exists a $\delta = \delta(\varepsilon)$ such that for any $z \in S$ with $\chi(z, z_0) < \delta$ we have*

$$(4.1.1) \qquad |f(z) - f(z_0)| < \varepsilon.$$

*Here $\chi(z, z_0)$ is the chordal distance defined by (2.5.12). We say that $f(z)$ is continuous in $S$ if it is continuous at all points of $S$.*

At the present juncture we are not admitting functions taking on the value infinity. On the other hand, it is necessary from the start to admit unbounded point sets and to allow $z = \infty$ as a point of $S$. This explains why we use the chordal metric for $S$ but the Euclidean one for $f(z)$. If $S$ is a bounded set, say $|z| < R$ for all $z$ in $S$, then we may replace the chordal metric by the Euclidean one, since in this case

$$\chi(z_1, z_2) \leqq |z_1 - z_2| \leqq (1 + R^2)\chi(z_1, z_2),$$

so that the two metrics are equivalent.

A function $f(z)$ may very well be continuous in a set $S$ without being bounded there. Thus $f(z) = z^2$ is continuous in the finite plane but not bounded. If $f(z)$ is bounded in $S$, the non-negative quantity

$$(4.1.2) \qquad \sup_{z \in S} |f(z)| \equiv \|f\|$$

is called the *sup-norm* of $f(z)$ in $S$, or, if $f(z)$ is also continuous in $S$, the *C-norm*. The terminology will be explained and elaborated in Section 4.7. Usually we cannot assert the existence of a $z_0 \in S$ such that

$$(4.1.3) \qquad |f(z_0)| = \|f\|.$$

Suppose now that $f(z)$ is defined in $S$ as a bounded continuous function of $z$ and that it is possible to extend the definition of $f(z)$ to the closure of $S$,

68

$\bar{S} = S \cup S'$, in such a manner that $f(z)$ becomes continuous in $\bar{S}$. We can then define a *C-norm* of $f(z)$ in $\bar{S}$, but this quantity obviously will coincide with the previously defined $C$-norm in $S$. For if $|f(z)| \leq \|f\|$ for all $z$ in $S$ and if $z_0 \in \bar{S} \ominus S$, then $|f(z_0)|$ as the limit of a suitably chosen sequence $\{|f(z_n)|\}$, with $z_n \in S$, $z_n \to z_0$, will satisfy the same inequality.

In this case we can assert the existence of a $z_0 \in \bar{S}$ such that (4.1.3) holds. Indeed, by the definition of the supremum there exists a sequence $\{z_n\}$ such that $z_n \in \bar{S}$ and

$$\|f\| - \frac{1}{n} < |f(z_n)| \leq \|f\|.$$

By the Bolzano-Weierstrass theorem the sequence $\{z_n\}$ has at least one limit point $z_0$ which belongs to $\bar{S}$ since $\bar{S}$ is closed. Without restricting the generality, we may assume that $z_n \to z_0$. Hence

$$|f(z_0)| = \lim_{n \to \infty} |f(z_n)| = \|f\|.$$

We come next to the notion of *uniform continuity*.

**THEOREM 4.1.1.**  *If $f(z)$ is defined and continuous in $\bar{S}$, then $f(z)$ is uniformly continuous there, that is, given any $\varepsilon > 0$, there exists a $\delta = \delta(\varepsilon)$ such that $z'$, $z'' \in \bar{S}$ and $\chi(z', z'') < \delta$ imply that*

(4.1.4)                     $$|f(z') - f(z'')| < \varepsilon.$$

*Proof.*  The argument is based on the Heine-Borel theorem, which is proved in Appendix A. Since $f(z)$ is continuous in the closed set $\bar{S}$, for every point $z_0 \in \bar{S}$ and every given $\varepsilon > 0$ there exists a $\delta = \delta(\varepsilon, z_0)$ such that

(4.1.5)          $$|f(z) - f(z_0)| < \varepsilon \quad \text{when} \quad \chi(z, z_0) < \delta(\varepsilon, z_0).$$

The spherical caps

$$C(z_0) = [z \mid \chi(z, z_0) < \tfrac{1}{2}\delta(\varepsilon, z_0), z_0 \in \bar{S}]$$

form an open covering of the closed set $\bar{S}$ on the sphere. By the Heine-Borel theorem there exists a finite subcovering by caps

(4.1.6)                $$C(z_1), C(z_2), \cdots, C(z_n)$$

also covering $\bar{S}$. Suppose that

$$\min_k \delta(\varepsilon, z_k) = \rho,$$

and consider any two points $z'$ and $z''$ of $\bar{S}$ such that

$$\chi(z', z'') < \tfrac{1}{2}\rho.$$

Then there exists a $k$ such that $z' \in C(z_k)$, whence

$$\chi(z', z_k) < \tfrac{1}{2}\delta(\varepsilon, z_k).$$

By the triangle inequality,

$$\chi(z_k, z'') \leq \chi(z_k, z') + \chi(z', z'') < \tfrac{1}{2}\delta(\varepsilon, z_k) + \tfrac{1}{2}\rho < \delta(\varepsilon, z_k),$$

and by (4.1.5) we then obtain

$$|f(z') - f(z'')| \leq |f(z') - f(z_k)| + |f(z_k) - f(z'')| < \varepsilon + \varepsilon = 2\varepsilon.$$

This completes the proof.

We denote the class of functions $f(z)$ continuous in a given set $S$ by $C[S]$ and write $CB[S]$ for the subset of bounded functions. We note that

$$(4.1.7) \qquad\qquad C[S] = CB[S] \quad \text{if} \quad S = \bar{S}.$$

Suppose that $f_1(z)$ and $f_2(z)$ belong to $C[S]$ and that $\alpha$ and $\beta$ are arbitrary complex numbers. Then

$$(4.1.8) \qquad\qquad \alpha f_1(z) + \beta f_2(z) \in C[S], \quad f_1(z)f_2(z) \in C[S].$$

The proof of this fact is the same as in the case of real-valued continuous functions of a real variable familiar to the student. A set of elements $a, b, \cdots$, containing sums, products, and scalar multiples of its elements, is known as an *algebra* (see Section 4.7 below). Thus $C[S]$ and $CB[S]$ are algebras for any choice of $S$. They are not fields, for the quotient of two continuous functions is not always continuous.

Suppose now that $\{f_n(z)\}$ is a sequence of elements of $C[S]$. It may happen that this sequence converges to a limit as $n \to \infty$ for every fixed $z_0 \in S$. By Theorem 1.3.1 this will be the case if and only if for each $\varepsilon > 0$ and each $z_0 \in S$ there exists an integer $M(\varepsilon, z_0)$ such that

$$|f_m(z_0) - f_n(z_0)| < \varepsilon \quad \text{for} \quad m, n > M(\varepsilon, z_0).$$

For the following, the most important case is that in which there exists an integer $M(\varepsilon)$ that will serve for *all* points of $S$.

DEFINITION 4.1.2. *The sequence $\{f_n(z)\}$ of elements in $C[S]$ is said to converge uniformly in $S$ if for each $\varepsilon > 0$ there exists an integer $M(\varepsilon)$, independent of $z$, such that*

$$(4.1.9) \qquad\qquad |f_m(z) - f_n(z)| < \varepsilon$$

*for $m, n > M(\varepsilon)$ and for all $z \in S$.*

THEOREM 4.1.2. *A uniformly convergent sequence of elements in $C[\bar{S}]$ converges to an element of $C[\bar{S}]$.*

*Proof.* Suppose the sequence $\{f_n(z)\}$ satisfies (4.1.9) for every $\varepsilon > 0$. Since we have pointwise convergence in $\bar{S}$, there exists a uniquely determined function $f(z)$ such that

$$\lim_{n \to \infty} f_n(z) = f(z)$$

for each $z \in \bar{S}$. It remains to show that $f(z) \in C[\bar{S}]$, that is, $f(z)$ is continuous and bounded. For any choice of points $z'$, $z''$ in $\bar{S}$ and any integer $m > 0$ we have

$$(4.1.10) \quad |f(z') - f(z'')| \leq |f(z') - f_m(z')| + |f_m(z') - f_m(z'')| + |f_m(z'') - f(z'')|.$$

Choose now a fixed $m$ exceeding $M(\varepsilon/3)$ in the notation of (4.1.9). Passing to the limit with $n$ in (4.1.9), we see that the first and the last terms on the right in (4.1.10) are then each $\leq \varepsilon/3$. Since $\bar{S}$ is closed and $f_m(z) \in C[\bar{S}]$, Theorem 4.1.1 asserts the existence of a quantity $\delta_m(\varepsilon/3)$ such that the second term in (4.1.10) is less than $\varepsilon/3$ for $\chi(z', z'') < \delta_m(\varepsilon/3)$. It follows that $|f(z') - f(z'')| < \varepsilon$ for $\chi(z', z'') < \delta_m(\varepsilon/3)$. Thus $f(z)$ is continuous and $f(z) \in C[\bar{S}]$.

Condition (4.1.9) may obviously be restated as follows:
*There exists an integer $M(\varepsilon)$ such that*

$$(4.1.11) \qquad\qquad \|f_m - f_n\| < \varepsilon \quad for \quad m, n > M(\varepsilon).$$

Such a sequence $\{f_n\}$ is said to be *fundamental*, or a *Cauchy sequence*.
This leads to reformulations of Theorem 4.1.2 which we state as corollaries.

COROLLARY 1.   *If $f_n(z) \in C[\bar{S}]$ and if*

$$(4.1.12) \qquad\qquad \lim_{m, n \to \infty} \|f_m - f_n\| = 0,$$

*then there exists an $f(z) \in C[\bar{S}]$ such that*

$$(4.1.13) \qquad\qquad \lim_{n \to \infty} \|f - f_n\| = 0.$$

COROLLARY 2.   *Cauchy's convergence principle holds in $C[\bar{S}]$.*

The restriction to closed sets is neither necessary nor suitable for our future needs. In particular, we shall need convergence theorems for functions defined in a domain $D$ of the complex plane.

THEOREM 4.1.3.   *If $\{f_n(z)\} \subset CB[D]$ and $\|f_m - f_n\| \to 0$ as $m, n \to \infty$, then there exists an $f(z) \in CB[D]$ such that $\|f_n - f\| \to 0$.*

*Proof.*   We know that $\lim\limits_{n \to \infty} f_n(z) \equiv f(z)$ exists for every $z \in D$. Let $S$ be a closed subset of $D$. Then $f_n(z) \in C[S]$ and converges uniformly in $S$ to an element of $C[S]$. This element must coincide with $f(z)$ in $S$. Since $S$ is arbitrary, it follows that $f(z) \in C[D]$. Now, given any $\varepsilon > 0$, there exists an $M(\varepsilon)$ such that

$$\|f_m - f_n\| = \sup_{z \in D} |f_m(z) - f_n(z)| < \varepsilon \quad for \quad m, n > M(\varepsilon).$$

Letting $m \to \infty$ we see that

$$\sup_{z \in D} |f(z) - f_n(z)| \leq \varepsilon, \quad n > M(\varepsilon),$$

and from this we conclude that $f(z) \in CB[D]$ and that $\|f - f_n\| \leq \varepsilon$.

## EXERCISE 4.1

**1.** Find an upper bound of $| (z')^2 - (z'')^2 |$ when $z'$ and $z''$ range over $| z | \leq 1$ in such a manner that $| z' - z'' | \leq \delta < 2$.

**2.** Given $f(z) = z^2$, $| z | < 1$, find a $\delta(\varepsilon)$ in the notation of Theorem 4.1.1 for this function, replacing chordal distance by the Euclidean one.

**3.** Answer the same two questions for $f(z) = z + \frac{1}{2}z^2$, $| z | \leq 1$.

**4.** Same questions for $f(z) = \dfrac{z}{2 - z}$, $| z | \leq 1$.

**5.** Same questions for $f(z) = \sqrt{z}$, $| z - 1 | \leq 1$.

**6.** If the roots are assumed to be real positive when $z$ is real positive, prove that the sequence $\{ \sqrt[n]{z} \}$ converges uniformly to a limit in $| z - 1 | \leq r < 1$. Is this statement true for $r = 1$? Determine an admissible $M(\varepsilon)$ as a function of $\varepsilon$ and $r$.

**7.** Let $r$ be fixed, $0 < r < 1$. Let $R_1$ and $R_2$ be the regions $| z | \leq r$ and $| z | \geq 1/r$ respectively. The sequence of functions

$$f_n(z) = \frac{1 - z^n}{1 + z^n}$$

belongs to $C[R_1]$ as well as to $C[R_2]$. It converges uniformly in $R_1$ and $R_2$. What are the limit functions? Determine an admissible $M(\varepsilon)$ in each case.

**4.2. Differentiability; holomorphic functions.** We shall restrict the class of functions to be considered by the further requirement of differentiability.

DEFINITION 4.2.1.    *Let $f(z)$ be a single-valued continuous function of $z$ in a domain $D$. We say that $f(z)$ is differentiable at the point $z_0$ of $D$ if*

(4.2.1)                      $$\lim_{h \to 0} \frac{1}{h} [f(z_0 + h) - f(z_0)]$$

*exists as a finite number and is independent of how the complex increment $h$ tends to zero. The limit, when it exists, will be denoted by $f'(z_0)$ and called the derivative of $f(z)$ at $z_0$.*

*The function $f(z)$ is differentiable in $D$ if it is differentiable at all points of $D$, and $f(z)$ is then said to be a holomorphic function of $z$ in $D$. Finally, $f(z)$ is holomorphic at $z = z_0$ if it is holomorphic in some $\varepsilon$-neighborhood of $z_0$.*

Thus a holomorphic function is single-valued, continuous, and differentiable in the domain under consideration. Here the property of being continuous is evidently implied by the stronger property of being differentiable. Instead of "holomorphic" many other terms are used, such as *analytic, monogenic, regular,* and *regular-analytic.* The old term *synectic* is rarely encountered in modern literature. We shall later on speak of *analytic functions*, but in a wider sense, allowing the functions to be many-valued and to have certain singularities in their domains of definition.

We shall denote the operation of differentiation with respect to $z$ by $d/dz$ and write equivalently

$$(4.2.2) \qquad \frac{d}{dz} f(z) = \frac{df}{dz} = f'(z).$$

The formal rules of differential calculus extend to differentiable functions of a complex variable. Thus

$$\frac{d}{dz}[f_1(z) + f_2(z)] = f_1'(z) + f_2'(z),$$

$$\frac{d}{dz}[Cf(z)] = Cf'(z),$$

$$(4.2.3)$$

$$\frac{d}{dz}[f_1(z)f_2(z)] = f_1'(z)f_2(z) + f_1(z)f_2'(z),$$

$$\frac{d}{dz}\frac{f_1(z)}{f_2(z)} = \frac{f_2(z)f_1'(z) - f_1(z)f_2'(z)}{[f_2(z)]^2},$$

provided the right-hand sides exist. We also note that

$$(4.2.4) \qquad \frac{d}{dz} z^n = nz^{n-1},$$

where $n$ is any integer positive, zero, or negative.

These formulas tell us that any polynomial in $z$ is holomorphic in the finite plane and that a rational function is holomorphic in any domain in the finite plane which does not contain any zero of the denominator. In particular, the linear fraction

$$\frac{az + b}{cz + d}, \quad c \neq 0,$$

is holomorphic except at $z = -d/c$. Of the functions considered in Chapter 3 only the roots present a problem. We take

$$f(z) = \sqrt[m]{z}$$

and restrict $z$ to a domain $D$ in which a single-valued determination of the root may be defined. We take one of these determinations, immaterial which one, and we use that throughout the discussion. We set

$$\sqrt[m]{z} = w, \quad \sqrt[m]{z_0} = w_0,$$

and form

$$\frac{\sqrt[m]{z} - \sqrt[m]{z_0}}{z - z_0} = \frac{w - w_0}{w^m - w_0^m} = [w^{m-1} + w^{m-2}w_0 + \cdots + w_0^{m-1}]^{-1}.$$

As $z \to z_0$, $w \to w_0$, and the last member in the display tends to

$$[mw_0^{m-1}]^{-1} = \frac{1}{m}[\sqrt[m]{z_0}]^{1-m} = \frac{1}{m}z_0^{\frac{1}{m}-1}.$$

We conclude that an $m$th root is holomorphic in any domain where it is single-valued. Further, we see that with the appropriate determination of the fractional power, formula (4.2.4) is also valid for $n = 1/m$. The extension to arbitrary fractional exponents $p/q$ is then immediate. If $p$ is a positive integer, we note that complete induction with respect to the number of factors, applied to the third formula under (4.2.3), gives

(4.2.5)
$$\frac{d}{dz}[f(z)]^p = p[f(z)]^{p-1}f'(z),$$

and this holds also for negative integers $p$ by the fourth formula. We then set $f(z) = z^{1/q}$ and get the desired result.

Formula (4.2.5) is of course a special case of the *chain rule*, the formula for the *derivative of a function of a function*. The general case of the latter involves a delicate point which must be postponed to a later occasion. The situation is the following: Given a non-constant function $g(z)$, holomorphic in a domain $D_1$, whose range will be denoted by $R_1$. Given a second function $f(z)$ holomorphic in a domain $D_2$ such that $D_2 \supset R_1$. Then $f[g(z)]$ is well defined, single-valued, and continuous in $D_1$. For the differentiability at $z = z_0 \in D_1$, we have to examine the difference quotient

$$\frac{f[g(z)] - f[g(z_0)]}{z - z_0} = \frac{f[g(z)] - f[g(z_0)]}{g(z) - g(z_0)} \frac{g(z) - g(z_0)}{z - z_0}$$

for values of $z$ near to $z_0$. Here, the right-hand side makes sense unless $g(z) = g(z_0)$. Now, if $g'(z_0) \neq 0$, we can certainly find an $\varepsilon$-neighborhood in which $g(z) \neq g(z_0)$ for $z \neq z_0$. When $z \to z_0$, $g(z) \to g(z_0)$, and we obtain

(4.2.6)
$$\frac{d}{dz}f[g(z)] = f'[g(z)]g'(z).$$

The important point in this argument is the existence of an $\varepsilon$-neighborhood where

$$g(z) \neq g(z_0), \quad |z - z_0| < \varepsilon, \quad z \neq z_0.$$

It will be shown in Chapter 8 that such a neighborhood always exists, regardless of the value of $g'(z_0)$, unless $g(z)$ is a constant. Anticipating this result we may accept (4.2.6) as always valid.

Our definition of being holomorphic at the point $z_0$ clearly presupposes that $z_0 \neq \infty$. The omitted case will now be covered.

DEFINITION 4.2.2.   *$f(z)$ is holomorphic at $z = \infty$ if $f(1/w)$ is holomorphic at $w = 0$.*

Rational functions of the form

$$f(z) = \frac{P(z)}{Q(z)},$$

where the degree of the polynomial $P(z)$ does not exceed that of $Q(z)$, give examples of functions holomorphic at infinity. We note that the function

$$\frac{1}{2}\left(z + \frac{1}{z}\right)$$

of Section 3.5 is holomorphic at every point except 0 and $\infty$.

We denote the class of all functions holomorphic in a given domain $D$ by $H[D]$ and use $HB[D]$ for the subset of functions of $H[D]$ which are also bounded in $D$. Formulas (4.2.3) give:

THEOREM 4.2.1.　　$H[D]$ and $HB[D]$ are algebras.

We note that the field $C$ of complex numbers is a subalgebra of every $HB[D]$, so these algebras are never void. If $D$ is the extended plane (= the complex sphere), then $H[D] = C$; if $D$ is the finite plane, $HB[D] = C$. These assertions are consequences of the theorem of Liouville, proved below in Chapter 8. A similar situation holds for $HB[D]$ whenever the complement of $D$ is a finite point set. On the other hand, if $D$ is not the extended plane, $H[D]$ can never reduce to $C$. Indeed, if $D$ does not contain the point at infinity, then the class of all polynomials in $z$ forms a subalgebra of $H[D]$. If $D$ contains $z = \infty$, and $D$ is not the extended plane, then $D$ must omit at least one finite point $z = a$, and the polynomials in $(z - a)^{-1}$ form a subalgebra of $H[D]$.

The set $H[D]$ is a proper subset of $C[D]$. In other words, there exist continuous functions of a complex variable which are not differentiable. This fact is much easier to prove than the corresponding result for functions of a real variable. Suppose, for instance, that $f(z)$ is a real-valued continuous function in $D$ and that $f(z)$ is not a constant. For the difference quotient

$$\frac{1}{h}[f(z + h) - f(z)] \equiv q(z, h)$$

we then find

$$\arg q(z, h) \equiv -\arg h \pmod{\pi}.$$

Hence, there are two possibilities: either $q(z, h)$ tends to zero with $h$ or it does not have a unique limit. If the first alternative holds in a square, it may be shown that $f(z)$ has a constant value there. Hence, if $f(z)$ is not a constant, there must be points $z$ in $D$ where the second alternative arises, so that $f(z)$ cannot belong to $H[D]$.

The argument also shows that $HB[D]$ is a proper subset of $CB[D]$. We observe in passing that a function in $HB[D]$ need not be uniformly continuous in $D$, but we cannot exhibit a counterexample from the limited set of holomorphic functions with which the student is familiar at this stage. If $D_0$ is any domain such that $\overline{D}_0 \subset D$ and $f(z) \in HB[D]$, then $f(z)$ is uniformly continuous in $\overline{D}_0$, because any function in $CB[D]$ has this property as observed in the proof of Theorem 4.1.3. It will be shown later that such a function is also *uniformly differentiable* in $\overline{D}_0$, that is, the difference quotient approaches its limit $f'(z)$ uniformly with respect to $z$ in $\overline{D}_0$. In general this is not true in the larger domain $D$.

If $f(z) \in HB[D] \subset CB[D]$, then the sup-norm

$$\|f\| = \sup |f(z)|$$

is well defined, and we may raise convergence questions. Given a sequence $\{f_n(z)\} \subset HB[D]$ such that $\lim_{m,\,n\to\infty} \|f_m - f_n\| = 0$, does the limit function $f(z)$, which exists and belongs to $CB[D]$ by Theorem 4.1.3, also belong to $HB[D]$? In other words, does Theorem 4.1.3 hold if we replace $CB[D]$ by $HB[D]$? The answer is in the affirmative, but we have to postpone the discussion of this important question until Chapter 7. For the general background of this type of question see also Section 4.7.

The central theme of this section is differentiation, but it is natural to ask at this juncture whether or not an inverse operation can be found. While the systematic study of integration in the complex plane is deferred to Chapter 7, we shall indicate here a simple type of integral that takes care of our needs in the meantime.

Given a function $f(z) \in C[D]$ and two points $a$ and $b$ such that the line segment $[a, b]$ lies in $D$, we want to define the integral

$$\int_a^b f(z)\,dz.$$

This we can do in two different but equivalent ways. The most direct is to set

$$z = a + r(b - a), \quad 0 \leqq r \leqq 1, \quad f(z) = U(r) + iV(r)$$

and to define

(4.2.7)
$$\int_a^b f(z)\,dz = (b - a)\left\{ \int_0^1 U(r)\,dr + i \int_0^1 V(r)\,dr \right\},$$

where the integrals on the right are the ordinary Riemann integrals of continuous functions. (Cf. Section C.1 of Appendix C.)

Secondly, we can choose two sets of points $z_{n,\,0} = a, z_{n,\,1}, \cdots, z_{n,\,n} = b$ and $\zeta_{n,\,1}, \zeta_{n,\,2}, \cdots, \zeta_{n,\,n}$ on $[a, b]$ in such a manner that $z_{n,\,j}$ precedes $z_{n,\,k}$ on $[a, b]$ if $j < k$, and that $\zeta_{n,\,j}$ is a point of $[z_{n,\,j-1}, z_{n,\,j}]$. We then form the sequence of Riemann sums

(4.2.8)
$$\sum_{j=1}^n f(\zeta_{n,\,j})(z_{n,\,j} - z_{n,\,j-1})$$

and prove in the usual manner that these sums approach a unique limit as

$\max\limits_{j} \mid z_{n,\,j} - z_{n,\,j-1} \mid \to 0$, and this limit is the desired integral. Since every sum (4.2.8) is of the form

$$(b - a) \left\{ \sum_{j=1}^{n} U(\rho_{n,\,j})(r_{n,\,j} - r_{n,\,j-1}) + i \sum_{j=1}^{n} V(\rho_{n,\,j})(r_{n,\,j} - r_{n,\,j-1}) \right\},$$

it is clear that the two definitions are equivalent.

This integral has a number of properties in common with the classical Riemann integral. We note in particular the following: *Integration is a linear operation*, that is, if $f(z)$, $g(z) \in C[D]$ and $\alpha$ and $\beta$ are arbitrary complex numbers, then

$$(4.2.9) \qquad \int_a^b [\alpha f(z) + \beta g(z)]\, dz = \alpha \int_a^b f(z)\, dz + \beta \int_a^b g(z)\, dz.$$

Further, if $\mid f(z) \mid \leq M$ on $[a, b]$, then

$$(4.2.10) \qquad \left| \int_a^b f(z)\, dz \right| \leq M \mid b - a \mid.$$

Finally we list a property showing that we are actually dealing with an operation which is the inverse of differentiation. Suppose that $f(z) \in H[D]$ and that $f'(z)$ is continuous on $[a, b]$. Then

$$(4.2.11) \qquad \int_a^b f'(z)\, dz = f(b) - f(a).$$

In fact we have

$$f'(z) = [U'(r) + iV'(r)](b - a)^{-1},$$

where the primes indicate differentiation with respect to $z$ on the left and with respect to $r$ on the right. Hence by (4.2.7)

$$\int_a^b f'(z)\, dz = \int_0^1 U'(r)\, dr + i \int_0^1 V'(r)\, dr$$
$$= U(1) + iV(1) - U(0) - iV(0)$$
$$= f(b) - f(a),$$

as asserted. We shall see later that the assumed continuity of $f'(z)$ on the line segment $[a, b]$ is actually implied by the fact that $f(z)$ is holomorphic in $D$.


## EXERCISE 4.2

**1.** At what points do the following functions fail to be holomorphic?

    **a.** $\dfrac{z^4 + 2z}{z^3 - 3z + 2}$      **b.** $\dfrac{z^4 + 1}{z^2 + 1}$      **c.** $\dfrac{\sqrt{z} + 1}{\sqrt{z} - 1}$      **d.** $\sqrt{1 - z^2}$

**2.** Find the largest circle with center at the origin in which the two determinations of $(1 - z^2)^{\frac{1}{2}}$ are holomorphic. What conditions should a domain $D$ satisfy to be a domain of holomorphy of one of the determinations of $(1 - z^2)^{\frac{1}{2}}$?

**3.** Same question for $[z + (1 - z^2)^{\frac{1}{2}}]^{\frac{1}{2}}$, which has four determinations. Where are the branch points?

**4.** If the domain $D$ does not contain the disk $|z - a| \leq r$, show that the polynomials in $(z - a)^{-1}$ form a subalgebra of $HB[D]$.

**5.** We define

$$z^i = e^{-\theta}[\cos{(\log r)} + i \sin{(\log r)}],$$

$$z = r(\cos \theta + i \sin \theta).$$

Let $D$ be the semicircle $|\theta| < \frac{1}{2}\pi$, $0 < r < 1$. Show that $z^i \in CB[D]$ but is not uniformly continuous in $D$. (Since $z^i$ actually is holomorphic in $D$, this is an example of a function in $HB[D]$ which is not uniformly continuous in $D$.)

**6.** Show that $f(z) = |z|$ does not have a derivative anywhere. Determine the limiting values of the difference quotient as $h \to 0$, and show that they form the circumference of a circle in the complex plane. Consider $z = 0$ and $z \neq 0$ separately. (This is a special case of a theorem on "polygenic" functions discovered by Edward Kasner (1878–1955) in 1936.)

**7.** Same question for $f(z) = x$.

**8.** Show that $f(z) = |z|^2$ has a derivative at $z = 0$ but nowhere else. Find the locus of the limits of the difference quotient for $z \neq 0$.

**9.** Suppose that $f(z) = f(x + iy) = F(x, y)$ is real-valued and continuous and that its derivative is zero in the rectangle $a < x < b$, $c < y < d$. Show that the partial derivatives of $F(x, y)$ with respect to $x$ and $y$ exist, that they are zero in the rectangle, and, hence, that $f(z)$ is constant there.

**10.** Find the value of the integral of $z^i$ taken along the real axis from 0 to 1.

**4.3. The Cauchy-Riemann equations.** The question of the existence of the derivative of a function of a complex variable may be reduced to a study of the existence and the properties of the partial derivatives of the real and imaginary parts considered as functions of $(x, y)$. We set

(4.3.1) $$f(z) = f(x + iy) = U(x, y) + iV(x, y).$$

THEOREM 4.3.1. *Suppose that $f(z)$ is defined and continuous in some neighborhood of $z = c = a + ib$. A necessary condition for the existence of $f'(c)$ is that $U(x, y)$ and $V(x, y)$ have first-order partials at $(x, y) = (a, b)$, and that*

(4.3.2) $$U_x(a, b) = V_y(a, b),$$
$$U_y(a, b) = -V_x(a, b).$$

*Proof.* Suppose that

$$\lim_{h \to 0} \frac{1}{h}[f(c+h) - f(c)] \equiv f'(c)$$

exists and is independent of direction. Taking $h = \alpha$ real, we have

$$\frac{1}{\alpha}[f(c+\alpha) - f(c)] = \frac{1}{\alpha}[U(a+\alpha, b) - U(a, b)] + i\frac{1}{\alpha}[V(a+\alpha, b) - V(a, b)],$$

and this quantity approaches a limit as $\alpha \to 0$ by assumption. The limit must clearly be

$$U_x(a, b) + iV_x(a, b).$$

Likewise,

$$\frac{1}{\beta i}[f(c+i\beta) - f(c)] = -i\frac{1}{\beta}[U(a, b+\beta) - U(a, b)]$$
$$+ \frac{1}{\beta}[V(a, b+\beta) - V(a, b)]$$

must have a limit as $\beta \to 0$, namely,

$$-iU_y(a, b) + V_y(a, b),$$

and these two limits must be equal. Separating reals and imaginaries, we get (4.3.2). We note the resulting expressions for the derivative

$$(4.3.3) \qquad f'(c) = U_x(a, b) + iV_x(a, b) = V_y(a, b) - iU_y(a, b).$$

COROLLARY. *If $f(z)$ is holomorphic in a domain $D$, then $U(x, y)$ and $V(x, y)$ have first-order partial derivatives in $D$ and these partials satisfy the Cauchy-Riemann equations in $D$:*

$$(4.3.4) \qquad \frac{\partial U}{\partial x} = \frac{\partial V}{\partial y}, \qquad \frac{\partial U}{\partial y} = -\frac{\partial V}{\partial x}.$$

We note the complex form of these equations:

$$(4.3.5) \qquad \frac{\partial f}{\partial x} = -i\frac{\partial f}{\partial y}.$$

The conditions of Theorem 4.3.1 are not sufficient for the existence of a derivative, as is shown by the following example due to D. Menchoff.[1] We take

$$f(z) = \begin{cases} z^5 \, |\, z\, |^{-4}, & z \neq 0, \\ 0, & z = 0. \end{cases}$$

---

[1] *Les Conditions de Monogénéité* (Actualités Scientifiques et Industrielles; Hermann et Cie, Paris, 1936), p. 329. We refer to this tract also for a discussion of sufficient conditions much weaker than those of Theorem 4.3.2, below.

**Here**

$$\frac{f(h)}{h} = \left(\frac{h}{|h|}\right)^4,$$

and this has the value 1 when $h$ is real or purely imaginary. We get

$$U_x(0, 0) = V_y(0, 0) = 1, \quad U_y(0, 0) = -V_x(0, 0) = 0,$$

so the Cauchy-Riemann equations are satisfied at the origin, but there is obviously not a unique derivative. It is clear then that additional conditions are needed for the existence of the derivative. A set of such conditions is given in the next theorem; less restrictive conditions are known, however.

THEOREM 4.3.2.    *Suppose that $f(z) = U(x, y) + iV(x, y)$ is defined in some ε-neighborhood of $z = c$ as a single-valued continuous function having continuous partials with respect to $x$ and $y$, and suppose that the partials satisfy (4.3.4) for $z = c$. Then $f(z)$ has a unique derivative at $z = c$.*

*Proof.*    The difference quotient at $z = c$ with increment $h = \alpha + i\beta$ equals

$$\frac{U(a + \alpha, b + \beta) - U(a, b)}{\alpha + i\beta} + i\frac{V(a + \alpha, b + \beta) - V(a, b)}{\alpha + i\beta}$$

$$= \frac{\alpha}{\alpha + i\beta}\left\{\frac{1}{\alpha}[U(a + \alpha, b + \beta) - U(a, b + \beta)]\right.$$

$$\left. + i\frac{1}{\alpha}[V(a + \alpha, b + \beta) - V(a, b + \beta)]\right\}$$

$$+ \frac{\beta}{\alpha + i\beta}\left\{\frac{1}{\beta}[U(a, b + \beta) - U(a, b)] + i\frac{1}{\beta}[V(a, b + \beta) - V(a, b)]\right\}$$

$$= \frac{\alpha}{\alpha + i\beta}[U_x(a + \theta_1\alpha, b + \beta) + iV_x(a + \theta_2\alpha, b + \beta)]$$

$$+ \frac{\beta}{\alpha + i\beta}[U_y(a, b + \theta_3\beta) + iV_y(a, b + \theta_4\beta)],$$

where $0 < \theta_k < 1$, by the mean value theorem of the differential calculus. Since the partials are continuous at $(x, y) = (a, b)$, we may replace the last expression by

$$\frac{\alpha}{\alpha + i\beta}[U_x(a, b) + iV_x(a, b) + \eta_1] + \frac{\beta}{\alpha + i\beta}[U_y(a, b) + iV_y(a, b) + \eta_2],$$

where $\eta_1$ and $\eta_2$ tend to zero with $h$. Since (4.3.4) holds at $(a, b)$, this expression reduces to

$$U_x(a, b) + iV_x(a, b) + \frac{\alpha\eta_1}{\alpha + i\beta} + \frac{\beta\eta_2}{\alpha + i\beta}.$$

The last two terms are dominated in absolute value by $|\eta_1|$ and $|\eta_2|$ and, hence, tend to zero as $\alpha \to 0$, $\beta \to 0$. Thus, a unique derivative exists.

COROLLARY. *If $f(z)$ is defined as a single-valued function having continuous partials with respect to $x$ and $y$ in a domain $D$, and if the partials satisfy the Cauchy-Riemann equations in $D$, then $f(z)$ is holomorphic in $D$.*

The system of partial differential equations (4.3.4) evidently has an abundance of solutions having remarkable properties. Thus, for a given domain $D$ any $f(z) \in H[D]$ will give a solution $U(x, y) = \Re[f(z)]$, $V(x, y) = \Im[f(z)]$. Conversely, if $U(x, y)$, $V(x, y)$ satisfy (4.3.4) and have continuous partials in $D$, then $U(x, y) + iV(x, y) = f(z) \in H[D]$. There exists a class of systems of first-order partial differential equations, similar to (4.3.4) but having variable coefficients, for which the composite function $U(x, y) + iV(x, y)$ has many basic properties in common with holomorphic functions.[1]

There are many useful variants of the Cauchy-Riemann equations. In Problem 2 of Exercise 4.3 the reader is invited to transform the equations into polar coordinates. At this juncture we shall merely take up the expressions resulting from the identities

$$2x = z + \bar{z}, \quad 2iy = z - \bar{z}.$$

Suppose that a pair of real-valued differentiable functions, $U(x, y)$ and $V(x, y)$, is given. Since

$$U(x, y) = U\left[\frac{1}{2}(z + \bar{z}), \frac{1}{2i}(z - \bar{z})\right],$$

$$V(x, y) = V\left[\frac{1}{2}(z + \bar{z}), \frac{1}{2i}(z - \bar{z})\right],$$

we may define

(4.3.6) $$U(x, y) + iV(x, y) \equiv f(z, \bar{z}).$$

Though $z$ and $\bar{z}$ are not independent variables, since $z$ determines $\bar{z}$, we shall nevertheless treat them as if they were and form partial derivatives with respect to $z$ and $\bar{z}$, using the formal rules of the calculus. Thus, we define

$$\frac{\partial f}{\partial z} = \frac{\partial f}{\partial x}\frac{\partial x}{\partial z} + \frac{\partial f}{\partial y}\frac{\partial y}{\partial z}$$

and a similar expression with $z$ replaced by $\bar{z}$. This leads to the formulas

(4.3.7)
$$\frac{\partial f}{\partial z} = \frac{1}{2}\left[\frac{\partial U}{\partial x} + \frac{\partial V}{\partial y}\right] + \frac{i}{2}\left[\frac{\partial V}{\partial x} - \frac{\partial U}{\partial y}\right],$$
$$\frac{\partial f}{\partial \bar{z}} = \frac{1}{2}\left[\frac{\partial U}{\partial x} - \frac{\partial V}{\partial y}\right] + \frac{i}{2}\left[\frac{\partial V}{\partial x} + \frac{\partial U}{\partial y}\right].$$

---

[1] A systematic study has been made by Lipman Bers and Abe Gelbart, "On a Class of Functions Defined by Partial Differential Equations," *Transactions of the American Mathematical Society*, Vol. 56 (1944), pp. 67–93.

Here, the right sides have a sense if $U$ and $V$ have partial derivatives, and we take them as definitions of the left sides. We see that the Cauchy-Riemann equations now assume the simple and condensed form

(4.3.8) $$\frac{\partial f}{\partial \bar{z}} = 0.$$

This condition sometimes provides a convenient test for holomorphism.

For future reference we list the formula

(4.3.9)   $|f'(z)|^2 = [U_x(x, y)]^2 + [V_x(x, y)]^2 = [U_y(x, y)]^2 + [V_y(x, y)]^2.$

This is an immediate consequence of (4.3.3).

## EXERCISE 4.3

**1.** Verify that the following pairs of functions satisfy the Cauchy-Riemann equations:

**a.** $U = x^2 - y^2, \quad V = 2xy.$

**d.** $U = x^3 - 3xy^2, \quad V = 3x^2y - y^3.$

**b.** $U = e^x \cos y, \quad V = e^x \sin y.$

**e.** $U = \dfrac{x}{x^2 + y^2}, \quad V = -\dfrac{y}{x^2 + y^2}.$

**c.** $U = \sin x \cosh y, \quad V = \cos x \sinh y.$

**2.** Transform the Cauchy-Riemann equations by the introduction of polar coordinates

$$x = r \cos \theta, \quad y = r \sin \theta, \quad U = R \cos \Theta, \quad V = R \sin \Theta,$$

and derive the following equations:

**a.** $r \dfrac{\partial U}{\partial r} = \dfrac{\partial V}{\partial \theta}, \quad r \dfrac{\partial V}{\partial r} = -\dfrac{\partial U}{\partial \theta}.$

**b.** $\dfrac{1}{R} \dfrac{\partial R}{\partial x} = \dfrac{\partial \Theta}{\partial y}, \quad \dfrac{1}{R} \dfrac{\partial R}{\partial y} = -\dfrac{\partial \Theta}{\partial x}.$

**c.** $\dfrac{r}{R} \dfrac{\partial R}{\partial r} = \dfrac{\partial \Theta}{\partial \theta}, \quad \dfrac{1}{R} \dfrac{\partial R}{\partial \theta} = -r \dfrac{\partial \Theta}{\partial r}.$

**3.** Show that the function defined by

$$f(z) = \log |z| + i \arg z$$

is holomorphic in the sector $0 < |z|, \; |\arg z| < \pi$. What is the value of its derivative?

**4.** Same questions for

$$f(z) = \sqrt{r} \, [\cos \tfrac{1}{2}\theta + i \sin \tfrac{1}{2}\theta], \quad r > 0, \quad |\theta| < \pi.$$

**5.** Same questions for the function $z^i$ defined in Problem 5 of Exercise 4.2.

**6.** Defining $U$ and $V$ as in Problem 1a above, express them in terms of $z$ and $\bar{z}$, and verify that $U + iV$ satisfies (4.3.8).

**7.** Use (4.3.8) to show that the functions $x$, $|z|$, and $|z|^2$ are not holomorphic.

**4.4. Laplace's equation.** Given $f(z) \in H[D]$ and

$$f(z) = U(x, y) + iV(x, y),$$

we know that the first-order partials of $U$ and $V$ exist and satisfy the Cauchy-Riemann equations. Supposing, in addition, that continuous second-order partials exist, we have

$$\frac{\partial^2 U}{\partial x^2} = \frac{\partial^2 V}{\partial x\, \partial y} = \frac{\partial^2 V}{\partial y\, \partial x} = -\frac{\partial^2 U}{\partial y^2},$$

and

$$\frac{\partial^2 V}{\partial x^2} = -\frac{\partial^2 U}{\partial x\, \partial y} = -\frac{\partial^2 U}{\partial y\, \partial x} = -\frac{\partial^2 V}{\partial y^2},$$

so that

(4.4.1)
$$\Delta U \equiv \frac{\partial^2 U}{\partial x^2} + \frac{\partial^2 U}{\partial y^2} = 0,$$

and the same relation holds for $V$. We shall see later that $U$ and $V$ have partial derivatives of all orders, so it is not necessary to assume the existence of the second-order partials.

Equation (4.4.1) is known as *Laplace's equation in two dimensions*. Its solutions are called *harmonic functions*, or *potential functions*. The same names are used for the solutions of Laplace's equation in three dimensions

(4.4.2)
$$\frac{\partial^2 U}{\partial x^2} + \frac{\partial^2 U}{\partial y^2} + \frac{\partial^2 U}{\partial z^2} = 0.$$

When it is necessary to make a distinction, $U(x, y)$ is called a *logarithmic potential*, $U(x, y, z)$ a *Newtonian potential*. They correspond to different laws of attraction; the force is inversely proportional to the distance in the first case and to the square of the distance in the second case. $\Delta U$ is called the *Laplacean* of $U$.

An elementary solution of (4.4.1) is given by

(4.4.3)
$$\log{[(x - a)^2 + (y - b)^2]}^{-\frac{1}{2}}.$$

This corresponds to the solution

(4.4.4)
$$[(x - a)^2 + (y - b)^2 + (z - c)^2]^{-\frac{1}{2}}$$

of (4.4.2). Both solutions are used for the purpose of building up more general solutions of the respective equations. With $z = x + iy$, it is clear that

$$(4.4.5) \qquad U(z) = \sum_{k=1}^{n} \alpha_k \log | z - z_k |^{-1}$$

is a solution of (4.4.1) for any choice of the real numbers $\alpha_k$ and the complex numbers $z_k$. We can think of $U(z)$ as the potential at the point $z$ produced by the masses $\alpha_k$ located at the points $z_k$. If we have a unit mass located at $z$, it is attracted by the mass $\alpha_k$ at $z_k$ if $\alpha_k > 0$; otherwise it is repelled. The force exerted by $z_k$ on $z$ is then a vector given in magnitude and direction by the complex number

$$\frac{\alpha_k}{\bar{z}_k - \bar{z}},$$

and the resulting force from all the masses is given by

$$(4.4.6) \qquad F(z) = \sum_{k=1}^{n} \frac{\alpha_k}{\bar{z}_k - \bar{z}}.$$

A point $z_0$ is a *point of equilibrium* of $z$ if $F(z_0) = 0$. This is a point where the system of the $n$ masses exerts no force on $z$ and where $z$, then, can stay at rest. We shall use this interpretation to prove the following theorem due to Gauss and F. Lucas:

THEOREM 4.4.1.  *Let $P(z)$ be a polynomial of degree $n$ having the zeros $z_1$, $z_2$, $\cdots$, $z_n$ (multiple roots repeated in this sequence according to their multiplicity), and let $\Pi$ be the least convex polygon containing the zeros. Then $P'(z)$ cannot vanish anywhere in the exterior of $\Pi$.*

*Proof.*  We have

$$(4.4.7) \qquad \frac{P'(z)}{P(z)} = \sum_{k=1}^{n} \frac{1}{z - z_k},$$

so that a zero $z_0$ of $P'(z)$ is a point of equilibrium with respect to a system of particles with unit mass at each of the points $z_k$. Physically it is fairly obvious that the force acting on $z$ can never be zero when $z$ is outside of $\Pi$. Analytically we argue as follows: The *convex hull* of the set $(z_1, z_2, \cdots, z_n)$, that is, $\Pi$ and its interior, is given by a formula similar to (2.3.4), namely,

$$(4.4.8) \qquad [z \mid z = \Sigma_1^n \alpha_k z_k, \; \alpha_k \geq 0, \; \Sigma_1^n \alpha_k = 1].$$

Suppose now that $P'(z_0) = 0$. We have then

$$0 = \sum_{k=1}^{n} \frac{1}{\bar{z}_0 - \bar{z}_k} = \sum_{k=1}^{n} \frac{z_0 - z_k}{| z_0 - z_k |^2},$$

so that

$$z_0 \sum_{k=1}^{n} \frac{1}{| z_0 - z_k |^2} = \sum_{k=1}^{n} \frac{z_k}{| z_0 - z_k |^2},$$

or $z_0$ belongs to the convex hull as asserted.

We do not have to restrict ourselves to point masses. We can also consider mass distributions of a more general nature. The finite sum in (4.4.5) is then replaced by a Stieltjes integral. Further consideration of this matter would take us beyond the scope of this text, however, so we shall not pursue the subject.

To every solution $U(x, y)$ of (4.4.1), defined and harmonic in a domain $D$, corresponds a *conjugate harmonic function* $V(x, y)$, in general infinitely many-valued in $D$, such that $U + iV$ is *locally holomorphic* in $D$; that is, each determination is holomorphic in some neighborhood of any point of $D$ where it is defined. Let $z_0 \in D$, let $N(z_0)$ be a circular neighborhood of $z_0$ such that $N(z_0) \subset D$. The Cauchy-Riemann equations give us the conditions

$$V_x = -U_y, \quad V_y = U_x.$$

Hence

(4.4.9) $$dV \equiv -U_y \, dx + U_x \, dy$$

is an exact differential in $N(z_0)$, and $V(x, y)$ may be found by evaluating the line integral

(4.4.10) $$\int_{(x_0, y_0)}^{(x, y)} [-U_t(s, t) \, ds + U_s(s, t) \, dt] \equiv V(x, y) - V(x_0, y_0)$$

along a line segment in $N(z_0)$. $V(x_0, y_0)$ may be chosen arbitrarily, but then $V(x, y)$ is uniquely determined in $N(z_0)$, and $U(x, y) + iV(x, y)$ is holomorphic in $N(z_0)$. The extension to all of $D$ is based on the method of analytic continuation presented in Chapter 10, Volume II.

For example, the conjugate harmonic function corresponding to the function $U(z)$ of (4.4.5) is given by the infinitely many-valued function

(4.4.11) $$-\sum_{k=1}^{n} \alpha_k \arg(z - z_k).$$

See Problem 3 of Exercise 4.3.

There are various important transformations of Laplace's equation to be noted. Suppose $U(z)$ is harmonic in a domain $D$ of the $z$-plane and $D$ is the one-to-one image of a domain $D_0$ under the mapping

$$z = f(\zeta), \quad \zeta = \xi + i\eta,$$

where $f(\zeta)$ is holomorphic in $D_0$. We have then

(4.4.12) $$\frac{\partial^2 U}{\partial \xi^2} + \frac{\partial^2 U}{\partial \eta^2} = |f'(\zeta)|^2 \left( \frac{\partial^2 U}{\partial x^2} + \frac{\partial^2 U}{\partial y^2} \right).$$

It follows that the function $U[f(\zeta)]$ is harmonic in $D_0$.

We also note the expression of the Laplacean in terms of $z$ and $\bar{z}$:

(4.4.13) $$\Delta U = 4 \frac{\partial^2 U}{\partial \bar{z} \, \partial z}.$$

The expression of the Laplacean in terms of polar coordinates is given in Problem 6 of Exercise 4.4.

## EXERCISE 4.4

**1.** Verify that $U = y(x^2 + y^2)^{-1}$ is harmonic except at the origin and find the conjugate harmonic function $V(z)$ if $V(\infty) = 0$.

**2.** If $U(z) = \arg z$, $|\arg z| < \pi$, find $V(z)$, given that $V(1) = 0$.

**3.** Prove (4.4.12).

**4.** Let $U(z)$ be defined by (4.4.5). The curve $U(z) = C$ is known as an equipotential line. Sketch these curves if $n = 2$, $z_1 = -1$, $z_2 = 1$, and
**a.** $\alpha_1 = \alpha_2 = 1$;
**b.** $\alpha_1 = -\alpha_2 = 1$.

**\*5.** If $F(z_0) \neq 0$, where $F(z)$ is defined by (4.4.6), prove that the vector $iF(z_0)$ is parallel to a tangent vector of $U(z) = U(z_0)$ at $z = z_0$. In other words, the force is normal to the equipotential line. (*Hint:* Take one fraction at a time in (4.4.6), multiply by $i$, and interpret. Show that when equipotential lines are "added," tangent vectors are also added.)

**6.** If $x = r \cos \theta$, $y = r \sin \theta$, show that

$$\Delta U = \frac{\partial^2 U}{\partial r^2} + \frac{1}{r} \frac{\partial U}{\partial r} + \frac{1}{r^2} \frac{\partial^2 U}{\partial \theta^2}.$$

**7.** Show that $U = r^n \cos n\theta$, $n$ a positive integer, is harmonic, and find the conjugate harmonic function.

**8.** The function $\log r$ is a logarithmic potential function. Does $r$ or any power of $r$ have this property?

**9.** Let $U(z)$ be harmonic in the domain $D$ and let its range be contained in the interval $(a, b)$. Let $F(u)$ be defined for $u$ in $(a, b)$ and have continuous first- and second-order derivatives. What further properties must $F(u)$ have in order that $F[U(z)]$ be harmonic in $D$?

**10.** Show that $U(x^2 - y^2, 2xy)$ is harmonic if and only if $U(x, y)$ is.

**4.5. The inverse function.** In earlier chapters we have studied some mappings defined by special holomorphic functions. Two features of this study may have been noticed by the student: the role played by the inverse function and the property of preserving angles that is possessed by the mapping, certain points excepted. The time has come to make a general study of these features, and we start with the inverse function. We shall have to postulate that the mapping function $f(z)$ has a *continuous* first derivative. It will be shown in Chapter 7 that a holomorphic function has derivatives of all orders, so that this assumption is clearly redundant. It is therefore put in brackets in the formulation of the theorems.

THEOREM 4.5.1.    *Suppose that $f(z)$ is holomorphic in the circle $| z - z_0 | < r_1$ and that $[f'(z)$ is continuous there and $] f'(z_0) \neq 0$. If $f(z_0) = w_0$, there exist a circle $| w - w_0 | < r_2$ and a function $g(w)$, defined and holomorphic for $| w - w_0 | < r_2$, such that (1) $g(w_0) = z_0$, (2) $| g(w) - z_0 | < r_1$, (3)*

$$(4.5.1) \qquad\qquad f[g(w)] = w,$$

*and (4) if $z$ and $w$ are corresponding points with $z = g(w)$, then*

$$(4.5.2) \qquad\qquad f'(z)g'(w) = 1.$$

*Moreover, $z = g(w)$ is the only solution of*

$$(4.5.3) \qquad\qquad w = f(z)$$

*satisfying (2) for small values of $| w - w_0 |$.*

Proof.    We shall base the argument upon a variant of the *method of successive approximations.*[1] We start by observing that since $f'(z)$ is continuous, we can find an $r_3 \leq r_1$ such that

$$(4.5.4) \qquad | f'(z) - f'(z_0) | < \tfrac{1}{2} | f'(z_0) | \quad \text{for} \quad | z - z_0 | < r_3.$$

For such values of $z$ we define

$$(4.5.5) \qquad\qquad R(z, z_0) \equiv f(z) - f(z_0) - (z - z_0)f'(z_0),$$

so that

$$(4.5.6) \qquad R(z_2, z_0) - R(z_1, z_0) = f(z_2) - f(z_1) - (z_2 - z_1)f'(z_0)$$

$$= \int_{z_1}^{z_2} [f'(z) - f'(z_0)] \, dz.$$

Here the integral is taken in the sense defined in Section 4.2. The line segment $[z_1, z_2]$ lies in the circle $| z - z_0 | < r_3$ where $f'(z)$ is continuous. We have used (4.2.9) and (4.2.11) plus the fact that 1 is the derivative of $z$. Using (4.2.10) and (4.5.4), we get the estimate

$$(4.5.7) \qquad | R(z_1, z_0) - R(z_2, z_0) | < \tfrac{1}{2} | f'(z_0) | \, | z_1 - z_2 |,$$

---

[1] This particular variant is due to Édouard Goursat (1858–1936) and was published in 1903. The method of successive approximations as a tool for proving existence and uniqueness theorems was created by Émile Picard (1856–1941) in 1890 for the study of differential equations. It was later extended to many other classes of functional equations. In principle, the method is the time-honored device of trial and error, but with emphasis on estimation of the error. Picard is famous for his theorem on exceptional values of entire and meromorphic functions (see Chapter 14, Volume II). That theorem has served as point of departure for a large part of the research in complex function theory during the three quarters of a century that have elapsed since its discovery. Much of Picard's work was devoted to algebraic geometry and functions of two complex variables: algebraic functions and their integrals, transformation groups and related automorphic functions.

which is basic for the following. Such an inequality is known as a *Lipschitz condition*, after Rudolf Lipschitz (1832–1903).

We now consider (4.5.5), which we write in the equivalent form

$$(4.5.8) \qquad z - z_0 = a(w - w_0) - aR(z, z_0), \quad \text{where} \quad a = [f'(z_0)]^{-1}.$$

We want to solve this equation for $z$ in terms of $w$. Since $| R(z, z_0) |$ is small in comparison with $| z - z_0 |$, we get a first crude approximation by neglecting $R(z, z_0)$ altogether. This gives

$$z_1 = z_0 + a(w - w_0).$$

We can probably get a better approximation by computing $R(z_1, z_0)$ and setting

$$z_2 = z_1 - aR(z_1, z_0).$$

We may iterate this procedure as often as we please, and we must now show that there exist values of $w$ near to $w_0$ for which the approximations actually become better and better and converge to the desired solution.

Let us restrict $w$ by the condition

$$(4.5.9) \qquad | w - w_0 | < \tfrac{1}{2} | f'(z_0) | \, r_3 \equiv r_2$$

and define

$$g_1(w) = z_0 + a(w - w_0),$$

$$(4.5.10)$$

$$g_n(w) = z_0 + a(w - w_0) - aR[g_{n-1}(w), z_0], \quad n > 1$$

for $| w - w_0 | < r_2$. It is clear that $g_1(w)$ is uniquely defined as a continuous function of $w$, and $| g_1(w) - z_0 | < \tfrac{1}{2}r_3$. Suppose that we have shown for $n \leq k$ that the functions $g_n(w)$ exist as continuous functions of $w$ for $| w - w_0 | < r_2$, satisfying the two inequalities

$$(4.5.11) \qquad | g_n(w) - g_{n-1}(w) | < 2^{-n}r_3,$$

$$(4.5.12) \qquad | g_n(w) - z_0 | < (1 - 2^{-n})r_3.$$

Here the second inequality is obviously a consequence of the first, since

$$| g_n(w) - z_0 | \leq | g_1(w) - z_0 | + \sum_{m=2}^{n} | g_m(w) - g_{m-1}(w) |.$$

The induction hypothesis is clearly verified for $k = 1$ if we set $g_0(w) \equiv z_0$.

Since $g_k(w)$ is well defined, continuous, and satisfies

$$| g_k(w) - z_0 | < (1 - 2^{-k})r_3 < r_1,$$

it follows that

$$g_{k+1}(w) = z_0 + a(w - w_0) - aR[g_k(w), z_0]$$

is also well defined and continuous, provided $| w - w_0 | < r_2$. Further, condition (4.5.7) shows that

$$
\begin{aligned}
| g_{k+1}(w) - g_k(w) | &= | a | \, | R[g_k(w), z_0] - R[g_{k-1}(w), z_0] | \\
&< 2^{-1} | g_k(w) - g_{k-1}(w) | \\
&< 2^{-k-1} r_3
\end{aligned}
$$

by (4.5.11). We have then also (4.5.12) holding for $n = k + 1$. It follows that all approximations $g_n(w)$ are well defined and continuous for $| w - w_0 | < r_2$, and that (4.5.11) and (4.5.12) hold for all $n$.

These estimates show that the series

$$
g_1(w) + \sum_{n=2}^{\infty} [g_n(w) - g_{n-1}(w)]
$$

converges absolutely and uniformly for $| w - w_0 | < r_2$. Its sum,

(4.5.13)
$$
g(w) = \lim_{n \to \infty} g_n(w),
$$

is consequently a continuous function of $w$, and

$$
| g(w) - z_0 | \leq r_3 \leq r_1.
$$

Further,

$$
\begin{aligned}
g(w) = \lim_{n \to \infty} g_n(w) &= z_0 + a(w - w_0) - \lim_{n \to \infty} a R[g_{n-1}(w), z_0] \\
&= z_0 + a(w - w_0) - a R[\lim_{n \to \infty} g_{n-1}(w), z_0] \\
&= z_0 + a(w - w_0) - a R[g(w), z_0].
\end{aligned}
$$

In view of (4.5.5), this relation reduces to (4.5.1). Thus $g(w)$ is a continuous solution of (4.5.3) satisfying (1) and (2) in $| w - w_0 | < r_2$.

To prove the uniqueness of the solution we use the Lipschitz condition once more. Suppose that $z = h(w)$ is any solution of (4.5.3) which is defined for $| w - w_0 | < r_0 \leq r_2$ and which satisfies $| h(w) - z_0 | < r_1$. It is not necessary to assume that $h(w)$ is continuous or that $h(w_0) = z_0$. From

$$
h(w) = z_1 - a R[h(w), z_0], \quad g_n(w) = z_1 - a R[g_{n-1}(w), z_0]
$$

we get

$$
\begin{aligned}
| h(w) - g_n(w) | &= | a | \, | R[h(w), z_0] - R[g_{n-1}(w), z_0] | \\
&< 2^{-1} | h(w) - g_{n-1}(w) | \\
&< 2^{-2} | h(w) - g_{n-2}(w) | \\
&\quad \cdot \; \cdot \; \cdot \; \cdot \; \cdot \; \cdot \; \cdot \; \cdot \; \cdot \; \cdot \; \cdot \; \cdot \\
&< 2^{-n+1} | h(w) - g_1(w) | \\
&< 2^{-n+1} [\, | h(w) - z_0 | + | g_1(w) - z_0 | \,] \\
&< 2^{-n+2} r_1.
\end{aligned}
$$

It follows that

$$h(w) = \lim_{n \to \infty} g_n(w),$$

or the solution is unique.

It remains to prove that $g(w)$ is actually holomorphic for $|w - w_0| < r_2$. Suppose that $z_1$, $w_1$ and $z_2$, $w_2$ are two pairs of associated values

$$z_1 = g(w_1),\ z_2 = g(w_2) \quad \text{or} \quad w_1 = f(z_1),\ w_2 = f(z_2),$$

where $|w_k - w_0| < r_2$, $k = 1, 2$. Then

$$\frac{f(z_1) - f(z_2)}{z_1 - z_2} = \frac{w_1 - w_2}{g(w_1) - g(w_2)}.$$

As $w_2 \to w_1$, $z_2 \to z_1$, and the left side tends to $f'(z_1)$. Formula (4.5.4) shows that $f'(z_1) \neq 0$, since $|z_1 - z_0| < r_3$. It follows that the right member also has a limit, that is, $g'(w)$ exists for every $w$ with $|w - w_0| < r_2$, and (4.5.2) holds. Thus $g(w)$ is holomorphic, and the theorem is proved.

We have seen that (4.5.1) holds for $|w - w_0| < r_2$. We have no guarantee that this is the largest circle in which $g(w)$ exists and satisfies (4.5.1). We have also the inverse relation

(4.5.14) $$\qquad\qquad\qquad g[f(z)] = z$$

in a neighborhood of $z = z_0$. The function $g(w)$ is known as the *inverse* of $f(z)$. When we want to emphasize this relationship we shall use the notation $\breve{f}(w)$ (read "$f$ inverse").[1]

For other proofs of the inverse function theorem see Section 9.4.

## EXERCISE 4.5

**1.** Determine $\breve{f}(w)$ and its branch points, if $f(z)$ is given by

     **a.** $\dfrac{1 + z^2}{1 - z^2}$.      **b.** $z + 1/z$.      **c.** $1 + z^2 + z^4$.      **d.** $z^2 + 1/z^2$.

**2.** Given

$$w = z + \tfrac{1}{2}z^2.$$

Find the inverse function. Let $z_0$ be given, $z_0 \neq -1$, and apply the method of successive approximations to find $g_3(w)$. Find the corresponding value of $r_2$. How does $r_2$ compare with the distance from $w_0$ to the branch point of $g(w)$?

---

[1] This notation is due to H. Behnke and F. Sommer. The student is urged to distinguish carefully between the *reciprocal* and the *inverse* of a function $f(z)$. We denote the former by $[f(z)]^{-1}$ or $1/f(z)$, the latter by $\breve{f}(w)$ but never by $f^{-1}(w)$. In consulting other treatises, verify the terminology. Note, in particular, that modern French usage, as represented by N. Bourbaki, is the opposite of ours.

**3.** When the method of successive approximations is applied to

$$w = z^2, \quad z_0 = w_0 = 1,$$

the $n$th approximation is a polynomial in $(w - 1)$ of degree $2^{n-1}$. How large must $n$ be chosen in order that these polynomials have the same terms of degree $\leq 3$ from that point on?

## 4.6. Conformal mapping.

We are now ready to study the mapping properties of a holomorphic function. We start by reformulating the results obtained in the preceding section.

THEOREM 4.6.1.    *Suppose that $f(z)$ is holomorphic in a domain $D$, and $f'(z)$ [is continuous in $D$ and] is $\neq 0$ in $D$. Then the mapping $z \to f(z)$ is locally one-to-one, that is, for every $z_0 \in D$ there exists a neighborhood $N(z_0) \subset D$ in which $f(z_1) \neq f(z_2)$ if $z_1 \neq z_2$. Moreover, $f[N(z_0)]$, the image of $N(z_0)$, contains a neighborhood of $w_0 = f(z_0)$, $N(w_0)$ say. If $R \equiv [w \mid w = f(z), z \in D]$, then $R$ is a domain.*

*Proof.*    Formulas (4.5.6) and (4.5.7) show that if $\mid z_k - z_0 \mid < r_3, k = 1, 2$, and $z_1 \neq z_2$, then $f(z_1) \neq f(z_2)$. Thus, we can take $N(z_0)$ as the open circular disk $\mid z - z_0 \mid < r_3$. Similarly, $N(w_0)$ may be taken as the corresponding circular disk $\mid w - w_0 \mid < r_2$.

It remains to prove that the range $R$ is a domain. Since $w_0 \in R$ implies $N(w_0) \subset R$, we see that $R$ is an open set. To show that $R$ is arcwise connected, we have to exhibit for any pair of points, $w' = f(z')$ and $w'' = f(z'')$, a polygonal line in $R$ joining these points (see Definition 2.3.1 (4) and subsequent comments). Now, $D$ is arcwise connected, so that $z'$ and $z''$ may be joined by a polygonal line $\Pi$ in $D$. The image $f[\Pi]$ is a path in $R$ joining $w'$ with $w''$, but it is normally not a polygonal line. Such a line may be constructed as follows: For every point $z_0$ on $\Pi$ there is a corresponding radius $r_2(z_0)$. Now, $f(\Pi)$ is a bounded closed set (Why?) and as such it has the Heine-Borel property (see Appendix A); that is, every covering of the set contains a finite subcovering. The union of the disks $\mid w - f(z_0) \mid < r_2(z_0)$ for $z_0 \in \Pi$ is a covering of $f(\Pi)$. There exists then a finite subcovering of $f(\Pi)$; that is, there are points $z_1, z_2, \cdots, z_n$ on $\Pi$ such that the set $S = \bigcup_{k=1}^{n} C_k$, with $C_k : \mid w - f(z_k) \mid < r_2(z_k)$, is a covering of $f(\Pi)$. Then $S$ as the union of open disks is also open and $f(\Pi) \subset S \subset f(D) = R$. We must show that $S$ is also arcwise connected.

If this were not the case, then $S$ would break up into a finite number of maximal components, each of which would be open and arcwise connected. Further, each component would clearly be the union of some of the disks $C_k$. Suppose that $w'$ belongs to the component $S_1$ and let $f(\Pi)$ be given by the equation $w = g(t)$, $0 \leq t \leq 1$, where $g(0) = w'$, $g(1) = w''$. If not all of $f(\Pi)$ belongs to $S_1$, then we can find a $t_1$, $0 < t_1 \leq 1$, such that $g(t) \in S_1$ for $0 \leq t < t_1$, but $g(t_1) \notin S_1$. Since $S$ covers $f(\Pi)$, there exists, however, a disk, $C_j$ say, such that $g(t_1) \in C_j$. Here $S_1 \cap C_j$ must be an open non-void set. It follows that $S_1 \cup C_j$ is arcwise connected; that is, $S_1$ cannot be a maximal component of $S$. Thus, $S$ consists of a single component and is, consequently, arcwise connected. It follows that we can connect $w'$ to $w''$ by a polygonal line in $S \subset R$. This completes the proof.

DEFINITION 4.6.1.    *A mapping $z \to f(z)$ is said to be an interior, or open, mapping if it takes open sets into open sets.*[1]

COROLLARY [INTERIOR MAPPING THEOREM].    *Under the assumptions of Theorem 4.6.1, $z \to f(z)$ is an interior mapping.*

It should be observed that the condition $f'(z) \neq 0$ in $D$ is not necessary for the validity of this theorem; it suffices that $f'(z) \not\equiv 0$, so that $f(z)$ is not a constant. At the present time we do not have the necessary means at our disposal to discuss the mapping in the neighborhood of a zero of $f'(z)$.

We turn now to questions of isogonality.

DEFINITION 4.6.2.    *Let $f(z)$ be a continuous function of domain $D$ and range $R$. The mapping $z \to f(z)$ is said to be conformal at $z = z_0$ if it is isogonal and a pure magnification at $z = z_0$. The mapping is conformal in $D$ if it is conformal at all points of $D$. $R$ is the conformal map of $D$ under this mapping if the correspondence is one-to-one and conformal in $D$.*

It should be observed that a mapping may be conformal with the sense of the angles either preserved or reversed. Further, it is assumed that $D$ is a domain, that is, an open arcwise connected set. We are not concerned here with the behavior of the mapping function on the boundary even if $f(z)$ has boundary values. Thus $w = z^2$ maps the interior of the first quadrant in the $z$-plane conformally on the interior of the upper half of the $w$-plane, although the angle at the origin is doubled.

THEOREM 4.6.2.    *If $f(z)$ is holomorphic in $D$, then the mapping $z \to f(z)$ is conformal, with the sense of angles preserved, at all points of $D$ where $f'(z) \neq 0$. The mapping is conformal in $D$ if $f'(z) \neq 0$ in $D$.*

*Proof.*    We shall see later that the mapping is actually not conformal at the points where $f'(z) = 0$. Now let $z_0 \in D$ and $f'(z_0) \neq 0$. Then

$$(4.6.1) \qquad \frac{f(z) - f(z_0)}{z - z_0} = f'(z_0) + r(z, z_0)$$

---

[1] For this notion see G. T. Whyburn, *Analytic Topology* (Colloquium Publications, Vol. 28; American Mathematical Society, New York, 1942). Interior transformations are one of the main themes of this treatise, where it is shown that they have many properties in common with the analytic transformations here considered.

where $r(z, z_0)$ tends to zero with $|z - z_0|$. Hence, modulo $2\pi$,

$$(4.6.2) \qquad \lim_{z \to z_0} \arg [f(z) - f(z_0)] \equiv \arg f'(z_0) + \lim_{z \to z_0} \arg (z - z_0),$$

provided the limit on the right exists. Suppose now that two arcs $C_1$ and $C_2$ in the $z$-plane end at $z = z_0$, where they have definite semitangents, $t_1$ and $t_2$ respectively, and suppose that the angle from $t_1$ to $t_2$ equals $\alpha$. These curves are mapped onto curves $\Gamma_1 = f[C_1]$ and $\Gamma_2 = f[C_2]$ in the $w$-plane. Formula (4.6.2) shows that $\Gamma_1$ and $\Gamma_2$ have semitangents, $T_1$ and $T_2$ respectively, at $w = w_0$, and the angle from $T_1$ to $T_2$ also equals $\alpha$. Hence sense and magnitude of angles are preserved by the mapping. As a further consequence of (4.6.1) we note that

$$(4.6.3) \qquad \lim_{z \to z_0} \frac{|f(z) - f(z_0)|}{|z - z_0|} = |f'(z_0)|.$$

Thus the mapping is also a pure magnification.

There are various consequences of these formulas together with the continuity of $f'(z)$ which are worth listing here even if further discussion is postponed until Chapters 14 and 17 of Volume II.

*Suppose that $f'(z) \neq 0$ in $D$ and that $C$ is a rectifiable arc in $D$. Then $\Gamma = f[C]$ is also rectifiable, and its length is*

$$(4.6.4) \qquad l[\Gamma] = \int_C |f'(z)| \, |dz| = \int_0^L |f'[z(s)]| \, ds$$

*if $z = z(s)$ is a parametric representation of $C$ in terms of its arc length and if $L = l[C]$.*

Another consequence is that *the mapping is approximately linear in the neighborhood of any point $z_0$ where $f'(z_0) \neq 0$.* To make this assertion more precise, let us consider the following simple case: $f(z)$ *is holomorphic in a domain $D$ such that* (i) $D$ *is bounded,* (ii) $f'(z) \neq 0$ *in $D$,* (iii) $f'(z)$ *is uniformly continuous in $D$, and* (iv) $f(z_1) \neq f(z_2)$ *for $z_1 \neq z_2$.* Thus $w = f(z)$ maps $D$ conformally on $f[D] \equiv R$. Condition (iii) allows us to introduce the *modulus of continuity* of $f'(z)$ defined for small $\delta \geq 0$ by

$$(4.6.5) \quad \mu(\delta; f') \equiv \sup |f'(z_1) - f'(z_2)|, \quad z_1, z_2 \in D, \quad |z_1 - z_2| \leq \delta.$$

Here $\mu(0; f') = 0$ and $\mu(\delta; f')$ is continuous and monotone increasing.

Suppose now that $z$ and $z_0$ in $D$ are so chosen that $[z_0, z] \subset D$. Since

$$(4.6.6) \qquad \int_{z_0}^z [f'(t) - f'(z_0)] \, dt = f(z) - f(z_0) - (z - z_0)f'(z_0),$$

we have

$$(4.6.7) \quad |f(z) - f(z_0) - (z - z_0)f'(z_0)| \leq |z - z_0| \, \mu(|z - z_0|; f').$$

This inequality measures the deviation of the given mapping $z \to f(z)$ from the local linear mapping at $z_0$:

$$z \to f(z_0) + (z - z_0)f'(z_0).$$

Consider in particular a small square region $S$ of side $\delta$ in $D$ having one vertex at $z_0$. The local linear mapping at $z_0$ maps $S$ onto a square region $S' \subset R$ having one vertex at $w = f(z_0)$ and side $|f'(z_0)| \delta$. The mapping $w = f(z)$ is one-to-one and approximately linear. This implies that the perimeter of $S$ is mapped onto a curvilinear quadrilateral $Q$ whose sides are approximately rectilinear, approximately of the same length, and meet at right angles. Formula (4.6.7) enables us to estimate the deviation of $Q$ from $S'$. Let $S_-$ and $S_+$ be two squares whose sides are parallel to those of $S'$ at the distance

$$\sqrt{2}\delta\, \mu(\sqrt{2}\delta; f') \equiv \delta\, \eta(\delta)$$

with $S_- \subset S' \subset S_+$. See Figure 12. If $\delta$ is sufficiently small, $\eta(\delta)$ is as small as we please. Using (4.6.7) we conclude that $Q$ lies in the interior of $S_+$ but exterior to $S_-$ and is tangent to $S'$ at $w = f(z_0)$.

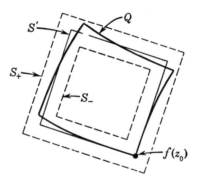

**Figure 12**

The interior of $S$, $S_i$ say, then maps onto the interior of $Q$, $Q_i$ say. We note first that $f[S_i] \equiv \Delta$ is a domain and that $Q$ is at least a portion of the boundary of $\Delta$. Further, $\Delta \supset Q_i \cap N(w_0)$ in the notation of Theorem 4.6.1. Since the mapping is one-to-one and continuous, $\Delta$ cannot straddle $Q$, and we must have $\Delta \subset Q_i \subset R$. For the same reason, every point of $Q_i$ must be the image of some point of $S_i$ so that $\Delta \supset Q_i$ or $f(S_i) = Q_i$, as asserted.

The area of the map $R = f(D)$ is a concept of considerable importance. The meaning and existence of such a quantity follow from the Lebesgue theory of planar measure, but this discussion will be postponed until Chapter 17, Volume II. At the present juncture we shall be content with the naïve idea of area inherited from the calculus. This leads to cogent results, at least if the region under consideration is bounded by a finite number of simple arcs each having a continuously turning tangent. In particular, if $\delta$ is sufficiently small, the domain $Q_i = f(S_i)$ considered above has a well-defined area $A[Q_i]$, and Figure 12 gives the inequality

$$(4.6.8) \qquad \delta^2 [\, |f'(z_0)\,| - \eta(\delta)]^2 \leq A[Q_i] \leq \delta^2 [\, |f'(z_0)\,| + \eta(\delta)]^2.$$

Let the domain $D_0 \subset D$ be the union of a finite number of squares $S_{j,n}$ belonging to a quadratic mesh $M_n$, where the lines forming the mesh are at the distance $2^{-n}$ from each other and $n$ is large. Set $f(D_0) = R_0$. Then $R_0$ has an area $A[R_0]$ and by (4.6.8)

$$(4.6.9) \qquad \begin{aligned} 2^{-2n} \Sigma_j [\, |f'(z_{j,n})\,| - \eta(2^{-n})]^2 &\leq A[R_0] \\ &\leq 2^{-2n} \Sigma_j [\, |f'(z_{j,n})\,| + \eta(2^{-n})]^2, \end{aligned}$$

where $z_{j,n}$ is a vertex of $S_{j,n}$. We now replace the mesh $M_n$ by a mesh $M_p$, $p > n$, obtained by subdividing each square of $M_n$ into $2^{2(p-n)}$ equal squares. $D_0$ will then still be the union of squares from $M_p$, only these squares will be smaller, and the approximation to $A[R_0]$ is better. Passing to the limit with $p$ and recalling the definition of the Riemann integral, we obtain

$$(4.6.10) \qquad A[R_0] = \iint_{D_0} |f'(z)|^2 \, dx \, dy.$$

Here, we can clearly drop the assumption that $D_0$ is the union of squares. $D_0$ may be any subdomain of $D$ to which the methods of the calculus apply. Actually we have also

$$(4.6.11) \qquad A[R] = \iint_D |f'(z)|^2 \, dx \, dy,$$

where, however, the integral may have to be taken in the sense of Lebesgue, and $A[R]$ is the planar measure of $R$. For the validity of the formula it is not necessary that $|f'(z)|$ be bounded; it suffices that the integral converges. When needed, new variables may be introduced in (4.6.11) according to the rules of the calculus. In particular, if $D$ is a circle, it pays to introduce polar coordinates. The surface element $dx \, dy$ is then replaced by $r \, dr \, d\theta$.

This concludes our discussion of the mapping properties of holomorphic functions. We shall now consider the converse question: If a mapping is isogonal or a pure magnification everywhere in the domain of definition, what limitations does this place on the mapping function?

THEOREM 4.6.3. *If* $w = f(z) = f(x + iy)$ *is defined and continuous together with its first-order partials with respect to* $x$ *and* $y$ *in a domain* $D$, *if* $|f_x(z)|^2 + |f_y(z)|^2 \neq 0$ *in* $D$, *and if the mapping is isogonal everywhere in* $D$, *then* $f(z)$ *is either holomorphic in* $D$ *or the conjugate of a holomorphic function. The same conclusion holds if the mapping defines a pure magnification in* $D$.

*Proof.* By the corollary to Theorem 4.3.2 it suffices to prove that either $f(z)$ or $\overline{f(z)}$ satisfies the Cauchy-Riemann equations. Consider a neighborhood of a point $z_0 \in D$. Since continuous partial derivatives exist, we may apply Taylor's theorem with remainder to the real and imaginary parts of $f(z)$. Combining, we obtain

$$(4.6.12) \qquad f(z) - f(z_0) = f_x(z_0)(x - x_0) + f_y(z_0)(y - y_0) + R(z, z_0),$$

where the remainder tends to zero faster than $z - z_0$ as $z \to z_0$. It follows that the local mapping properties at $z = z_0$ are determined by the linear terms, whose sum does not vanish identically by assumption. Set

$$z = z_0 + r(\cos \theta + i \sin \theta),$$

$$P(z_0, \theta) \equiv f_x(z_0) \cos \theta + f_y(z_0) \sin \theta.$$

Then (4.6.12) may be written

$$(4.6.13) \qquad\qquad f(z) - f(z_0) = r[P(z_0, \theta) + \eta(r)],$$

where $\eta(r)$ tends to zero with $r$.

Suppose now that the mapping is isogonal in $D$. From the continuity of $f(z)$ and its partials we conclude that the sense of angles is either preserved everywhere or reversed everywhere. Suppose the sense is preserved. A necessary and sufficient condition that magnitude and sense of angles be preserved is that

$$(4.6.14) \qquad\qquad P(z_0, \theta) = P(z_0)[\cos \theta + i \sin \theta],$$

where $P(z_0)$ is independent of $\theta$, since this expresses that

$$\arg P(z_0, \theta_1) - \arg P(z_0, \theta_2) \equiv \theta_1 - \theta_2 \quad (\mathrm{mod}\ 2\pi).$$

Since (4.6.14) must hold identically in $\theta$, we see that

$$P(z_0) = f_x(z_0) = -if_y(z_0).$$

Hence we must have everywhere in $D$

$$(4.6.15) \qquad\qquad f_x(z) = -if_y(z),$$

which is formula (4.3.5), the complex form of the Cauchy-Riemann equations. In this case $f(z)$ is holomorphic in $D$.

If angles are to be reversed instead, we must have

$$(4.6.16) \qquad\qquad P(z_0, \theta) = P(z_0)[\cos \theta - i \sin \theta],$$

and this gives

$$(4.6.17) \qquad\qquad f_x(z) = if_y(z).$$

Hence the conjugate function is holomorphic in $D$.

Suppose now, instead, that the mapping is a pure magnification. This requires that

(4.6.18) $$| P(z_0, \theta) | \equiv Q(z_0)$$

where $Q(z_0)$ is independent of $\theta$. But the square of the left-hand side equals

$$| f_x |^2 \cos^2 \theta + [f_x \bar{f}_y + \bar{f}_x f_y] \sin \theta \cos \theta + | f_y |^2 \sin^2 \theta,$$

where the partials are evaluated at $z = z_0$. If this is to be independent of $\theta$ we must have

$$| f_x |^2 - | f_y |^2 = 0,$$
$$f_x \bar{f}_y + \bar{f}_x f_y = 0.$$

Multiplying the second relation by $i$ and adding it to the first we get

$$[f_x + i f_y][\bar{f}_x + i \bar{f}_y] = 0,$$

that is, either (4.6.15) or (4.6.17) must hold everywhere in $D$. For reasons of continuity we conclude that if one of these relations holds at $z = z_0$, the same relation holds everywhere in $D$. Thus either $f(z)$ or $\overline{f(z)}$ is holomorphic in $D$. This completes the proof.

## EXERCISE 4.6

**1.** Given $f(z) = 3z + z^3$. Where does the mapping cease to be locally conformal? Prove that $f(z_1) \neq f(z_2)$ if $z_1 \neq z_2$ and $| z_1 | < 1, | z_2 | < 1$.

**2.** Show that $w = 3z + z^3$ maps the disk $| z | < 1$ conformally onto a domain $\Delta$ of the $w$-plane, and determine the boundary of $\Delta$.

**3.** By Problem 3, Exercise 3.4, $f(z) = z + \frac{1}{2}z^2$ maps $| z | < 1$ in a one-to-one manner on the interior of a certain cardioid. Verify that the mapping is conformal.

**4.** Use formula (4.6.4) to find the length of the cardioid.

**5.** Find the area of the image of the disk $| z | < 1$ under the same mapping using (4.6.11) transformed to polar coordinates.

**6.** If $f(z) = c_1 z + c_2 z^2$, find the area of the image of $| z | < R$ as a function of $c_1$ and $c_2$. Interpret your result geometrically. Here it is supposed that $R \leq R_0 = | c_1 |/(2 | c_2 |)$. What does the formula represent if $R > R_0$?

**7.** Generalize Problem 6 to a cubic polynomial.

**8.** What is the length of the boundary of $\Delta$ in Problem 2?

**9.** Let $f(z) \in C[\overline{D}]$ and define the modulus of continuity of $f(z)$ as in (4.6.5), replacing $f'(z)$ by $f(z)$. Show that this function is continuous, monotone increasing, and subadditive.

**4.7. Function spaces.** We shall return briefly to the ideas of the first two sections of this chapter. Though the notion of a linear vector space was first introduced in Section 1.3, and that of an algebra in Section 4.1, it will be appropriate to recapitulate the definitions here and to add some further concepts.

DEFINITION 4.7.1.   *X is a linear vector space over the complex field $C$ [over the real field $R$] if the elements of $X$ admit of the two operations of addition and scalar multiplication subject to the following conditions:*
  *Addition satisfies Postulates $A_1$–$A_5$ of Section 1.1.*
  *Scalar multiplication satisfies:*

  $S_1$.   *For every scalar $\alpha \in C$ [$\alpha \in R$] and every element $x \in X$ there is a uniquely defined scalar product $\alpha x = x\alpha$ in $X$;*

  $S_2$.   $(\alpha + \beta)x = \alpha x + \beta x$;

  $S_3$.   $\alpha(x + y) = \alpha x + \alpha y$;

  $S_4$.   $\alpha(\beta x) = (\alpha\beta)x$;

  $S_5$.   $1 \cdot x = x$.

We recall that the postulates for addition imply the existence of a unique zero element and of a unique negative. Combining $A_3$, $S_2$, and $S_5$, we see that $-x = (-1) \cdot x$ and $0 \cdot x = 0$. In the last relation we have on the left the zero element of $C$ or of $R$, and on the right the zero element of $X$, for which we use the same notation.

A real-valued function $p(x)$ defined everywhere in $X$ is said to be a *norm* if:

  $N_1$.   *$p(x) \geq 0$ and $= 0$ if and only if $x = 0$;*

  $N_2$.   *$p(\alpha x) = | \alpha | \, p(x)$;*

  $N_3$.   *$p(x + y) \leq p(x) + p(y)$.*

Usually we write $\| x \|$ instead of $p(x)$.
  We introduce a *metric* in $X$ by defining

$$(4.7.1) \qquad\qquad d(x_1, x_2) = \| x_1 - x_2 \|.$$

That this is a *distance function* in the sense of Section 1.2 follows from $N_1$, $N_2$, and $N_3$. The resulting metric space is known as a *normed linear vector space*. We recall that a set $\{x_n\} \subset X$ is called a *Cauchy sequence* if

$$(4.7.2) \qquad\qquad \lim_{m,\, n \to \infty} \| x_m - x_n \| = 0.$$

DEFINITION 4.7.2.   *A metric space is said to be complete if for each Cauchy sequence $\{x_n\} \subset X$ there exists a (unique) element $x_0 \in X$ such that*

$$(4.7.3) \qquad\qquad \lim_{n \to \infty} \| x_n - x_0 \| = 0.$$

DEFINITION 4.7.3.  *A complete normed linear vector space is called a Banach space, or (B)-space for short.*[1]

If elements may also be multiplied, we obtain a richer theory.

DEFINITION 4.7.4.  *A is an algebra over the scalar field C (or R) if its elements admit of the three operations of addition, multiplication, and scalar multiplication subject to the following conditions:*
*A is a linear vector space in the sense of Definition* 4.7.1.
*The multiplication satisfies:*

$M_1$.  *Every ordered pair of elements x, y has a unique product xy;*

$M_2$.  *Multiplication is associative:* $(xy)z = x(yz)$;

D.  *Addition and multiplication are distributive:*

$$x(y + z) = xy + xz, \quad (y + z)x = yx + zx.$$

$S_6$.  *Multiplication and scalar multiplication commute:*

$$(\alpha x)(\beta y) = (\alpha\beta)(xy).$$

*A is said to have the unit element e if*
$M_3$.  $ex = xe = x$ *for each* $x \in X$.
*A is said to be commutative if*
$M_4$.  $xy = yx$ *for* $x, y \in X$.

DEFINITION 4.7.5.  *A is a normed algebra if A is an algebra as well as a normed linear vector space and if, in addition,*

$N_4$.  $\| xy \| \leq \| x \| \| y \|$.

*A is a (B)-algebra if it is a normed algebra as well as a (B)-space.*

The reader has already encountered a number of linear vector spaces and algebras in this treatise.  Thus, the rational field $Q$ (over $Q$), the real field $R$, and the complex field $C$ are normed algebras under the absolute value as norm. $R$ and $C$ are (B)-algebras, but $Q$ is not complete and hence not a (B)-algebra.

We say that a linear vector space $X$ is a *function space* if its elements are functions of a real or a complex variable or of several such variables.  The continuous functions $f(t)$ on a closed interval $[a, b]$ form a function space $C[a, b]$ which is a (B)-algebra under the sup-norm.  The functions of bounded variation on $[a, b]$ form the space $BV[a, b]$.  This is also a (B)-algebra under a

---

[1] Stefan Banach (1892–1945), Polish mathematician, professor at Lwów.  Together with Maurice Fréchet (1878—) and Frigyes Riesz (1880–1956) he is considered one of the founders of modern "functional analysis," to which the earlier "general analysis" of Eliakim Hastings Moore (1862–1932) is loosely related.

suitably chosen norm. Passing over to functions of a complex variable, we have seen that the function spaces $CB[D]$ and $C[\overline{D}]$ of Section 4.1 are (B)-algebras. The function space $HB[D]$ of Section 4.2 was shown to be a normed algebra, and one of our main problems in later chapters will be to show that $HB[D]$ is actually a (B)-algebra. All these algebras are commutative and have a unit element.

Not all function spaces, however, are algebras. As an example, let $X$ be the set of all functions $f(z)$ holomorphic in the open unit disk such that

$$(4.7.4) \qquad \qquad \| f \| = \sup_{|z|<1} |\, (1-z)f(z)\,| < \infty.$$

Then $X$ is a (B)-space under this norm; it is not an algebra since $(1-z)^{-1} \in X$, and its square does not belong to $X$.

## EXERCISE 4.7

**1.** Verify that the sup-norm is actually a norm.

**2.** Suppose that $f(t) \in BV[a,\ b]$, and let $V[f]$ be the total variation of $f(t)$ in $[a,\ b]$. Show that $V[f]$ cannot be used as a norm in this function space, but that $\sup |f| + V[f]$ can.

**3.** A function $f(z)$ is said to be *univalent* (German *schlicht*) in the domain $D$ if $z_1 \neq z_2$ implies $f(z_1) \neq f(z_2)$. Show that the set of functions $f(z)$, holomorphic and univalent in the open unit disk, is not closed under addition or multiplication.

**4.** Verify that the space $X$ of holomorphic functions with norm (4.7.4) is a (B)-space.

## COLLATERAL READING

For Sections 4.1–4.3 see

BEHNKE, H., and SOMMER, F. *Theorie der analytischen Funktionen einer komplexen Veränderlichen*, Chap. 1, Sections 5–7. Springer-Verlag, Berlin, 1955.

For the theory of harmonic functions see

AHLFORS, L. V. *Complex Analysis, An Introduction to the Theory of Analytic Functions of One Complex Variable*, Chap. 5. McGraw-Hill Book Company, Inc., New York, 1953.

HURWITZ, A., and COURANT, R. *Vorlesungen über allgemeine Funktionentheorie und elliptische Funktionen*, Second Edition, Part III, Chap. 3, Sections 5–10. Springer-Verlag, Berlin, 1925.

The Gauss-Lucas theorem 4.4.1 and its many generalizations are the theme of

MARDEN, MORRIS. *The Geometry of the Zeros of a Polynomial in a Complex Variable*, Mathematical Surveys, Vol. III. American Mathematical Society, Providence, R.I., 1949.

WALSH, J. L. *The Location of Critical Points of Analytic and Harmonic Functions*, Colloquium Publications, Vol. 34. American Mathematical Society, Providence, R.I., 1950.

The proof of Theorem 4.5.1 is an adaptation of

PICARD, E. *Traité d'Analyse*, Vol. 2, Note IV. Gauthier-Villars, Paris, 1893.

For different aspects of the theory of conformal mapping see

CHURCHILL, R. V. *Introduction to Complex Variables and Applications*, Chaps. 8–10, Appendix 2. McGraw-Hill Book Company, Inc., New York, 1948.

KOBER, H. *Dictionary of Conformal Representations*. Dover Publications, Inc., New York, 1952.

For Section 4.7 see

HILLE, E., and PHILLIPS, R. S. *Functional Analysis and Semi-Groups*, Revised Edition, Colloquium Publications, Vol. 31, Chap. 1. American Mathematical Society, Providence, R.I., 1957.

# 5

# POWER SERIES

**5.1. Infinite series.**[1] We start with a brief review of the theory of infinite series needed for a fairly detailed study of power series.

Let $\{w_n \mid n = 0, 1, 2, \cdots\}$ be a given infinite sequence of complex numbers and define

(5.1.1) $$W_n = \sum_{k=0}^{n} w_k, \quad w_n = W_n - W_{n-1}, \quad W_{-1} = 0,$$

(5.1.2) $$|w_n| = a_n, \quad A_n = \sum_{k=0}^{n} a_k.$$

Then

(5.1.3) $$\sum_{n=0}^{\infty} w_n$$

is the formal infinite series whose $n$th term equals $w_n$, and $W_n$ is called the $n$th *partial sum* of the series.

DEFINITION 5.1.1.    *The series (5.1.3) is said to be convergent, or to converge, if*

(5.1.4) $$\lim_{n \to \infty} W_n \equiv W$$

*exists as a finite quantity. $W$ is then called the sum of the series. The series is said to be divergent, or to diverge, if $\lim W_n$ either does not exist or is infinite. The series is absolutely convergent if*

(5.1.5) $$\lim_{n \to \infty} A_n \equiv A$$

*exists as a finite quantity.*

It is clear that *a series with complex terms converges if and only if the two series formed by the real parts and by the imaginary parts of the terms converge.* Cauchy's convergence principle gives:

THEOREM 5.1.1.    *A necessary and sufficient condition that (5.1.3) be convergent [absolutely convergent] is that given any $\varepsilon > 0$ there exists an $N = N(\varepsilon)$ such that*

$$|W_{n+p} - W_n| < \varepsilon \qquad [A_{n+p} - A_n < \varepsilon]$$

*for $n > N(\varepsilon)$ and every $p > 0$. An absolutely convergent series is convergent, but the converse is not always true.*

---

[1] A student who is well versed in the theory of infinite series may start with Section 5.4 and refer to the earlier sections only as needed.

The first part of the final assertion follows from

$$| W_{n+p} - W_n | \leqq A_{n+p} - A_n.$$

Examples of convergent series which are not absolutely convergent will be found below.

The theory of absolutely convergent series is essentially that of series with positive terms, for which a number of convergence criteria and comparison theorems exist. A series of positive terms $p_n$ is obviously convergent if there exists a convergent series with positive terms $c_n$ such that

$$p_n \leqq c_n$$

for every large $n$. It is divergent if there exists a divergent series with positive terms $d_n$ such that

$$p_n \geqq d_n$$

for every large $n$.

The following general principle enables us to construct comparison series with positive terms. Let $\{M_n\}$ be a monotone increasing sequence of positive numbers such that $M_n \to \infty$. Then

(5.1.6)
$$\sum_{n=1}^{\infty} [M_n - M_{n-1}]$$

is divergent, while

(5.1.7)
$$\sum_{n=0}^{\infty} \left[ \frac{1}{M_n} - \frac{1}{M_{n+1}} \right]$$

is convergent.

Choosing $M_n = a^n$, $a > 1$, we see that the geometric series

(5.1.8)
$$\sum_{n=0}^{\infty} q^n$$

converges for $q < 1$ and diverges for $q \geqq 1$. In the second case the terms do not tend to zero, and this is obviously a necessary condition for convergence. In the present case $M_n - M_{n-1}$ is easily handled. In more complicated cases the mean value theorem is used in the form

$$f(n) - f(n - 1) = f'(n - \theta_n), \quad 0 < \theta_n < 1,$$

in the discussion of the implications of (5.1.6) and (5.1.7).

Choosing $M_n = n^p$, $p > 0$, we may prove that

(5.1.9)
$$\sum_{n=1}^{\infty} n^{-s}$$

converges for $s > 1$ and diverges for $s \leqq 1$. This series defines the Riemann zeta function, $\zeta(s)$, for $s > 1$. It is a special Dirichlet series.[1] An immediate generalization is furnished by the series

$$(5.1.10) \qquad \sum_{n=1}^{\infty} e^{-\lambda_n s}, \quad 0 < \lambda_n < \lambda_{n+1}, \quad \lambda_n \to \infty.$$

Assuming

$$(5.1.11) \qquad \lambda_n/(\log n) \to +\infty,$$

we can prove that (5.1.10) converges for $s > 0$. It obviously diverges for $s \leqq 0$. The series in the so-called *logarithmic scale*, of which

$$(5.1.12) \qquad \sum_{n=4}^{\infty} n^{-1}(\log n)^{-\alpha}(\log \log n)^{-\beta}$$

is a sample, may be discussed by a suitable choice of $M_n$ as a product of powers of $\log n$, $\log \log n$, etc. The series (5.1.12) converges for $\alpha > 1$ regardless of the value of $\beta$, converges for $\alpha = 1$ only when $\beta > 1$, and diverges for $\alpha < 1$.

We shall now give some convergence criteria for series with positive terms.

THEOREM 5.1.2. *If $p_n > 0$, set*

$$(5.1.13) \qquad \limsup_{n \to \infty} (p_n)^{1/n} = \mu.$$

*The series whose nth term is $p_n$ converges if $\mu < 1$ and diverges if $\mu > 1$. No assertion can be made when $\mu = 1$.*

*Proof.* Suppose $\mu < 1$ and choose $q$, $\mu < q < 1$. By the definition of the superior limit we can then find an integer $N(\varepsilon)$ such that

$$(p_n)^{1/n} < q \quad \text{or} \quad p_n < q^n, \quad n > N(\varepsilon).$$

Thus the series $\Sigma\, p_n$ converges by comparison with a geometric series. Similarly, if $\mu > 1$ the inequality

$$(p_n)^{1/n} \geqq 1, \quad \text{or} \quad p_n \geqq 1,$$

must hold for infinitely many values of $n$, so that the series diverges.

For the series (5.1.9) we have $\mu = 1$ regardless of the value of $s$, and the series converges for $s > 1$ and diverges for $s \leqq 1$.

In Theorem 5.1.2 we may replace (5.1.13) by

$$(5.1.14) \qquad \limsup (p_n)^{1/\lambda_n} = \mu,$$

where $\{\lambda_n\}$ is a monotone increasing sequence of numbers satisfying (5.1.11). Again we have convergence for $\mu < 1$, divergence for $\mu > 1$, and no information when $\mu = 1$.

---

[1] Peter Gustav Lejeune Dirichlet (1805–1859), German mathematician of Huguenot family, Gauss's successor in Göttingen in 1855, previously professor in Berlin. He contributed to number theory and analysis. He examined the primes in an arithmetic progression and gave the first rigorous discussion of the convergence of a Fourier series.

THEOREM 5.1.3.    *Set*

(5.1.15) $$\liminf_{n\to\infty}\frac{p_{n+1}}{p_n}=\nu,\qquad \limsup_{n\to\infty}\frac{p_{n+1}}{p_n}=\mu.$$

*The series* $\Sigma\, p_n$ *converges if* $\mu < 1$. *It diverges if* $\nu > 1$. *It may converge even though* $\mu = \infty$ *and diverge even though* $\nu = 0$.

*Proof.*    If $\mu < 1$ and $\mu < q < 1$, we can find an $N$ such that

(5.1.16) $$p_{N+m} < q^m p_N$$

for all $m > 0$. For $N$ we take the least integer such that $p_{n+1}/p_n < q$ for all $n \geqq N$. The product of these inequalities with $n = N,\, N+1,\, \cdots,\, N+m-1$ gives (5.1.16). Convergence then follows by comparison with a geometric series. If $\nu > 1$ we have $p_n \geqq p_N$ for some $N$ and all large $n$, and divergence is immediate. An example to illustrate the remaining cases is to be found in Exercise 5.1.

It is interesting to compare the last two theorems. Both apply to series converging or diverging as a geometric series, but while the first theorem applies to all such series, the second becomes practically useless as soon as the terms fail to form an ultimately monotone sequence. On the other hand, when the second theorem does apply, it is normally ever so much easier to use than the first. There is a vast class of power series with coefficients which are rational functions of the index to which Theorem 5.1.3 applies and where it would be a waste of effort to use Theorem 5.1.2. For this special class of series additional information is furnished by

THEOREM 5.1.4.    *If*

(5.1.17) $$\frac{p_{n+1}}{p_n}=1+\frac{\alpha}{n}+\frac{\beta_n}{n^2},\qquad |\,\beta_n\,| \leqq M,$$

*for all large* $n$, *then the series* $\Sigma\, p_n$ *converges for* $\alpha < -1$ *but diverges for* $\alpha \geqq -1$.

*Proof.*    By Taylor's theorem with remainder

$$\log\left(1+\frac{\alpha}{n}\right)-\alpha\log\left(1+\frac{1}{n}\right)=\frac{\gamma_n}{n^2},\qquad |\,\gamma_n\,| \leqq M_1.$$

Further, if $|\,\delta_n\,| \leqq M_2$, then

$$\log\left(1+\frac{\delta_n}{n^2}\right)=\frac{\eta_n}{n^2}\quad\text{where}\quad |\,\eta_n\,| \leqq M_3.$$

But

$$1+\frac{\alpha}{n}+\frac{\beta_n}{n^2}=\left(1+\frac{\alpha}{n}\right)\left(1+\frac{\delta_n}{n^2}\right)$$

$$=\left(\frac{n+1}{n}\right)^{\alpha}\exp\left(\frac{\sigma_n}{n^2}\right),\qquad \sigma_n=\gamma_n+\eta_n,\quad |\,\sigma_n\,| \leqq M_4.$$

Without restricting the generality, we may assume that this is true for $n \geq 1$. Hence

$$p_n = p_1 \prod_{k=1}^{n-1} \left\{ \left( \frac{k+1}{k} \right)^\alpha \exp \left( \frac{\sigma_k}{k^2} \right) \right\} = p_1 n^\alpha \exp \left[ \sum_{k=1}^{n-1} \frac{\sigma_k}{k^2} \right].$$

Since $\Sigma \, n^{-2}$ converges, the multiplier of $n^\alpha$ in the last member is bounded away from zero and infinity. It follows that the series $\Sigma \, p_n$ converges if and only if the series $\Sigma \, n^\alpha$ does, that is, if and only if $\alpha < -1$.

This theorem can be extended to series with complex terms. If

$$(5.1.18) \qquad \frac{w_{n+1}}{w_n} = 1 + \frac{\alpha}{n} + \frac{\beta_n}{n^2}, \quad |\beta_n| \leq M$$

for all large $n$, then the series (5.1.3) converges for $\Re(\alpha) < -1$ and diverges for $\Re(\alpha) \geq -1$. The same proof applies; we have merely to anticipate some of the properties of the exponential function and the logarithm for complex variables.

THEOREM 5.1.5.  *Suppose $p_n \geq p_{n+1}$, $\lim p_n = 0$.  Then the series*

$$(5.1.19) \qquad \sum_{n=0}^{\infty} (-1)^n p_n$$

*is convergent, and its sum lies between $p_0$ and $p_0 - p_1$.*

*Proof.*  If $P_n$ is the $n$th partial sum, it is seen that

$$P_1 \leq P_{2k+1} \leq P_{2k+3} \leq \cdots \leq P_{2k+2} \leq P_{2k} \leq P_0,$$

$$P_{2k+1} - P_{2k} = - p_{2k+1} \to 0,$$

and these relations imply convergence and also the stated inequalities.

In particular, we see that

$$\sum_{n=0}^{\infty} (-1)^n (n+1)^{-s} \quad \text{and} \quad \sum_{n=0}^{\infty} (n+1)^{-s}$$

converge for $s > 0$ and $s > 1$ respectively. Thus there exist convergent series which are not absolutely convergent.

Just as integration by parts is a very useful device in the discussion of integrals, *summation by parts* is very useful in the theory of infinite series. Given a series whose $n$th term is of the form $a_n b_n$, the basic identity reads

$$(5.1.20) \qquad \sum_{k=0}^{n} a_k b_k = \sum_{k=0}^{n-1} (a_k - a_{k+1}) B_k + a_n B_n, \quad B_k = \sum_{m=0}^{k} b_m.$$

This leads to the following theorem, one of the many to which the name of Abel is attached.[1]

THEOREM 5.1.6.    *The series*

$$(5.1.21) \qquad \sum_{n=0}^{\infty} a_n b_n$$

*converges if* $\Sigma \mid a_n - a_{n+1} \mid$ *converges,* $\lim a_n = 0$, *and* $\mid B_n \mid$ *is bounded.*

The proof follows immediately from (5.1.20).

In the following the terms of the series to be considered are frequently functions of a complex variable, and we shall need the notion of *uniform convergence*. This concept is related to Definitions 4.1.2 and 5.1.1.

DEFINITION 5.1.2.    *Let the sequence* $\{w_n(z)\}$ *be defined for z belonging to the set S of the complex plane. Then the series*

$$(5.1.22) \qquad \sum_{n=0}^{\infty} w_n(z), \quad \text{with partial sums} \quad W_n(z) = \sum_{k=0}^{n} w_k(z),$$

*is said to converge uniformly with respect to z in S if the sequence* $\{W_n(z)\}$ *converges uniformly in S.*

The best-known condition for uniform convergence is the so-called *M-test* of Weierstrass, which is an immediate consequence of the definitions:

THEOREM 5.1.7.    *If* $\mid w_n(z) \mid \leq M_n$ *for all z in S, and if* $\Sigma M_n$ *converges, then the series* (5.1.22) *converges uniformly in S.*

This test presupposes that the series is absolutely convergent for all $z$ in $S$ and uniformly convergent with respect to $z$. It is consequently of rather limited applicability. More refined criteria may be obtained by partial summation. The following is an example:

THEOREM 5.1.8.    *The series*

$$(5.1.23) \qquad \sum_{n=0}^{\infty} a_n b_n(z)$$

*converges uniformly in S if* $\sum_{n=0}^{\infty} \mid a_n - a_{n+1} \mid$ *converges,* $\lim_{n \to \infty} a_n = 0$, *and the partial sums* $B_n(z)$ *of the series* $\sum_{k=0}^{\infty} b_k(z)$ *are uniformly bounded in S.*

---

[1] Niels Henrik Abel (1802–1829), a Norwegian genius who succumbed to tuberculosis before he became famous. His collected works, many of them posthumous, form two volumes of close to 1000 pages. Abel solved two of the outstanding problems of his day. He showed that the general algebraic equation of degree greater than four cannot be solved algebraically, and he determined which equations could be so solved. He inverted elliptic integrals and, in competition with Carl Gustav Jacob Jacobi (1804–1851), he laid the foundation of the theory of elliptic functions. His most profound research was devoted to what later came to be known as the theory of Abelian integrals.

*Proof.*   From (5.1.20) we obtain

$$\sum_{k=n+1}^{n+p} a_k b_k(z) = \sum_{k=n+1}^{n+p-1} (a_k - a_{k+1}) B_k(z) - a_{n+1} B_n(z) + a_{n+p} B_{n+p}(z).$$

The absolute value of this expression does not exceed

$$B \left\{ \sum_{n+1}^{n+p-1} | a_k - a_{k+1} | + | a_{n+1} | + | a_{n+p} | \right\} \quad \text{if} \quad | B_k(z) | \leq B,$$

and this clearly tends to zero when $n \to \infty$.

We return now to absolutely convergent series.

THEOREM 5.1.9.  *An absolutely convergent series may be rearranged arbitrarily without affecting the absolute convergence or the sum of the series.*

*Proof.*   Suppose that

$$\Sigma\, w_n \quad \text{and} \quad \Sigma\, \omega_n, \quad \text{with} \quad W_n = \sum_{k=0}^{n} w_k, \quad \Omega_n = \sum_{k=0}^{n} \omega_k,$$

are the original and the rearranged series, respectively. Thus, every $\omega_n$ is a $w_k$, say with $k = k(n)$, and every $w_n$ is an $\omega_j$, say with $j = j(n)$. Then $k(n)$ and $j(n)$ are positive integers tending to infinity with $n$. Let $\varepsilon > 0$ be given, and suppose that $N = N(\varepsilon)$ is so large that in the notation of (5.1.2)

$$A_{n+p} - A_n < \varepsilon, \quad n > N, \quad p > 0 \text{ arbitrary}.$$

We fix such an $n$ and observe that $| W - W_n | < \varepsilon$. Set

$$j = \max j(m), \quad 0 \leq m \leq n.$$

Then $\Omega_j$ contains every single term occurring in $W_n$ and normally also certain other terms $w_\alpha$ with $\alpha > n$. It follows that

$$| \Omega_j - W_n | \leq A_{n+p} - A_n < \varepsilon \quad \text{if all } \alpha \leq n + p.$$

In the first member of this inequality we may replace $W_n$ by $A_n$ and $\Omega_j$ by the sum of the absolute values of its terms without affecting the validity of the inequality. This implies the truth of the theorem.

An absolutely convergent series may be split into mutually exclusive subseries, finite or infinite in number. The sum of the sums of these subseries is still equal to the sum of the original series. In the case of nonabsolutely convergent series the situation is entirely different. If the terms are real, the positive as well as the negative terms form divergent subseries. Further, it is possible to rearrange the series in such a manner that the partial sums of the new series converge to a preassigned real number, or diverge to $+ \infty$ or to $- \infty$, or oscillate between finite or infinite bounds (theorem of Riemann). For series with complex terms there are two possibilities:  the limit points of the partial sums of all possible rearrangements fill either a line in the complex plane or the whole plane. We shall not need these results and will not take time to prove them.

Finally we shall have occasion to use some elementary methods of summability, in particular the $(C, 1)$-*means*.[1] We set

$$(5.1.24) \qquad W_n{}^1 = W_0 + W_1 + \cdots + W_n$$

and say that *the series* (5.1.3) *is summable* $(C, 1)$ *to the sum* $W$ if

$$(5.1.25) \qquad \lim_{n \to \infty} \frac{W_n{}^1}{n+1} = W.$$

We shall prove that $(C, 1)$-summability is *consistent* with convergence, which in this context is sometimes referred to as $(C, 0)$-summability.

THEOREM 5.1.10.  *A convergent series of sum* $W$ *is always summable* $(C, 1)$ *to the same sum.*

*Proof.*  Set

$$W_n = W + \eta_n, \quad \eta_n \to 0.$$

Suppose $\varepsilon > 0$ is given. Then there exists an integer $N$ and a positive quantity $M$ such that

$$|\eta_k| < \varepsilon \text{ for } k > N, \quad |\eta_k| < M \text{ for all } k.$$

Hence for $n > N$

$$\left| \frac{W_n{}^1}{n+1} - W \right| \le \frac{1}{n+1} \sum_{k=0}^{n} |\eta_k| \le \frac{1}{n+1} [(N+1)M + (n - N - 1)\varepsilon].$$

The superior limit as $n \to \infty$ of the last member is $\varepsilon$. It follows that the first member has a limit and that this limit must be zero, as asserted.

There are series which are summable $(C, 1)$ without being convergent: $w_n = (-1)^n$ is a case in point.

## EXERCISE 5.1

**1.** Complete the discussion of the geometric series started in the text. What is the relation between $a$ and $q$?

**2.** Prove the assertions concerning the series (5.1.9).

**3.** Same question for the series (5.1.12).

**4.** Discuss the series (5.1.10) under the assumption (5.1.11).

---

[1] The "C" refers to Ernesto Cesàro (1859–1906), who generalized these means. They were earlier known as arithmetic means of order 1.

**5.** Prove the analogue of Theorem 5.1.2 under the assumption (5.1.14).

**6.** In the series (5.1.9) replace $(k^2)^{-s}$ by $k^{-s}$, $k = 2, 3, 4, \cdots$, leaving the other terms alone. Prove that the new series converges or diverges for the same values of $s$ for which the original series does. For the new series show that the quantities $\nu$ and $\mu$ of (5.1.15) are 0 and $+\infty$ respectively if $s > 0$.

**7.** For what complex values of $\alpha$ does the binomial series

$$\sum_{n=0}^{\infty} (-1)^n \binom{\alpha}{n}$$

converge?

**8.** Prove by induction that the $n$th partial sum of the series in Problem 7 equals

$$(-1)^n \binom{\alpha - 1}{n}.$$

**9.** Show that the geometric series $\sum_{n=0}^{\infty} z^n$ converges uniformly in the set $|z| \leqq r < 1$.

**10.** Show that the partial sums of the geometric series are uniformly bounded in $R_\delta$: $|z| \leqq 1$, $|z - 1| \geqq \delta > 0$.

**11.** Suppose that $a_n$ is positive and decreases steadily to 0 as $n \to \infty$. Prove that the series $\sum_{n=0}^{\infty} a_n z^n$ is uniformly convergent in $R_\delta$.

**12.** Given that

$$\tfrac{1}{2} + \cos \theta + \cos 2\theta + \cdots + \cos n\theta = \frac{\sin (n + \tfrac{1}{2})\theta}{2 \sin \tfrac{1}{2}\theta},$$

and that $a_n$ is positive and steadily decreasing to 0, prove that the cosine series

$$\tfrac{1}{2}a_0 + \sum_{n=1}^{\infty} a_n \cos n\theta$$

converges uniformly with respect to $\theta$ if $0 < \delta \leqq \theta \leqq \pi$.

**13.** Find the region of absolute convergence of the series

    **a.** $\displaystyle\sum_{n=1}^{\infty} \left(\frac{z-1}{z+1}\right)^n.$         **b.** $\displaystyle\sum_{n=1}^{\infty} (1 - z^2)^n.$

**14.** What is the $(C, 1)$-sum of the series $\sum_{n=1}^{\infty} (-1)^n$?

**15.** Show that the geometric series is $(C, 1)$-summable for $|z| \leqq 1, z \neq 1$.

**16.** Is the series $\sum_{n=1}^{\infty} (-1)^n n$ summable $(C, 1)$?

**5.2. Operations on series.** We shall be concerned with the operations of addition, scalar multiplication, and multiplication of convergent infinite series. For such series we shall use the same symbol for the series and for its sum whenever the meaning is clear from the context.

Given two convergent series

$$U \equiv \Sigma_0^\infty u_n, \quad V \equiv \Sigma_0^\infty v_n,$$

we define their *sum* as the series

(5.2.1) $$\Sigma_0^\infty (u_n + v_n).$$

This is clearly a convergent series, and its sum is $U + V$. *Scalar multiplication* is defined by

(5.2.2) $$\alpha \, \Sigma_0^\infty u_n = \Sigma_0^\infty \alpha u_n.$$

The resulting series is convergent, and its sum is $\alpha U$. It follows that the set of convergent series forms a linear vector space. A suitable norm is given by

(5.2.3) $$\| U \| = \sup_n | U_n |, \quad U_n = \sum_{k=0}^n u_k.$$

The resulting normed linear vector space is commonly denoted by $c$. It is easily shown to be complete.

Multiplication of convergent series leads to less trivial questions. Formally, the product leads to a double series

$$\sum_{j=0}^\infty \sum_{k=0}^\infty u_j v_k.$$

We shall study such series in the next section. Here we shall be concerned with a special way of grouping the terms of the formal double series which leads to a simple series. Let us define

(5.2.4) $$w_n = \sum_{j=0}^n u_j v_{n-j}, \quad W_n = \sum_{k=0}^n w_k, \quad (W =) \sum_{n=0}^\infty w_n.$$

The series whose general term is $w_n$ is known as the *Cauchy product series* where, for the time being, the convergence of the series is undecided. This is the natural definition of the product when one works with power series, since the Cauchy product of two power series will again be a power series. It should be observed, however, that for other types of series a different definition of the product may be more convenient. Thus, in the theory of (special) Dirichlet series of the form

$$\sum_{n=1}^\infty a_n n^{-s},$$

the Dirichlet product series

(5.2.5) $$(D =) \sum_{n=1}^\infty d_n, \quad d_n = \sum_{jk=n} u_j v_k$$

is the natural tool.

THEOREM 5.2.1.    *The Cauchy product of two convergent series $U$ and $V$ need not be convergent, but the product series is always summable $(C, 1)$ to the sum $UV$.*

*Proof.*    To prove the first point, we form the square of the alternating series

$$\sum_{n=1}^{\infty} (-1)^{n-1} n^{-1/2}.$$

Here

$$w_n = (-1)^{n-1} \sum_{j=1}^{n-1} j^{-1/2}(n-j)^{-1/2}.$$

Since the minimum value of the function

$$x^{-1/2}(n-x)^{-1/2}$$

in the interval $(0, n)$ is $2/n$, it follows that

$$(-1)^{n-1} w_n > (n-1)\frac{2}{n} > 1 \quad \text{for} \quad n > 1,$$

so that the terms of the product series do not tend to zero. Hence the series diverges.

To prove $(C, 1)$-summability we form

$$W_n{}^1 = \sum_{m=0}^{n} W_m = \sum_{m=0}^{n} \sum_{k=0}^{m} w_k = \sum_{k=0}^{n} (n-k+1)w_k$$

$$= \sum_{k=0}^{n} (n-k+1) \sum_{j=0}^{k} u_j v_{k-j} = \sum_{j=0}^{n} u_j \sum_{k=j}^{n} (n-k+1)v_{k-j}$$

$$= \sum_{j=0}^{n} u_j \sum_{m=0}^{n-j} (n-j-m+1)v_m = \sum_{j=0}^{n} u_j \sum_{m=0}^{n-j} V_m$$

$$= \sum_{m=0}^{n} V_m \sum_{j=0}^{n-m} u_j = \sum_{m=0}^{n} V_m U_{n-m}.$$

Here we set

$$U_n = U + \delta_n, \quad V_n = V + \varepsilon_n.$$

It follows that

$$(5.2.6) \qquad W_n{}^1 - (n+1)UV = U \sum_{m=0}^{n} \varepsilon_{n-m} + V \sum_{m=0}^{n} \delta_m + \sum_{m=0}^{n} \delta_m \varepsilon_{n-m}.$$

Since $\delta_m \to 0$ and $\varepsilon_m \to 0$, Theorem 5.1.10 shows that the first two terms in the right member are $o(n+1)$. Thus, if we set

$$\sum_{m=0}^{n} \delta_m \varepsilon_{n-m} \equiv (n+1)\zeta_n,$$

the problem is reduced to showing that $\zeta_n \to 0$. Given an $\eta > 0$, we can find an integer $N = N(\eta)$ and a positive quantity $M$ such that

$$| \delta_k | < \eta, \quad | \varepsilon_k | < \eta \quad \text{for } k > N \text{ and}$$
$$| \delta_k | < M, \quad | \varepsilon_k | < M \quad \text{for all } k.$$

Thus, for $n > 2N$

$$(n + 1) | \zeta_n | \leq \left\{ \sum_{m=0}^{N} + \sum_{m=N+1}^{n-N} + \sum_{m=n-N+1}^{n} \right\} | \delta_m | | \varepsilon_{n-m} |$$
$$< (2N + 1)M\eta + (n - 2N)\eta^2.$$

Dividing through by $n + 1$ and passing to the limit with $n$, we obtain

$$\limsup_{n \to \infty} | \zeta_n | \leq \eta^2.$$

This quantity being arbitrarily small, it follows that $\zeta_n \to 0$, as asserted. This result combined with Theorem 5.1.8 gives the

COROLLARY. *If the Cauchy product series converges, its sum is the product of the sums of the factor series.*

The convergence of the product series may be proved under various assumptions on the factor series. The following theorem relates to the most important case:

THEOREM 5.2.2. *The Cauchy product series of two absolutely convergent series is absolutely convergent.*

*Proof.* We set
$$| u_n | = a_n, \quad | v_n | = b_n, \quad | w_n | = c_n$$
and use the corresponding capital letters for partial sums and sums. We have then

$$C_n \leq \sum_{m=0}^{n} \sum_{k=0}^{m} a_k b_{m-k} \leq \sum_{k=0}^{n} a_k \cdot \sum_{k=0}^{n} b_k = A_n B_n \leq AB,$$

whence it follows that the product series is absolutely convergent and

(5.2.7) $$C \leq AB.$$

The absolutely convergent series consequently form an algebra. We introduce the norm

(5.2.8) $$\| W \| = C = \sum_{n=0}^{\infty} | w_n |$$

and denote the resulting normed algebra by $l$. It is a (B)-algebra.

## EXERCISE 5.2

**1.** Find the square of the series in Problem 7 of Exercise 5.1.

**2.** Find the square of the series for $e$.

**3.** Verify that the metric space $c$ is complete under the norm (5.2.3).

**4.** Verify that (5.2.8) defines a norm and that the algebra $l$ is complete under this norm.

**5.** Prove that

$$\lim_{n \to \infty} \sum_{j=1}^{n-1} j^{-1/2}(n-j)^{-1/2} = \pi.$$

**6.** THEOREM OF MERTENS: *The Cauchy product series converges if one factor series, U say, is absolutely convergent while the other factor V is merely convergent.* (*Hint:* Prove that

$$W_n = \sum_{k=0}^{n} u_k V_{n-k} = V U_n + \sum_{k=0}^{n} u_k \eta_{n-k}, \quad V_m = V + \eta_m,$$

where $\eta_m \to 0$ as $m \to \infty$, and imitate the argument used in proving Theorem 5.2.1.)

**7.** Prove that the square of the alternating series $\sum_{n=1}^{\infty} (-1)^{n-1} n^{-\alpha}$ converges for $\alpha > \frac{1}{2}$.

**5.3. Double series.** Let $w_{jk}$ be a complex-valued function of the two subscripts $j$, $k$, defined for $j$, $k = 0, 1, 2, \cdots$. We imagine the values of this function listed in an infinite table with double entries, containing infinitely many rows and infinitely many columns. The rows and columns are numbered $0, 1, 2, \cdots$. We list $w_{jk}$ in the $j$th row and the $k$th column. The resulting table $T$ could also be called an *infinite matrix*. We now consider the elements of this matrix as the terms of an infinite double series

$$(5.3.1) \qquad\qquad \sum_{j=0}^{\infty} \sum_{k=0}^{\infty} w_{jk}$$

and ask what significance may be given to this symbol and, in particular, if it is possible to define the notions of convergence and sum. To simplify matters, we shall in the main restrict ourselves to absolutely convergent series.

This means that we have to consider an auxiliary table $T_a$ in which the entries are

$$(5.3.2) \qquad\qquad a_{jk} = |\,w_{jk}\,|,$$

so that the new matrix has real non-negative elements. We now introduce an exhaustion process defined by means of a function $\varphi(j, k)$ having the following properties:

(1) $\varphi(j, k)$ is a non-negative integer defined for $j$, $k = 0, 1, 2, \cdots$;
(2) $\varphi(0, 0) = 0$;
(3) $\varphi(j, k) \leq \varphi(j + 1, k)$, $\quad \varphi(j, k) \leq \varphi(j, k + 1)$;
(4) $\varphi(j, k) \to \infty$ with $j + k$.

Examples of such functions are

(5.3.3)                $\varphi(j, k) = \max(j, k), \quad \varphi(j, k) = j + k.$

Any such function will be called an *admissible function* $\varphi(j, k)$. Let such a $\varphi(j, k)$ be chosen, and denote by $A = A_n(\varphi)$ the sum

(5.3.4)                $A_n(\varphi) = \sum_{j, k} a_{jk}, \quad \varphi(j, k) \leq n.$

The sequence $\{A_n(\varphi)\}$ is monotone increasing. Suppose that it is bounded and denote its limit by $A$. Let $\psi(j, k)$ be another admissible function and form the corresponding sequence $\{A_n(\psi)\}$. It is to be proved that this sequence is also bounded and that its limit equals $A$. We note, first, that

$$A_n(\varphi) \leq A$$

for all $n$, and we want to prove the same inequality for $A_n(\psi)$. For this purpose let

(5.3.5)                $\max \varphi(j, k) = m(n) \quad \text{for} \quad \psi(j, k) \leq n.$

But then the sum $A_{m(n)}(\varphi)$ contains all the $a_{jk}$ occurring in $A_n(\psi)$ and, in addition, possibly other terms, so that

$$A_n(\psi) \leq A_{m(n)}(\varphi) \leq A,$$

as asserted. It follows that

$$\lim_{n \to \infty} A_n(\psi) \equiv A^* \leq A$$

exists. Interchanging the roles of $\varphi$ and $\psi$ in this argument, we conclude also that $A \leq A^*$, so that $A = A^*$. Further we see that if one sequence is actually unbounded, then every sequence will have the same property. This justifies

DEFINITION 5.3.1. *The double series* $\Sigma\Sigma\, a_{jk}, a_{jk} \geq 0$, *is said to be convergent if for a particular choice of an admissible function* $\varphi(j, k)$ *the corresponding sequence* $\{A_n(\varphi)\}$ *is bounded. In this case* $\lim A_n(\varphi) \equiv A$ *is called the sum of the series. The series is divergent if one such sequence is unbounded. The series* $\Sigma\Sigma\, w_{jk}$ *is said to be absolutely convergent if* $\Sigma\Sigma\, a_{jk}, a_{jk} = |\,w_{jk}\,|$, *is convergent.*

This procedure of exhaustion may be applied to double series with complex terms. If $\Sigma\Sigma\, w_{jk}$ is the given series, and $\varphi(j, k)$ is an admissible function, we form

$$W_n(\varphi) \equiv \sum_{j, k} w_{jk}, \quad \varphi(j, k) \leq n.$$

It may very well happen that

$$\lim W_n(\varphi)$$

exists, but this no longer ensures the existence of

$$\lim W_n(\psi)$$

or the equality of the two limits, if they both exist. There is one case, however, where the procedure leads to a non-ambiguous result.

**THEOREM 5.3.1.** *If $\Sigma\Sigma\, w_{jk}$ is absolutely convergent, then*

$$(5.3.6) \qquad \lim_{n\to\infty} W_n(\varphi) \equiv W$$

*exists for every admissible choice of $\varphi(j, k)$, and the limit is independent of $\varphi$.*

*Proof.* We have

$$| \, W_{n+k}(\varphi) - W_n(\varphi) \, | \leq A_{n+k}(\varphi) - A_n(\varphi) \to 0$$

as $n \to \infty$. It follows that (5.3.6) exists. Take another function $\psi(j, k)$, let $\varepsilon > 0$ be preassigned, and choose $n$ so large that

$$A - A_n(\psi) < \varepsilon.$$

If $m(n)$ is given by (5.3.5) we have

$$A_{m(n)}(\varphi) - A_n(\psi) < \varepsilon.$$

But

$$| \, W_{m(n)}(\varphi) - W_n(\psi) \, | \leq A_{m(n)}(\varphi) - A_n(\psi),$$

whence it follows that also

$$\lim W_n(\psi) = W.$$

**THEOREM 5.3.2.** *If the double series is absolutely convergent, then*

$$(5.3.7) \qquad \sum_{j=0}^{\infty}\left\{ \sum_{k=0}^{\infty} w_{jk} \right\} = \sum_{k=0}^{\infty}\left\{ \sum_{j=0}^{\infty} w_{jk} \right\} = W.$$

*Proof.* We have to prove that each of the series

$$(5.3.8) \qquad S_j = \sum_{k=0}^{\infty} w_{jk}$$

is convergent and that

$$(5.3.9) \qquad \sum_{j=0}^{\infty} S_j = W$$

and similar results for the columns. But the absolute convergence of the double series clearly implies the absolute convergence of each of the series $S_j$, since for any choice of $m$ and $n$

$$\sum_{j=0}^{m}\left\{ \sum_{k=0}^{n} a_{jk} \right\} < A.$$

Letting $n \to \infty$, we see that the first $m$ series $S_j$ are absolutely convergent, and

$$\sum_{j=0}^{m} | \, S_j \, | < A$$

for every $m$, so that the left side of (5.3.9) is an absolutely convergent series whose sum does not exceed $A$ in absolute value. To prove that the sum equals $W$, we take $\varphi(j, k) = \max(j, k)$ and for a given $\varepsilon > 0$ we choose $n$ so large that

$$| W - W_n(\varphi) | \leq A - A_n(\varphi) < \varepsilon.$$

Here

$$W_n(\varphi) = \sum_{j=0}^{n} \left\{ \sum_{k=0}^{n} w_{jk} \right\}$$

and

$$\left| \sum_{j=0}^{n} S_j - W_n(\varphi) \right| < A - A_n(\varphi) < \varepsilon,$$

so that

$$\left| \sum_{j=0}^{n} S_j - W \right| < 2\varepsilon,$$

and the assertion follows. Columns are handled in the same manner.

Thus, an absolutely convergent double series may be summed by rows or by columns or by an exhaustion process based on an admissible function $\varphi(j, k)$, the sum being the same in each case. To illustrate what may happen when absolute convergence is lost, let us consider the following trivial example. Take

$$w_{j,k} = 1 \text{ if } k = j + 1, \quad w_{j,k} = -1 \text{ if } k = j - 1, \quad w_{j,k} = 0 \text{ otherwise.}$$

Here the sum by rows is $+1$, the sum by columns $-1$, and using either of the functions $\varphi(j, k)$ defined by (5.3.3) we get the sum 0.

Theorem 5.1.9 holds for double series. We state the result, but leave the proof to the reader.

THEOREM 5.3.3.    *An absolutely convergent double series may be rearranged in an arbitrary manner without affecting the absolute convergence or the sum of the series.*

## EXERCISE 5.3

**1.** For what values of $\alpha$ does the series $\sum\limits_{m=1}^{\infty} \sum\limits_{n=1}^{\infty} | m + ni |^{-\alpha}$ converge?

**2.** When is $\sum\limits_{m=1}^{\infty} \sum\limits_{n=1}^{\infty} z^m w^n$ absolutely convergent and what is its sum?

**3.** If the formal product of two convergent simple series is written as a double series, what choice of $\varphi(j, k)$ leads to the Cauchy product? to the Dirichlet product?

**4.** Prove Theorem 5.3.3.

**5.4. Convergence of power series.** We turn now to the main topic of this chapter: the theory of power series. Suppose that $\{a_n \mid n = 0, 1, 2, \cdots\}$ is a given sequence of complex numbers and form the series

$$(5.4.1) \qquad\qquad \sum_{n=0}^{\infty} a_n z^n.$$

This is a power series in $z$. We shall also have occasion to consider power series in the variable $(z - a)$ and, later, in other variables, but for the time being (5.4.1) will suffice.

This series certainly converges for $z = 0$, but it may possibly converge for other values of $z$. The following theorem due to Weierstrass is basic:

**THEOREM 5.4.1.** *If there exists a $z_0 \neq 0$ such that the terms of the series (5.4.1) are bounded for $z = z_0$:*

$$|a_n z_0^n| \leq M \text{ for all } n,$$

*then the series converges absolutely for every $z$ with $|z| < |z_0|$ and uniformly with respect to $z$ in $|z| \leq (1 - \varepsilon)|z_0|$, $\varepsilon > 0$.*

The proof is immediate. For $|z| < |z_0|$ we have

$$|a_n z^n| = |a_n z_0^n| \left|\frac{z}{z_0}\right|^n \leq M \left|\frac{z}{z_0}\right|^n,$$

so the series converges absolutely by comparison with a geometric series. Moreover, if $|z| \leq (1 - \varepsilon)|z_0|$, then

$$|a_n z^n| \leq (1 - \varepsilon)^n M,$$

and the uniform convergence follows from Theorem 5.1.7.

This theorem shows that for (5.4.1) one of the following mutually exclusive possibilities must hold: (i) $z = 0$ is the only point for which the terms are bounded. (ii) Other points with this property exist, but not every finite $z_0$ has it. (iii) Every finite $z_0$ has this property. The theorem shows that in case (ii) there exists a circle $|z| = R$ such that all interior points are admissible while every point in the exterior gives unbounded terms. We define

$$(5.4.2) \qquad\qquad \limsup_{n \to \infty} |a_n|^{\frac{1}{n}} = \alpha, \quad R = \frac{1}{\alpha},$$

where $1/\infty$ and $1/0$ are to be read as $0$ and $\infty$ respectively.

**THEOREM 5.4.2.** *If $R = 0$, the series (5.4.1) diverges for all $z \neq 0$. If $0 < R < \infty$, the series converges absolutely for $|z| < R$ and diverges for $|z| > R$. It converges uniformly for $|z| \leq (1 - \varepsilon)R$, $\varepsilon > 0$. If $R = \infty$, the series converges absolutely for any finite $z$ and uniformly in any bounded set.*

This is an immediate consequence of Theorems 5.1.2 and 5.4.1, since

$$\limsup_{n \to \infty} |a_n z^n|^{\frac{1}{n}} = \alpha |z|.$$

$R$ is called the *radius of convergence* of the power series, and $|z| = R$ is known as the *circle of convergence*. The series may converge for some or all points on this circle. The cases $R = 0$ and $R = \infty$ are not excluded, as shown by, respectively,

$$a_n = n^n \quad \text{and} \quad a_n = n^{-n}.$$

THEOREM 5.4.3. $R = \lim |a_n/a_{n+1}|$ *if this limit exists.*

This follows from Theorem 5.1.3.

For $|z| < R$ the formula

$$(5.4.3) \qquad \sum_{n=0}^{\infty} a_n z^n \equiv f(z)$$

defines a function $f(z)$. The partial sums of the series are polynomials in $z$, hence continuous functions of $z$ in any bounded domain. The series being uniformly convergent in $|z| \leq (1 - \varepsilon)R$ for any fixed $\varepsilon > 0$, it follows from Theorem 4.1.2 that $f(z)$ is continuous in this region and, since $\varepsilon$ is arbitrary, this implies that

$$(5.4.4) \qquad f(z) \in C\,[\,|z| < R].$$

If $R$ is finite, the behavior of the power series on its circle of convergence is highly arbitrary. We shall consider a few of the possibilities, taking $R = 1$ for the sake of simplicity.

*Absolute convergence on the circle of convergence holds either everywhere or nowhere.* If $R = 1$,

$$(5.4.5) \qquad \sum_k |a_k| < \infty$$

is a necessary and sufficient condition for absolute convergence on $|z| = 1$. In this case we have uniform convergence in $|z| \leq 1$ by the M-test so that

$$(5.4.6) \qquad f(z) \in C\,[\,|z| \leq 1].[1]$$

Nonabsolute convergence raises various questions, one of which is the following: Suppose the series (5.4.3) converges for $z = z_0$ with $|z_0| = 1$. The function $f(z)$ defined by (5.4.3) will then be defined also for $z = z_0$. We know that $f(z)$ is continuous for $|z| < 1$. How does it behave for $z = z_0$? The answer is given by a theorem due to Abel.

---

[1] The remainder of this section may be omitted in a first reading.

THEOREM 5.4.4.   *If the series (5.4.3) with $R = 1$ converges for $z = z_0$,
$|z_0| = 1$, then the series*

$$\sum_{n=0}^{\infty} a_n z_0{}^n r^n, \quad 0 \leq r \leq 1,$$

*converges uniformly with respect to $r$, so that*

(5.4.7)               $$f(z_0) = \lim_{r \to 1} f(r z_0).$$

*Proof.* Without restricting the generality we may assume that $z_0 = 1$.
As usual, we write

$$\sum_{k=0}^{n} a_k = A_n, \quad \lim A_n = A.$$

The desired result suggests the use of Theorem 5.1.8, but the assumptions do
not fit. Partial summation does apply, however, with a suitable variation.
Since

$$a_k = (A - A_{k-1}) - (A - A_k),$$

we obtain

$$\sum_{k=n+1}^{n+p} a_k r^k = -\sum_{k=n+1}^{n+p-1} (A - A_k)(r^k - r^{k+1}) + (A - A_n)r^{n+1} - (A - A_{n+p})r^{n+p},$$

the absolute value of which does not exceed

$$M_n \left[ \sum_{k=n+1}^{n+p-1} (r^k - r^{k+1}) + r^{n+1} + r^{n+p} \right] = 2M_n r^{n+1} \leq 2M_n,$$

where

$$M_n = \max_{k \geq n} |A - A_k|.$$

Since this tends to zero when $n \to \infty$, the uniform convergence follows. Thus
$f(r z_0)$ is a continuous function of $r$, $0 \leq r \leq 1$, and (5.4.7) holds.

This theorem shows that $f(z)$ *is continuous for radial approach at all points
of the circle of convergence where the power series actually converges.* An extension
to sectorial approach is given in Problem 8 of Exercise 5.4. The following
interpretation of Theorem 5.4.4 should be compared with Theorem 5.1.10:

*A series*

$$\Sigma_0^{\infty} w_k$$

*is said to be summable Abel to the sum $W$ if* (i) *the series $\Sigma_0^{\infty} w_k r^k$ converges for
$0 < r < 1$ and* (ii)

(5.4.8)               $$\lim_{r \to 1} \Sigma_0^{\infty} w_k r^k = W.$$

Theorem 5.4.4 expresses that *a convergent series is Abel summable to its Cauchy
sum.*

We note the following corollary of Theorem 5.4.4:

COROLLARY.　*Suppose that the series*

$$(5.4.9) \qquad \sum_{n=0}^{\infty} a_n(\cos n\theta + i \sin n\theta)$$

*converges uniformly in θ for* $\alpha \leq \theta \leq \beta$. *Then the power series* (5.4.3) *converges uniformly with respect to z in the closed sector* $0 \leq |z| \leq 1$, $\alpha \leq \arg z \leq \beta$. *In particular, if the series* (5.4.3) *converges uniformly in* $|z| = 1$, *then it converges uniformly in* $|z| \leq 1$, *and* (5.4.6) *holds.*

We have merely to replace $A$ and $A_n$ in the above estimates by the sum $A(\theta)$ and the partial sums $A_n(\theta)$ of (5.4.9). Since $A(\theta) - A_n(\theta)$ converges to zero uniformly with respect to $\theta$ for $\alpha \leq \theta \leq \beta$, we see that

$$M_n = \max_{n \leq k} \max_{\alpha \leq \theta \leq \beta} |A(\theta) - A_k(\theta)|$$

tends to zero as $n \to \infty$. This ensures uniform convergence with respect to $r$ and $\theta$.

We shall see later that the last assertion of the Corollary is a simple consequence of the so-called principle of the maximum. See Chapter 8.

We turn now to an important case in which we have convergence on the circle of convergence except possibly for $z = 1$.

THEOREM 5.4.5.　*If* $\Sigma |a_n - a_{n+1}|$ *converges and* $\lim a_n = 0$, *then the power series* (5.4.3) *converges for* $|z| \leq 1$, *except possibly for* $z = 1$, *and the convergence is uniform in* $|z| \leq 1$, $|z - 1| \geq \delta > 0$. *If* $a_n > 0$ *and* $\Sigma a_n$ *is divergent, the power series does not converge uniformly in* $0 < |z - 1| < \delta$, $|z| \leq 1$.

*Proof.*　Since for $|z| \leq 1$, $|1 - z| \geq \delta > 0$

$$|1 + z + z^2 + \cdots + z^n| = \left| \frac{1 - z^{n+1}}{1 - z} \right|$$

$$\leq \frac{2}{|1 - z|} \leq \frac{2}{\delta},$$

the first assertion is an immediate consequence of Theorem 5.1.8. If $a_n > 0$, and $\Sigma a_n$ diverges, the partial sums of the power series are unbounded on the line segment $[1 - \delta, 1)$, since for any choice of $N$

$$\sum_{n=0}^{\infty} a_n r^n > \sum_{n=0}^{N} a_n r^n > r^N \sum_{n=0}^{N} a_n.$$

Taking $N$ as the least integer $> (1 - r)^{-1}$, we have $N \to \infty$ as $r \to 1$, and $r^N > \frac{1}{4}$ for $1 > r > \frac{1}{2}$. Thus, since $A_N = \Sigma_0^N a_n \to \infty$ as $N \to \infty$, uniform convergence in $0 < |1 - z| \leq \delta$, $|z| \leq 1$, is out of the question.

COROLLARY.    *Let $a_n \geqq a_{n+1} \rightarrow 0$ and consider the trigonometric series*

$$(5.4.10) \qquad \sum_{n=1}^{\infty} a_n \cos n\theta \quad and \quad \sum_{n=1}^{\infty} a_n \sin n\theta.$$

*The first series converges uniformly in $\theta$ for $0 < \delta \leqq |\theta| \leqq \pi$. It converges for $\theta = 0$, and then uniformly for all $\theta$, if and only if $\Sigma a_n$ converges. The second series converges for all $\theta$ and uniformly for $0 < \delta \leqq |\theta| \leqq \pi$.*

Since the two series are the real and imaginary parts of a power series on the unit circle to which Theorem 5.4.5 applies, the assertions are obvious. For the more delicate convergence properties of the sine series near the origin, see Problem 9 of Exercise 5.4.

In conclusion we shall give a power series which converges uniformly but not absolutely on the unit circle. This particular construction is due to Lipot Fejér (1880–1959), who has used this device to produce power series and trigonometric series having selected convergency defects. We use the partial sums of the logarithmic series to form the polynomials

$$(5.4.11) \qquad L(z, n) = -\frac{z}{n} - \frac{z^2}{n-1} - \cdots - \frac{z^n}{1} + \frac{z^{n+1}}{1} + \frac{z^{n+2}}{2} + \cdots + \frac{z^{2n}}{n}.$$

The basic property of these polynomials is that they are uniformly bounded for $|z| \leqq 1$. Further, $L(1, n) = 0$ for all $n$. We now form

$$(5.4.12) \qquad f(z) \equiv \sum_{n=1}^{\infty} \frac{1}{n^2} z^{2^{n^2}} L(z, 2^n).$$

As written, this is a series of polynomials converging absolutely and uniformly in $|z| \leqq 1$. Thus $f(z) \in C[|z| \leqq 1]$. On the other hand, substituting the expression for $L(z, 2^n)$, multiplying out, and writing the terms consecutively, we see that (5.4.12) may also be regarded as a power series. The coefficients of the power series are bounded, so $R \geqq 1$. It is easily seen that the power series is not absolutely convergent on $|z| = 1$ since the harmonic series is divergent. Thus $R = 1$. On the other hand, we do have convergence everywhere on the unit circle and, indeed, uniform convergence. We can modify this example in many ways, by taking "longer" polynomials or by changing the basic sequence $\{1/n\}$. We shall not pursue the problem further.

## EXERCISE 5.4

**1.** Determine the radii of convergence of the power series whose $n$th coefficients are given by

    **a.** $n^p$.    **b.** $\log (n + 2)$.    **c.** $\cosh n$.    **d.** $e^{-n^2}$.    **e.** $\sin n$.

Determine the radii of convergence of the following series:

    **2.** The binomial series $\sum_{n=0}^{\infty} \binom{\alpha}{n} z^n$.    **3.** The logarithmic series $\sum_{n=1}^{\infty} \frac{z^n}{n}$.

**4.** The exponential series $1 + \sum\limits_{n=1}^{\infty} \dfrac{z^n}{n!}$.

**5.** The hypergeometric series (where $\gamma$ is neither zero nor a negative integer)

$$F(\alpha, \beta, \gamma;\ z) = 1 + \sum_{n=1}^{\infty} \frac{\alpha(\alpha+1)\cdots(\alpha+n-1)\beta(\beta+1)\cdots(\beta+n-1)}{1\cdot2\cdots n\ \gamma(\gamma+1)\cdots(\gamma+n-1)}\, z^n.$$

**6.** Examine the convergence on the circle of convergence in Problem 5 under various assumptions on $\alpha$, $\beta$, $\gamma$. If these parameters are real, when does Theorem 5.4.5 apply? (*Hint:* Estimate the coefficients as in the proof of Theorem 5.1.4.)

**7.** Show that the power series

$$\sum_{n=1}^{\infty} \frac{(-1)^n}{n}\, z^{n(n+1)}$$

has the radius of convergence 1. Examine convergence at $z = 1$, $-1$, and $i$.

**\*8.** [THEOREM OF OTTO STOLZ (1842–1905)] Prove the following extension of Theorem 5.4.4: *If* $\sum\limits_{n=0}^{\infty} a_n = A$ *is convergent, then*

$$\lim_{z\to1} \sum_{n=0}^{\infty} a_n z^n = A$$

*when* $z$ *approaches* 1 *in any sector* $|z| < 1$, $|\arg(1-z)| \leq \dfrac{\pi}{2} - \delta$, $0 < \delta$.

(*Hint:* Use summation by parts and estimate $\sum\limits_{k=0}^{n} |z^k - z^{k+1}|$ in the sector.)

**\*9.** The sine series $\sum\limits_{n=1}^{\infty} a_n \sin n\theta$ with $a_n \geq a_{n+1} \to 0$ converges uniformly for all $\theta$ if and only if $na_n \to 0$. The partial sums are bounded near the origin (and hence everywhere) if and only if $na_n$ is bounded. (*Hint:* If the $n$th partial sum is $S_n(\theta)$, examine $S_n(\theta_n) - S_m(\theta_n)$ with $\theta_n = 2\pi/n$ and show that it is positive and exceeds a constant (depending on $m$) times $na_n$. From this conclude the necessity of the stated conditions. For the sufficiency use summation by parts.)

**\*10.** Assume the uniform boundedness of the polynomials $L(z, n)$ and prove the assertions regarding the series (5.4.12) made in the text. (*Hint:* How fast do the partial sums of the harmonic series grow? A partial sum of the power series is not always a partial sum of the polynomial series. How do they differ? Estimate the difference.)

**11.** Prove that for a power series with bounded coefficients the radius of convergence is at least 1.

Find the radius of convergence of the power series $\sum\limits_{n=1}^{\infty} a_n z^n$ if

**12.** $a_{2m-1} = 1/m, \quad a_{2m} = 1.$

**13.** $a_{2m-1} = (1 - 1/m)^{m^2}, \quad a_{2m} = 2^m.$

**14.** $a_n = m$ for $n = m^2, \quad m = 1, 2, 3, \cdots$, otherwise 0.

**15.** $a_n = 1/m!$ when $n = 2^m, \quad m = 0, 1, 2, \cdots$, otherwise 0.

**16.** Prove that the power series specified in Problem 15 converges everywhere on its circle of convergence.

**17.** Prove that the power series specified in Problems 12–14 diverge everywhere on their circles of convergence.

**18.** If $\sum\limits_{n=0}^{\infty} a_n z^n \equiv f(z)$ converges for $|z| < 1$ and if $A_n = \sum\limits_{k=0}^{n} a_k$, $A_n^1 = \sum\limits_{m=0}^{n} A_m$, verify that for $|z| < 1$

$$f(z) = (1 - z) \sum_{n=0}^{\infty} A_n z^n = (1 - z)^2 \sum_{n=0}^{\infty} A_n^1 z^n.$$

### 5.5. Power series as holomorphic functions.

Given a power series

$$(5.5.1) \qquad \sum_{n=0}^{\infty} a_n z^n \equiv f(z)$$

with radius of convergence $R > 0$. We have seen that $f(z) \in C[|z| < R]$. We are now ready to prove the much stronger result that $f(z) \in H[|z| < R]$. In other words we have

THEOREM 5.5.1. *The sum of a power series is a holomorphic function in the interior of its circle of convergence.*

In Chapter 8 we shall prove the converse theorem to the effect that a function $f(z)$ holomorphic in the circle $|z| < R$ admits of a convergent power series expansion (5.5.1) in this circle. This fact adds to the importance of Theorem 5.5.1. The proof will be given in several steps.

Together with (5.5.1) we consider the infinite set of power series

$$\sum_{n=1}^{\infty} n a_n z^{n-1} \equiv f_1(z),$$

$$\sum_{n=2}^{\infty} n(n-1) a_n z^{n-2} \equiv f_2(z),$$

$$(5.5.2) \qquad \cdots\cdots\cdots\cdots\cdots\cdots\cdots$$

$$\sum_{n=k}^{\infty} n(n-1)(n-2)\cdots(n-k+1) a_n z^{n-k} \equiv f_k(z),$$

$$\cdots\cdots\cdots\cdots\cdots\cdots\cdots$$

For the time being these are referred to as the first, second, $\cdots$, $k$th, $\cdots$ *derived series*. The reader will note that these series are obtained by formal term-by-term differentiation of (5.5.1). For the moment, they are just other power series; we know neither that they converge nor how they are related to $f(z)$.

**LEMMA 5.5.1.** *All the series (5.5.2) have the same radius of convergence, namely, that of (5.5.1).*

*Proof.* It suffices to give the proof for $f_1(z)$, for if the assertion is valid for $f_1(z)$ then we have shown that the derived series has the same radius of convergence as the given series, and the general statement follows by induction since $f_{k+1}(z)$ is the derived series of $f_k(z)$. If the radius of convergence of $f_1(z)$ is $R_1$, then on the one hand $R_1 \leq R$, since the series for $zf_1(z)$ has larger terms than the series for $f(z)$. On the other hand, for positive sequences we have

$$(5.5.3) \qquad \limsup (a_n b_n) \leq (\limsup a_n)(\limsup b_n)$$

as long as the right-hand side does not take on the indeterminate form $0 \cdot \infty$. This gives

$$\frac{1}{R_1} = \limsup (n \mid a_n \mid)^{\frac{1}{n}} \leq \lim n^{\frac{1}{n}} \cdot \limsup \mid a_n \mid^{\frac{1}{n}} = \frac{1}{R},$$

or $R \leq R_1$. Hence $R = R_1$ as asserted.

**LEMMA 5.5.2.** *For $n \geq 2$ we have the identity*

$$\frac{b^n - a^n}{b - a} - na^{n-1} = (b - a)[b^{n-2} + 2ab^{n-3} + 3a^2b^{n-4} + \cdots + (n - 1)a^{n-2}].$$

*Proof.* This is evidently true for $n = 2$. Suppose that the identity holds for $n = k$. Denoting the quantity inside the bracket by $S_n(a, b)$, we have

$$bS_n(a, b) + na^{n-1} = S_{n+1}(a, b).$$

It follows that

$$
\begin{aligned}
\frac{b^{k+1} - a^{k+1}}{b - a} &= \frac{b^{k+1} - ba^k + ba^k - a^{k+1}}{b - a} = b\,\frac{b^k - a^k}{b - a} + a^k \\
&= b[(b - a)S_k(a, b) + ka^{k-1}] + a^k \\
&= (b - a)bS_k(a, b) + ka^{k-1}b + a^k \\
&= (b - a)[S_{k+1}(a, b) - ka^{k-1}] + ka^{k-1}b + a^k \\
&= (b - a)S_{k+1}(a, b) + (k + 1)a^k,
\end{aligned}
$$

and this is the desired identity for $n = k + 1$. Thus the relation is true for all $n$.

*Proof of Theorem 5.5.1.* It suffices to prove that the difference quotient converges to $f_1(z)$. Let $R_0 < R$ be given, and consider values of $z$ and $z + h$ such that $|z| \leq R_0$, $|z + h| \leq R_0$. The two lemmas give

$$\frac{1}{h}[f(z + h) - f(z)] - f_1(z) = \sum_{n=2}^{\infty} a_n \left\{ \frac{1}{h}[(z + h)^n - z^n] - nz^{n-1} \right\}$$

$$= h \sum_{n=2}^{\infty} a_n[(z + h)^{n-2} + 2z(z + h)^{n-3} + \cdots + (n - 1)z^{n-2}],$$

and the absolute value of the last member does not exceed

$$\tfrac{1}{2}|h| \sum_{n=2}^{\infty} n(n - 1)|a_n| R_0^{n-2} \leq \tfrac{1}{2}|h| F_2(R_0).$$

Here $F_2(z)$ is the second derived series of

$$F(z) \equiv \sum_{n=0}^{\infty} |a_n| z^n,$$

the radius of convergence of which also equals $R$. Thus

$$(5.5.4) \qquad \left| \frac{1}{h}[f(z + h) - f(z)] - f_1(z) \right| \leq \tfrac{1}{2}|h| F_2(R_0), \quad |z|, |z + h| \leq R_0.$$

As $h \to 0$ the right member tends to zero uniformly in $z$ for $|z| \leq R_0 < R$. From this result we conclude successively that (i) $f(z)$ is differentiable, (ii) $f'(z) = f_1(z)$, and (iii) the difference quotient of $f(z)$ tends uniformly to the derivative in any fixed circular disk interior to the circle of convergence. This proves Theorem 5.5.1 and much more than was asserted in that theorem. But there are several other conclusions that we can draw now. In view of their importance they will be stated as separate theorems.

THEOREM 5.5.2. *In the interior of its circle of convergence a power series may be differentiated term by term. The derived series converges and represents the derivative of the sum of the original power series.*

THEOREM 5.5.3. *A power series has derivatives of all orders, and for* $|z| < R$ *we have*

$$(5.5.5) \qquad f^{(k)}(z) = \sum_{n=k}^{\infty} n(n - 1) \cdots (n - k + 1)a_n z^{n-k}.$$

*In particular*

$$(5.5.6) \qquad f^{(k)}(0) = k!\, a_k.$$

*Hence the power series* (5.5.1) *is the Maclaurin series of its sum*

$$(5.5.7) \qquad f(z) = \sum_{k=0}^{\infty} \frac{f^{(k)}(0)}{k!} z^k.$$

We can also integrate a power series term by term where the integral is defined as in Section 4.1. Since

$$(5.5.8) \qquad\qquad \int_0^z t^k \, dt = \frac{z^{k+1}}{k+1} \, ,$$

we are justified in expecting that

$$(5.5.9) \qquad\qquad \int_0^z \left\{ \sum_{n=0}^{\infty} a_n t^n \right\} dt = \sum_{n=0}^{\infty} \frac{a_n}{n+1} z^{n+1}.$$

The integrated series obviously has the same radius of convergence as (5.5.1), but may possibly have better convergence properties on the circle of convergence. Formula (4.2.7) gives

$$\int_0^z \left\{ \sum_{n=0}^{\infty} a_n t^n \right\} dt = \int_0^r \left\{ \sum_{n=0}^{\infty} a_n \cos{(n+1)\theta} \, s^n \right\} ds + i \int_0^r \left\{ \sum_{n=0}^{\infty} a_n \sin{(n+1)\theta} \, s^n \right\} ds,$$

if $z = r(\cos\theta + i \sin\theta)$, $r < R$, and the coefficients are supposed to be real. Here $\theta$ is fixed and the series converge uniformly with respect to $s$ and $\theta$, so we can integrate termwise, obtaining (5.5.9) upon combining the real and the imaginary parts. The general case of complex coefficients can obviously be handled by setting

$$f(z) = \Sigma_0^{\infty} a_n z^n = \Sigma_0^{\infty} (\alpha_n + i\beta_n)z^n = \Sigma_0^{\infty} \alpha_n z^n + i \, \Sigma_0^{\infty} \beta_n z^n = f_r(z) + if_i(z)$$

and using the linearity of the integral. This proves (5.5.9).

## EXERCISE 5.5

**1.** Prove (5.5.3).

**2.** Compute and sum in closed form the first derived series of the logarithmic series (Problem 3 of Exercise 5.4).

**3.** Same question for the arc tangent series

$$\sum_{n=0}^{\infty} (-1)^n \frac{z^{2n+1}}{2n+1}.$$

**4.** Find the first derived series of a hypergeometric series and verify the differential equation

$$z(1-z)F'' + [\gamma - (\alpha + \beta + 1)z]F' - \alpha\beta F = 0.$$

**5.** Let $\triangle$ be a triangle with vertices $z_1$, $z_2$, $z_3$ located inside the circle of convergence $|z| = R$ of the power series (5.5.1). Evaluate

$$\left[ \int_{z_1}^{z_2} + \int_{z_2}^{z_3} + \int_{z_3}^{z_1} \right] f(z) \, dz.$$

**6.** What is the sum of the series $\sum_{n=1}^{\infty} n^2 z^n$, $\;|z| < 1$?

**5.6. Taylor's series.** We shall be concerned with an elementary transformation of power series which is of fundamental importance for the study of holomorphic functions. It is based on the binomial theorem in the form

$$(5.6.1) \qquad z^n = (z - a + a)^n = \sum_{k=0}^{n} \binom{n}{k} (z - a)^k a^{n-k}.$$

Suppose that the series (5.5.1) has the radius of convergence $R > 0$, and let $|a| < R$. Then for $|z| < R$,

$$(5.6.2) \qquad \sum_{n=0}^{\infty} a_n z^n = \sum_{n=0}^{\infty} a_n \sum_{k=0}^{n} \binom{n}{k} (z - a)^k a^{n-k}.$$

This may be regarded as a double series

$$\sum_{n=0}^{\infty} \sum_{k=0}^{\infty} w_{nk}$$

with

$$w_{nk} = \binom{n}{k} a_n (z - a)^k a^{n-k}, \quad k \leq n,$$

$$w_{nk} = 0, \qquad\qquad\qquad k > n.$$

With this interpretation (5.6.2) represents the double series summed by rows. In this case it would be of interest to sum the series by columns instead, and Theorem 5.3.2 shows that this is permitted without risk of changing the sum of the series if the latter is absolutely convergent. This will be the case if and only if

$$\sum_{n=0}^{\infty} |a_n| \sum_{k=0}^{n} \binom{n}{k} |z - a|^k |a|^{n-k} = \sum_{n=0}^{\infty} |a_n| (|z - a| + |a|)^n$$

converges, that is, if

$$(5.6.3) \qquad\qquad\qquad |z - a| < R - |a|.$$

Supposing this condition to be satisfied and summing by columns instead, we obtain

$$f(z) = \sum_{k=0}^{\infty} (z - a)^k \frac{1}{k!} \sum_{n=k}^{\infty} n(n - 1) \cdots (n - k + 1) a_n a^{n-k}.$$

Since $|a| < R$, formula (5.5.5) applies and we have

$$(5.6.4) \qquad\qquad f(z) = \sum_{k=0}^{\infty} \frac{f^{(k)}(a)}{k!} (z - a)^k.$$

This is Taylor's series for $f(z)$.[1]

---

[1] Brook Taylor (1685–1731) discovered the theorem named after him around 1712. He was also aware of the special case $a = 0$ usually called Maclaurin's theorem after Colin Maclaurin (1698–1746), who later rediscovered it.

Formula (5.6.4) represents $f(z)$ in the circle (5.6.3). But the right-hand side is a power series in $z - a$; as such it has a radius of convergence $R_a$, and the sum of the series is a holomorphic function $f(z;\ a)$ in the circle

$$(5.6.5) \qquad\qquad |z - a| < R_a.$$

The relationship between these two holomorphic functions $f(z)$ and $f(z;\ a)$ is basic for the theory. We have of course

$$(5.6.6) \qquad f(z;\ a) = f(z) \quad \text{for} \quad |z - a| < R - |a|.$$

Next we prove

THEOREM 5.6.1.    $R - |a| \leq R_a \leq R + |a|$.

*Proof.* The first part of this inequality has already been noticed. In order to prove the second half, suppose that $R_a > |a|$. In the opposite case we have nothing to prove. The assumption implies that the point $z = 0$ belongs to the common region of convergence of the two series where (5.6.6) holds. In particular we have then the relations

$$(5.6.7) \qquad\qquad f^{(n)}(0) = \sum_{k=n}^{\infty} \frac{f^{(k)}(a)}{(k - n)!} (-a)^{k-n}$$

valid for every $n$. Now we got the power series $f(z;\ a)$ from the power series $f(z)$ by setting $z = (z - a) + a$, using (5.6.1), and rearranging the resulting double series according to powers of $(z - a)$. But we may apply the same procedure to the power series $f(z;\ a)$, expanding $(z - a)^n$ by the binomial theorem and rearranging the resulting double series according to powers of $z$. This is permitted when the double series is absolutely convergent, that is, for $|z| + |a| < R_a$ or

$$|z| < R_a - |a|.$$

The rearranged series is

$$\sum_{n=0}^{\infty} \frac{z^n}{n!} \sum_{k=n}^{\infty} \frac{f^{(k)}(a)}{(k - n)!} (-a)^{k-n} = \sum_{n=0}^{\infty} \frac{f^{(n)}(0)}{n!} z^n = f(z)$$

by (5.6.7). Thus we are back at the starting point: the rearranged series at $z = a$, rearranged again at $z = 0$, gives the original series. The radius of convergence of the latter being $R$, we must have

$$R_a - |a| \leq R \quad \text{or} \quad R_a \leq R + |a|,$$

as asserted. This proves the theorem. We can reformulate the latter in geometrical language as follows:

Let $C_1(a)$ and $C_2(a)$ be the two circles with their centers at $z = a$ and tangent to $C\colon |z| = R$. Then the circle of convergence $C(a)$ of $f(z;\ a)$ contains the inscribed circle $C_1(a)$ and is contained in the circumscribed circle $C_2(a)$.

The two bounds found for $R_a$ are the best possible of their kind. This is shown by the situation for the geometric series

$$\frac{1}{1-z} = \sum_{n=0}^{\infty} z^n.$$

Here

$$(5.6.8) \qquad f(z;\ a) = \sum_{n=0}^{\infty} \frac{(z-a)^n}{(1-a)^{n+1}}, \quad R_a = |\,1-a\,|,$$

and $R_a = R - |\,a\,|$ if $a$ is real, $0 \leq a < 1$, while $R_a = R + |\,a\,|$ if $-1 < a < 0$.

THEOREM 5.6.2.    *The functions $f(z)$ and $f(z;\ a)$, defined by (5.5.1) and (5.6.4) respectively, coincide in their common domain of definition.*

*Proof.*    We know that $f(z)$ and $f(z;\ a)$ coincide in $|\,z-a\,| < R - |\,a\,|$. Suppose that we want to prove coincidence in a neighborhood of $z = b$ where

$$|\,b\,| < R, \quad R - |\,a\,| \leq |\,b-a\,| < R_a.$$

For this purpose we choose points

$$b_0 = a, b_1, b_2, \cdots, b_n = b$$

on the line segment $[a, b]$ in such a manner that

$$|\,b_k - b_{k-1}\,| < \delta, \quad k = 1, 2, \cdots, n$$

where

$$\delta = \min\left[\min_z\, (R - |\,z\,|),\ \min_z\, (R_a - |\,z-a\,|)\right]$$

for $z \in [a, b]$.

Since, in particular, $|\,b_1 - a\,| < \delta \leq R - |\,a\,|$, we see that $f(z)$ and $f(z;\ a)$ coincide in some neighborhood of $z = b_1$. We take the corresponding power series (5.5.1) and (5.6.4) and rearrange about the point $z = b_1$, obtaining

$$\sum_{k=0}^{\infty} \frac{f^{(k)}(b_1)}{k!}\, (z - b_1)^k \quad \text{and} \quad \sum_{k=0}^{\infty} \frac{f^{(k)}(b_1;\ a)}{k!}\, (z - b_1)^k$$

respectively. The first series converges and represents $f(z)$ for

$$|\,z - b_1\,| < R - |\,b_1\,|,$$

while the second converges and represents $f(z;\ a)$ for

$$|\,z - b_1\,| < R_a - |\,b_1 - a\,|.$$

On the other hand, the two power series are identical, that is, they have the same coefficients. Since $f(z)$ and $f(z;\ a)$ coincide in some neighborhood of $z = b_1$ and have derivatives of all orders, we must have

$$f^{(k)}(b_1) = f^{(k)}(b_1;\ a)$$

as asserted. This power series represents $f(z)$ in one circular disk and $f(z;\ a)$ in another concentric disk. These two functions must then coincide in the smaller of these disks, which we denote by $C_1$. Its radius is at least $\delta$, and $b_2 \in C_1$. We can then rearrange the power series for $f(z)$ and for $f(z;\ a)$ about $z = b_2$. The rearranged series have identical coefficients and enable us to conclude that $f(z)$ and $f(z;\ a)$ coincide also in a disk $C_2$ about $z = b_2$. After $n$ steps we find that $f(z)$ and $f(z;\ a)$ also coincide in a disk $C_n$ about $z = b_n = b$. Since $b$ was an arbitrary point in the common domain of definition of the two series, the theorem is proved.

Suppose now that $R_a > R - |a|$. Then $f(z;\ a)$ is defined in the disk

$$C_a:\quad |z - a| < R_a,$$

while $f(z)$ is defined in the disk

$$C_0:\quad |z| < R.$$

These two disks overlap, and neither is contained in the other. We can then define a function

$$F(z) = \begin{cases} f(z), & z \in C_0, \\ f(z;\ a), & z \in C_a \ominus C_0 \cap C_a. \end{cases}$$

From its definition this function is holomorphic in $C_0 \cup C_a$. We say that $f(z;\ a)$ gives an *analytic continuation* of $f(z)$ in $C_a \ominus C_0 \cap C_a$. Conversely, $f(z)$ gives an analytic continuation of $f(z;\ a)$ in $C_0 \ominus C_0 \cap C_a$.

We defer a closer study of analytic continuation to Volume II, Chapter 10, but in the next section we shall have something to say about the obstacles encountered in this process.

## EXERCISE 5.6

**1.** If $\dot{E}(z) \equiv 1 + \sum\limits_{n=1}^{\infty} \dfrac{z^n}{n!}$ with $R = \infty$, compute the derivatives and Taylor's series. From the latter or by Cauchy multiplication show that for all $z$ and $a$

$$E(z) = E(z - a)E(a).$$

**2.** Express $\sum\limits_{n=0}^{\infty} \dfrac{P(n)}{n!} z^n$ in terms of $E(z)$ if $P(t) = a_0 + a_1 t + a_2 t^2$.

**3.** Given

$$C(z) = \sum_{n=0}^{\infty} \frac{(-1)^n}{(2n)!} z^{2n}, \quad S(z) = \sum_{n=0}^{\infty} \frac{(-1)^n}{(2n+1)!} z^{2n+1}.$$

Compute the derivatives of $C(z)$ and $S(z)$ and the corresponding Taylor's series.

**5.7. Singularities; noncontinuable power series.** We have reached a stage where it is desirable to give a definition of singular points suitable for the study of power series.

DEFINITION 5.7.1.    *If $f(z)$ is defined by (5.5.1) with $R > 0$, if $|z_0| = R$, and if for all $\alpha$, $0 < \alpha < 1$*

$$(5.7.1) \qquad\qquad R_a = (1 - \alpha)R \quad where \quad a = \alpha z_0,$$

*then $z = z_0$ is a singular point of $f(z)$.*

We shall see later that if (5.7.1) holds for one $\alpha$ between 0 and 1, then it holds for all. The condition expresses that none of the power series $f(z;\ \alpha z_0)$, $0 < \alpha < 1$, obtained by rearranging the power series for $f(z)$ about a point $z = \alpha z_0$ on the radius from 0 to $z_0$, is defined outside of $C_0$: $|z| < R$. Thus none of these power series gives an analytic continuation of $f(z)$. Here we have restricted ourselves to radial approach. It will be seen later that for any value of $a$ with $|a| < R$ we have

$$(5.7.2) \qquad\qquad R_a \leq |z_0 - a|.$$

This implies that no function $F(z)$, defined and holomorphic in a domain which is the union of $C_0$ with an $\varepsilon$-neighborhood of $z = z_0$, and coinciding with $f(z)$ in $C_0$, can exist, no matter how small $\varepsilon$ is.

We shall also see later that *there is at least one singular point of the power series on its circle of convergence.*

The behavior of $f(z)$ for radial approach to a singular point differs from one case to the next. The simplest is that in which $f(\alpha z_0)$ does not tend to a limit or $|f(\alpha z_0)|$ becomes infinite as $\alpha \to 1$. The geometric series gives an example of this possibility. It is clear that $f(z)$ cannot be holomorphic at such a point $z_0$ since a necessary condition for holomorphy is that the function be defined and continuous. But $f(\alpha z_0)$ may very well have a finite limit for $\alpha \to 1$ without being holomorphic. The existence of a finite definite radial derivative is certainly a necessary condition, so that if $\lim_{\alpha \to 1} f'(\alpha z_0)$ does not exist, $z = z_0$ must be a singular point. We shall see later that a holomorphic function has derivatives of all orders. This shows that

$$(5.7.3) \qquad\qquad \lim_{\alpha \to 1} f^{(k)}(\alpha z_0)$$

must exist for every $k$ if $f(z)$ is to be holomorphic at $z = z_0$. If this condition is violated for some $k$, and hence for all larger values of $k$, then $z = z_0$ is certainly singular. But the existence of radial derivatives of all orders is only a necessary condition for holomorphy; it is by no means sufficient. Examples illustrating these various possibilities are given below.

We shall start our study of singular points by proving a theorem on power series with positive coefficients due to Alfred Pringsheim (1850–1941).

THEOREM 5.7.1.    *If $0 < R < \infty$ and $a_n \geq 0$ for all large $n$, then $z = R$ is a singular point of $f(z)$.*

*Proof.*    Without restricting the generality we may assume $a_n \geq 0$ for all $n$. If this is not true from the outset, we have merely to add a suitably chosen polynomial to $f(z)$. This does not affect convergence or singularities. If $a_n \geq 0$ for all $n$, and if for some $\alpha$, $0 < \alpha < 1$, $R_a > (1 - \alpha)R$, where $a = \alpha R$, then the series

$$f(z;\ \alpha R) = \sum_{k=0}^{\infty} (z - \alpha R)^k \sum_{n=k}^{\infty} \binom{n}{k} a_n (\alpha R)^{n-k}$$

converges for a real positive $z$ greater than $R$, say for $z = R(1 + \delta)$, $\delta > 0$. Since the supposedly convergent double series

$$\sum_{k=0}^{\infty} (1 - \alpha + \delta)^k R^k \sum_{n=k}^{\infty} \binom{n}{k} a_n (\alpha R)^{n-k}$$

has positive terms, we can interchange the order of the two summations without loss of convergence. Hence the series

$$\sum_{n=0}^{\infty} a_n \sum_{k=0}^{n} \binom{n}{k} (\alpha R)^{n-k} [(1 - \alpha + \delta)R]^k = \sum_{n=0}^{\infty} a_n [(1 + \delta)R]^n$$

must converge. But this is absurd. Hence (5.7.1) holds for $z_0 = R$, and $z = R$ is a singular point of $f(z)$.

This theorem has been generalized in many ways, and similar theorems have later been proved for Dirichlet series and other expansions. One of the simplest extensions occurs in Problem 1 of Exercise 5.7.

As an illustration consider the series

(5.7.4)             $$f_p(z) = \sum_{n=1}^{\infty} n^{-p} z^n, \quad p \text{ real.}$$

Here $R = 1$, and the theorem applies and shows that $z = 1$ is a singular point of $f_p(z)$. Actually, it is the only singularity on $|z| = 1$. In this case $f_p(\alpha)$ becomes infinite as $\alpha \to 1$ as long as $p \leq 1$ but not for $p > 1$. Similarly, the $k$th derivative $f_p^{(k)}(\alpha)$ becomes infinite for $p \leq k + 1$ but not for $p > k + 1$.[1]

With a view to several applications we shall work out a much more complicated illustration. We take

(5.7.5)             $$f(z) = \sum_{n=0}^{\infty} \frac{1}{n!} z^{2^n}.$$

Here again $R = 1$, and $z = 1$ is a singular point. But in this case radial

---

[1] The remainder of this section may be omitted in a first reading.

derivatives of all orders exist at $z = 1$. Further, every point on $|z| = 1$ is singular. A simple computation shows that

$$(5.7.6) \qquad f^{(k)}(\alpha) \to A_k \equiv \sum_{n=0}^{\infty} \frac{1}{n!} 2^n (2^n - 1) \cdots (2^n - k + 1).$$

All these series can be summed in closed form. We have

$$A_0 = e, \; A_1 = e^2, \; A_2 = e^4 - e^2, \cdots.$$

To get the general expression of $A_k$ we introduce the *factorial coefficients* $C_{m,k}$, also known as *Stirling numbers*, defined by

$$(5.7.7) \qquad \begin{aligned} t(t-1)(t-2) \cdots (t - k + 1) &= C_{0,k} t^k - C_{1,k} t^{k-1} + C_{2,k} t^{k-2} \\ &\quad - \cdots + (-1)^{k-1} C_{k-1,k} t. \end{aligned}$$

These are obviously positive integers, $C_{0,k} = 1$, $C_{k-1,k} = (k-1)!$ and

$$(5.7.8) \qquad \sum_{m=0}^{k-1} C_{m,k} = k!,$$

as we see by setting $t = -1$ in (5.7.7). In terms of these coefficients we have

$$(5.7.9) \qquad A_k = \sum_{m=0}^{k-1} (-1)^m C_{m,k} E_{k-m}, \quad E_m = e^{2^m}.$$

Here the first term $E_k$ dominates even for small values of $k$. In fact, we have

$$(5.7.10) \qquad \left(1 - \frac{k!}{E_{k-1}}\right) E_k < A_k < \left(1 + \frac{k!}{E_{k-1}}\right) E_k.$$

The quotient $k!/E_{k-1}$ has its largest value $1/e$ for $k = 0$ and tends very rapidly to zero as $k \to \infty$. To prove (5.7.10) we note that

$$A_k > E_k \left\{1 - \sum_{m=1}^{k-1} C_{m,k} \frac{E_{k-m}}{E_k}\right\} > E_k \left\{1 - \max_m \frac{E_{k-m}}{E_k} \sum_{m=0}^{k-1} C_{m,k}\right\}.$$

Here the maximum of $E_{k-m}/E_k$ is $1/E_{k-1}$ and occurs for $m = 1$, and the sum of the $C_{m,k}$ is given by (5.7.8). This gives the lower bound for $A_k$, and the upper bound is obtained in the same manner.

These estimates enable us to explain why $f(z)$ cannot be holomorphic at $z = 1$, in spite of the existence of all radial derivatives. The reason is simple enough: *the sequence of derivatives grows too fast*. We shall determine in Chapter 8 a permissible rate of growth for the derivatives of a holomorphic function. At this stage of the game we can argue as follows: Suppose that $f(z)$ were holomorphic at $z = 1$, and let us anticipate the result that a function holomorphic in $|z - a| < \delta$ may be expanded in Taylor's series about $z = a$ convergent for $|z - a| < \delta$. In the present case the series would have to be

$$\sum_{k=0}^{\infty} \frac{A_k}{k!} (z - 1)^k.$$

This series, however, has the radius of convergence 0 since

$$(E_k/k!)^{1/k} \to \infty.$$

Thus $f(z)$ cannot be holomorphic at $z = 1$.

Actually $f(z)$ is not holomorphic at any point of $|z| = 1$. Every point on the unit circle is a singular point of $f(z)$, the unit circle is said to be a *natural boundary* of $f(z)$, and the power series is *noncontinuable*. We shall prove something which on the face of it looks like a weaker statement, namely that every $2^k$th root of unity, $k = 0, 1, 2, \cdots$, is a singularity. These points, however, are dense on the unit circle, and singular points on the circle of convergence form a closed set, so the two assertions are actually equivalent. Let $\omega$ be a $2^k$th root of unity and form the function

$$f(\omega z) = \sum_{n=0}^{k-1} \frac{1}{n!} (\omega z)^{2^n} + \sum_{n=k}^{\infty} \frac{1}{n!} z^{2^n}.$$

Here we have used the fact that the $2^n$th power of $\omega$ equals 1 as soon as $n \geq k$. But the result of this transformation is a power series having positive coefficients for $n \geq k$, and by Theorem 5.7.1 $z = 1$ is a singular point of the function. But this means that $z = \omega$ is a singular point of $f(z)$. This argument holds for each of the $2^k$ unit roots involved and for each $k$.

The series (5.7.5) is an example of a so-called *gap series* or *lacunary series*. Most of the powers of $z$ are obviously missing and there are longer and longer gaps between the terms. Such a power series normally has its circle of convergence as a *natural boundary*. A sufficient condition for this to be the case is that the length of the gaps tends to infinity. A simple example is given by the power series in Problem 7 of Exercise 5.4.

All the properties of a function defined by a power series are determined by the sequence of coefficients $\{a_n\}$, and the finer properties such as distribution and nature of the singularities are regulated by the infinitary behavior of $a_n$. Suppose $R = 1$. If the sequence $\{a_n\}$ is very smooth, $z = 1$ is apt to be the only singularity on the unit circle. For this to be the case it suffices that $a_n = g(n)$ where $g(w)$ is, for instance, a rational function of $w$ or holomorphic at infinity or a Laplace transform. Numerous other conditions are also known.

Any irregularities in the sequence $\{a_n\}$ are apt to introduce other singularities. A periodic disturbance favors the appearance of singularities at unit roots. A trivial case is

$$\sum_{n=0}^{\infty} a_{kn} z^{kn}$$

where $k$ is a fixed integer and the coefficients are positive. Here, the $k$th roots of unity are singular. The disturbance referred to is the presence of gaps. They are of the same length and repeated periodically. A recurrent nonperiodic disturbance tends to spread the singularities all over the unit circle. The gap series (5.7.5) belongs to this category. It is by no means necessary to use gaps in order to produce noncontinuable power series. In a certain sense, which can be made precise, noncontinuability is the rule and continuability the exception, only in this case we are more familiar with the exceptions. We shall return to these questions in Volume II, Chapter 11.

## EXERCISE 5.7

**1.** Show that the conclusion of Theorem 5.7.1 holds if $\Re(a_n) \geq 0$ for all large $n$.

**2.** Verify that $R = 1$ for (5.7.5) and prove (5.7.9).

**3.** Prove that $\sum\limits_{n=0}^{\infty} z^{n!}$ has $|z| = 1$ as natural boundary by showing that every root of unity is a singular point.

**4.** Why do the singular points on the circle of convergence of a power series form a closed set?

**5.** If $f(z) = z + z^2 + z^4 + \cdots + z^{2^n} + \cdots$, $|z| < 1$, show that $f(z) = z + f(z^2)$ and use this functional equation to show that $|z| = 1$ is the natural boundary of $f(z)$.

**6.** Assuming $\alpha$ to be real, show that $z = -1$ is a singular point of the function defined by the binomial series

$$\sum_{n=0}^{\infty} \binom{\alpha}{n} z^n.$$

**7.** If $F_k(t) = t(t-1)(t-2)\cdots(t-k+1)$, show by induction or otherwise that

$$t^m = F_m(t) + A_{1,m}F_{m-1}(t) + \cdots + A_{m,m}F_1(t),$$

where the coefficients are positive integers.

**8.** Use the formula in Problem 7 to show that

$$\sum_{n=1}^{\infty} n^k z^n = P_k(z)(1-z)^{-k-1}, \quad |z| < 1,$$

where $P_k(z)$ is a polynomial in $z$ of degree $k$ having positive integral coefficients. Show that $z = 1$ is the only singularity of the function defined by the series.

## COLLATERAL READING

As general reference for infinite series consult

BROMWICH, T. J. I'A. *An Introduction to the Theory of Infinite Series*, Second Edition. Cambridge University Press, London, 1929.

KNOPP, K.　*Theory and Application of Infinite Series*, translated by R. C. H. YOUNG.  Blackie & Son, Ltd., Glasgow and London, 1951.

In connection with the series (5.4.12) see

FEJÉR, L.　"Ueber Potenzreihen deren Summen im abgeschlossenen Konvergenzkreise überall stetig sind," *Sitzungsberichte der Bayerischen Akademie der Wissenschaften, Mathematisch-physikalische Klasse*, 1917, pp. 33–50.

For Section 5.7 consult

MANDELBROJT, S.　*Les Singularités des Fonctions Analytiques Représentées par une Série de Taylor*.  Mémorial des Sciences Mathématiques, Vol. 54. Gauthier-Villars, Paris, 1932.

# 6

# SOME ELEMENTARY FUNCTIONS

**6.1. The exponential function.** The problem of extending to complex variables the elementary transcendental functions studied in the calculus poses many questions. The functions defined by the symbols $e^x$, $\log x$ $(x > 0)$, $\sin x$, $\cos x$, etc. are well defined for real values of $x$, but in most cases the usual definitions do not make sense for complex variables. In trying to make the extension we have to use properties or representations of the functions which remain meaningful in the complex plane.

We could, somewhat naïvely, pose the following problem: Given a function $f(x)$ defined on a real interval $(a, b)$, find a function $F(z)$ defined in some domain $D$ of the complex plane containing $(a, b)$ and such that $F(x) = f(x)$ for $a < x < b$. A moment's consideration shows that this can always be done and in infinitely many ways, most of which lead to results without interest. If we require, however, that $F(z)$ be holomorphic in $D$, then the problem has at most one solution. Thus *an analytic extension is unique if it exists*. To find out whether or not such an extension will exist, we need a test for holomorphy that can be applied to a real-valued function of a real variable. It would seem sufficient if $f(x)$ has derivatives of all orders and the corresponding Taylor's series

$$\sum_{n=0}^{\infty} \frac{f^{(n)}(c)}{n!} (x - c)^n, \quad a < c < b,$$

converges and represents $f(x)$. Actually this condition is necessary as well as sufficient. In the Taylor series we can replace $x$ by $z$ and thus obtain a function having the desired properties in some domain containing $(a, b)$.

Our program then will be to use Taylor's series to define the functions to be studied. We shall use this procedure for the exponential function and for the sine and the cosine. The inverse functions can then be handled directly.

We start with the exponential function defined by the power series

(6.1.1)
$$e^z \equiv 1 + \sum_{n=1}^{\infty} \frac{z^n}{n!},$$

where the series converges for all finite $z$. Since the series may be differentiated termwise, we have

(6.1.2)
$$\frac{d}{dz}[e^z] = e^z,$$

so that
$$w = e^z$$

138

is a solution of the differential system (i.e., differential equation with initial condition)

$$(6.1.3) \qquad \frac{dw}{dz} = w, \quad w(0) = 1;$$

in fact, it is the only solution. Taylor's series now takes the form

$$e^z = e^a \left\{ 1 + \sum_{n=1}^{\infty} \frac{(z-a)^n}{n!} \right\},$$

that is (see Problem 1 of Exercise 5.6),

$$(6.1.4) \qquad e^z = e^a \, e^{z-a}$$

or

$$(6.1.5) \qquad e^{z_1 + z_2} = e^{z_1} e^{z_2}.$$

Thus $e^z$ is a solution of the important functional equation

$$(6.1.6) \qquad f(z_1 + z_2) = f(z_1)f(z_2).$$

For real values of the variables the theory of this functional equation goes back to Cauchy's *Cours d'Analyse* of 1821, where it was shown that $e^{\alpha x}$ is the only continuous solution. Later research has shown that this is the only solution bounded on a finite interval as well as the only measurable solution. The situation is less satisfactory for complex values of the variables. Thus

$$f(x + iy) = e^{\alpha x + \beta y}$$

is a solution for any choice of $\alpha$ and $\beta$. It is obviously continuous, but holomorphic if and only if $\beta = \alpha i$.

THEOREM 6.1.1.    *$e^z$ is the only solution of (6.1.6) that is holomorphic at $z = 0$ and whose derivative equals 1 there.*

*Proof.*    We shall show that such a solution satisfies (6.1.3) and that the latter has a unique solution. First, it is clear from (6.1.6) that $f(0) = 0$ or 1 (set $z_1 = z_2 = 0$) and that the former alternative gives $f(z) \equiv 0$. Thus we must have $f(0) = 1$. Again, from (6.1.6) we get

$$f(z + h) - f(z) = f(z)f(h) - f(z)$$

or

$$\frac{1}{h}[f(z + h) - f(z)] = f(z)\frac{1}{h}[f(h) - 1].$$

By assumption the right-hand side has the limit $f(z)$ as $h \to 0$. It follows that $f'(z)$ exists everywhere and equals $f(z)$, that is, $f(z)$ satisfies (6.1.3). With the integral defined as in Section 4.2, we then have

$$f(z) = 1 + \int_0^z f(s) \, ds.$$

On the other hand, if $E_n(z)$ is the $n$th partial sum of the exponential series (6.1.1), we have

$$E_n(z) = 1 + \int_0^z E_{n-1}(s)\,ds.$$

Hence

(6.1.7) $$f(z) - E_n(z) = \int_0^z [f(s) - E_{n-1}(s)]\,ds.$$

The holomorphic function $f(z) - E_0(z)$ is bounded when $z$ is bounded. Suppose that

$$|f(z) - E_0(z)| \leq M \quad \text{for} \quad |z| \leq R.$$

Combining (6.1.7) for $n = 1$ with (4.2.10) we get

$$|f(z) - E_1(z)| \leq M\,|z| \quad \text{for} \quad |z| \leq R.$$

Suppose that we know that

(6.1.8) $$|f(z) - E_k(z)| \leq M\,\frac{|z|^k}{k!} \quad \text{for} \quad |z| \leq R$$

for some integer $k$. Setting $s = zr$, $0 \leq r \leq 1$, we get

$$\left| \int_0^z [f(s) - E_k(s)]\,ds \right| = \left| z \int_0^1 [f(rz) - E_k(rz)]\,dr \right| \leq |z| \int_0^1 |f(rz) - E_k(rz)|\,dr$$

$$\leq M\,\frac{|z|^{k+1}}{k!} \int_0^1 r^k\,dr = M\,\frac{|z|^{k+1}}{(k+1)!}\,,$$

where the first inequality follows from (4.2.8) together with the definition of the Riemann integral. In view of (6.1.7) we see that the inequality (6.1.8) holds for all values of $k$. But this implies that

$$f(z) = \lim_{n \to \infty} E_n(z) = e^z,$$

and the theorem is proved.

From (6.1.1) we get

$$e^{iy} = \sum_{n=0}^{\infty} \frac{1}{n!}(iy)^n = \sum_{k=0}^{\infty} \frac{(-1)^k}{(2k)!}\,y^{2k} + i \sum_{k=0}^{\infty} \frac{(-1)^k}{(2k+1)!}\,y^{2k+1}$$

or

(6.1.9) $$e^{iy} = \cos y + i \sin y,$$

a relation due to Euler.[1] Combining this with (6.1.5) we get

(6.1.10) $$e^{x+iy} = e^x \cos y + i\,e^x \sin y.$$

---

[1] Leonhard Euler (1707–1783), a native of Switzerland, was active in St. Petersburg 1727–1741, 1766–1783, with an intermission of twenty-five years spent in Berlin. Euler made fundamental contributions to all branches of mathematics and was a prolific writer. The Academies of Sciences of Prussia and Russia could not publish as fast as he wrote, and some of his papers are still unpublished.

We encountered this pair of conjugate harmonic functions in Problem 1$b$ of Exercise 4.3. Incidentally, this formula is sometimes taken as the definition of the exponential function in the complex plane.

From these formulas we can draw many conclusions. First, we note that

$$(6.1.11) \qquad\qquad | e^z | = e^x,$$

that is, the absolute value of $e^z$ is constant on vertical lines. This implies that

$$(6.1.12) \qquad\qquad e^z \neq 0 \quad \text{for every } z,$$

since this property holds for real values of $z$. From (6.1.10) it also follows that

$$(6.1.13) \qquad\qquad \arg e^z \equiv y \quad (\text{mod } 2\pi),$$

that is, the argument of $e^z$ is constant on horizontal lines. In particular, $e^z$ is real positive when $y$ is an even multiple of $\pi$, real negative when $y$ is an odd multiple of $\pi$, and purely imaginary when $y$ is an odd multiple of $\pi/2$.

*The exponential function is periodic with period $2\pi i$:*

$$(6.1.14) \qquad\qquad e^{z+2\pi i} = e^z.$$

This means of course that all integral multiples of $2\pi i$ are periods. These are the only periods, for if $p = \alpha + \beta i$ is such that

$$e^{z+p} = e^z$$

for all $z$, then necessarily $e^p = 1$, $\alpha = 0$, and $\beta = 2n\pi$.

The equation

$$(6.1.15) \qquad\qquad e^z = a, \quad a \neq 0,$$

z-plane

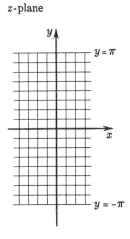

w-plane

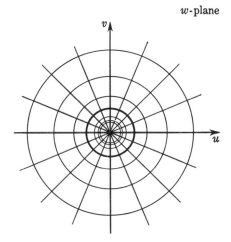

Figure 13                                      Figure 14

has infinitely many solutions. Formula (6.1.10) shows that

(6.1.16) $$z = \log | a | + i \arg a,$$

where $\arg a$ is given all admissible values. The straight lines $y = (2k + 1)\pi$ divide the plane into regions

(6.1.17) $$(2k - 1)\pi < y \leq (2k + 1)\pi, \quad k = 0, \pm 1, \pm 2, \cdots,$$

referred to as *period strips* in this connection. Formula (6.1.16) shows that the exponential function takes on every value except zero once and only once in each period strip.

It follows from this fact together with (6.1.2) and (6.1.12) that the exponential function maps the interior of each period strip conformally on the (finite) complex plane slit along the negative real axis. See Figures 13 and 14 above.

The exponential function is an example of a *transcendental entire function* (British terminology: *integral function*). This name is given to any function holomorphic in the finite plane which does not reduce to a polynomial. Such a function has an essential singular point at infinity. No value can be assigned to the function at $z = \infty$; in fact, if $a$ is a preassigned complex number or infinity, we can find a sequence $\{z_n\}$, $| z_n | \to \infty$, such that $f(z_n) \to a$. In the case of the exponential function, every value $a$, $a \neq 0, \infty$, is actually taken on infinitely often in any neighborhood of $w = \infty$ on the sphere, while $e^z \to 0$ if $\Re(z) \to - \infty$ and $| e^z | \to \infty$ as $\Re(z) \to + \infty$. This is entirely different from the behavior of a polynomial which is a *rational* entire function: a polynomial takes on large values only in the neighborhood of $z = \infty$ and may be assigned the value $w = \infty$ for $z = \infty$.

## EXERCISE 6.1

**1.** An alternative definition of the exponential function is

$$e^z = \lim_{n \to \infty} \left\{1 + \frac{z}{n}\right\}^n.$$

Expand by the binomial theorem and carry through the convergence proof. The convergence is uniform in $| z | \leq R < \infty$ for every $R$.

**2.** Show that

$$\lim_{n \to \infty} \left| 1 + \frac{z}{n} \right|^n = e^x.$$

**3.** Suppose $\delta > 0$ but arbitrarily small. Show that $e^z$ takes on every value except zero infinitely often in each of the two sectors $\left| \arg z \pm \frac{\pi}{2} \right| < \delta$.

**4.** Discuss the function $e^{1/z}$. Where is it real, where purely imaginary, where does it take on a given value $a \neq 0$?

**5.** The equation

$$e^z = z$$

has infinitely many roots, one in each period strip except the one containing the real axis. These roots may be found as the intersections of suitably chosen curves, for instance

$$e^{2x} = x^2 + y^2, \quad x = y \cot y.$$

Carry through the discussion and try to get approximate formulas for the roots remote from 0.

**6.** If $F_k(t) = t(t-1) \cdots (t-k+1)$, find the sum of the series

$$\sum_{n=0}^{\infty} \frac{F_k(n)}{n!} z^n.$$

**7.** If $P_k(t)$ is a polynomial in $t$ of degree $k$, show that

$$\sum_{n=0}^{\infty} \frac{P_k(n)}{n!} z^n = Q_k(z) e^z$$

where $Q_k(z)$ is a polynomial of degree $k$.

**8.** What is the absolute value of $e^{e^z}$, $z = x + iy$? Discuss the behavior of this function on horizontal and vertical lines.

**9.** For what values of $z$ does $e^{e^z} = 1$? If $z_m$ and $z_n$ range over distinct roots of this equation, is the set of distances $d(z_m, z_n)$ bounded away from zero?

**10.** Where are the zeros of the function

$$C(z) \equiv \tfrac{1}{2}(e^{iz} + e^{-iz})?$$

**11.** Find the derivatives of $C(z)$ and the corresponding Taylor series.

### 6.2. The logarithm.

We have seen that the equation

$$(6.2.1) \qquad\qquad e^w = z, \quad z \neq 0, \infty,$$

has infinitely many solutions. Each of these solutions is called a *logarithm* of $z$ and defines a determination of the infinitely many-valued function $\log z$. Thus

$$(6.2.2) \qquad\qquad \log z = \log |z| + i \arg z,$$

where on the right-hand side $\log |z|$ denotes the real logarithm and $\arg z$ is given all admissible values. Since the latter differ by multiples of $2\pi$, we see that the various determinations of $\log z$ differ by multiples of $2\pi i$. The *principal determination* of the logarithm corresponds to the principal determination of the argument. If the latter is chosen as $-\pi < \arg z \leqq \pi$ we have

$$-\pi < \Im[\log z] \leqq \pi$$

for the principal determination of the logarithm.

We note that

$$(6.2.3) \qquad \log(z_1 z_2) - \log z_1 - \log z_2 \equiv 0 \quad (\mathrm{mod}\ 2\pi i)$$

is always true, but the relation

$$(6.2.4) \qquad \log(z_1 z_2) = \log z_1 + \log z_2$$

does not hold for arbitrary determinations of the logarithms: two of the logarithms are at our disposal, the third is then uniquely determined.

We have obviously

$$(6.2.5) \qquad e^{\log z} = z$$

for any determination of $\log z$, while in the opposite direction

$$(6.2.6) \qquad \log[e^z] - z \equiv 0 \quad (\mathrm{mod}\ 2\pi i)$$

is the best that can be asserted. The logarithm and the exponential function are thus inverse functions, and Theorem 4.5.1 asserts that $\log z$ is locally holomorphic. Formula (4.5.2) combined with (6.1.2) shows that

$$(6.2.7) \qquad \frac{d}{dz}\log z = \frac{1}{z}.$$

Thus $\log z$ is actually holomorphic in any domain, omitting 0 and $\infty$, in which this function is single-valued. We can, of course, draw the same conclusions from the Cauchy-Riemann equations (see Problem 3 of Exercise 4.3).

Upon integration, formula (6.2.7) gives us a representation of the principal determination of $\log z$ as a definite integral:

$$(6.2.8) \qquad \log z = \int_1^z \frac{dt}{t}, \quad |\arg z| < \pi.$$

Here the integral is taken along the line segment $[1, z]$, and the result follows from formula (4.2.11). We note that $\log z$ is holomorphic in the finite plane cut along the negative real axis from 0 to $-\infty$, and that the line segment $[1, z]$ lies in this domain.

From (6.2.7) it follows that $\log z$ has derivatives of all orders:

$$(6.2.9) \qquad \frac{d^n}{dz^n}\log z = (-1)^{n-1}(n-1)!\, z^{-n}.$$

This leads formally to the Taylor series

$$(6.2.10) \qquad \log z = \log a + \sum_{n=1}^{\infty} \frac{(-1)^{n-1}}{n}\left(\frac{z-a}{a}\right)^n.$$

This series obviously converges for $|z - a| < |a|$, that is, inside the circle with center at $a$ passing through the origin. But it is not obvious at all that the series represents the function. At the present stage we do not know that a holomorphic function has derivatives of all orders and is represented by its Taylor series. If $z$ lies on the line segment $(0, 2a)$, the ratio $z/a$ is real positive and the series in the right member of (6.2.10) is known to represent the (real) logarithm of $z/a$. We have thus two holomorphic functions, $\log z - \log a$ and the power series, which coincide along a line segment. We shall see later that this makes them identical in their common domain of existence, but this theorem is not at our disposal now. Without anticipating later results, we can still prove that the power series represents the principal determination of $\log (z/a)$ when it converges. This is clearly all that is needed to justify (6.2.10). Moreover we can restrict ourselves to the case $a = 1$. Thus we want to prove that

$$(6.2.11) \qquad \log z = \sum_{n=1}^{\infty} \frac{(-1)^{n-1}}{n} (z - 1)^n, \quad |z - 1| < 1,$$

for the principal determination.

For this purpose we use the methods employed in the calculus in proving Taylor's theorem with remainder. We apply this to formula (6.2.8). Since integration by parts may be open to objection, we achieve the same result in a roundabout manner. For any positive integer $k$

$$\frac{1}{k} \frac{d}{dt} \frac{(z - t)^k}{t^k} = -\frac{(z - t)^{k-1}}{t^k} - \frac{(z - t)^k}{t^{k+1}}.$$

Here we integrate with respect to $t$ along the line segment from 1 to $z$ and use (4.2.11), obtaining after transposition of terms and multiplication by $(-1)^{k-1}$

$$(-1)^{k-1} \int_1^z \frac{(z - t)^{k-1}}{t^k} dt = \frac{(-1)^{k-1}}{k} (z - 1)^k + (-1)^k \int_1^z \frac{(z - t)^k}{t^{k+1}} dt.$$

Adding these relations for $k = 1, 2, \cdots, n$, we see that all integrals cancel except the first one on the left, which is $\log z$, and the last one on the right, leaving

$$(6.2.12) \qquad \log z = \sum_{k=1}^{n} \frac{(-1)^{k-1}}{k} (z - 1)^k + (-1)^n \int_1^z \frac{(z - t)^n}{t^{n+1}} dt.$$

This is the desired formula. It holds for any $z$ in the plane cut from 0 to $-\infty$ and for any $n$. Now we assume $|z - 1| < 1$. Then

$$(6.2.13) \qquad \max \left| \frac{z - t}{t} \right| = |z - 1|, \quad t \in [1, z],$$

so that

$$(6.2.14) \qquad \left| \int_1^z \frac{(z - t)^n}{t^{n+1}} dt \right| \leq |z - 1|^{n+1} M(z),$$

where $M(z) = \max |t|^{-1}$ for $t \in [1, z]$. Since the right-hand side of (6.2.14) tends to zero as $n \to \infty$, (6.2.11) and (6.2.10) are valid.

Our final topic is the Riemann surface of the logarithm. It follows from the discussion at the end of the preceding section that each determination of $\log z = w$ maps the domain

$$S: \quad 0 < |z| < \infty, \quad -\pi < \arg z < \pi$$

conformally onto one of the period strips

$$(2k - 1)\pi < \Im(w) < (2k + 1)\pi$$

of the exponential function. To cover the whole $w$-plane, we need infinitely many copies of $S$ with adjunction and identification of boundary points and a suitable definition of neighborhoods to define a Riemann surface. Consider a collection $\{S_k \mid k = 0, \pm 1, \pm 2, \cdots\}$, each element of which is a copy of the finite complex plane punctured at the origin and slit along the negative real axis, which serves as a branch line. Neighborhoods are defined as follows: A point $z_0$ in $S_k$ not on the negative real axis has an ordinary circular neighborhood $|z - z_0| < \varepsilon$ also in $S_k$. Neither zero nor infinity belongs to the surface, and no neighborhoods are defined for them. They are branch points of infinite order. The student should observe the difference between this situation and the one that holds for $\sqrt{z}$, where the branch points are points of the surface and have neighborhoods defined for them. If $z_0$ lies on the negative real axis and is the limit point of a set of points in the upper half plane of $S_k$ ($z_0$ is then said to lie on the "upper edge" of the cut or branch line), then a neighborhood of $z_0$ consists of a semicircular disk $|z - z_0| < \varepsilon$, $\Im(z) \geq 0$ in $S_k$ plus a semicircular disk $|z - z_0| < \varepsilon$, $\Im(z) < 0$ in $S_{k+1}$. If $z_0$ lies on the "lower edge" of the cut in $S_k$, then we have again two semicircular disks, the lower one in $S_k$, the upper one in $S_{k-1}$. Thus if $z$ crosses the branch line in such a manner that $\Im(z)$ decreases, then $z$ moves from $S_k$ to $S_{k+1}$, while a passage in the opposite direction takes $z$ from $S_k$ to $S_{k-1}$. The collection $\{S_k\}$ with the above conventions of neighborhoods and identification of boundary points constitutes the Riemann surface of the logarithm.

## EXERCISE 6.2

**1.** If $|\arg z| < \pi$ and the $n$th root is given its principal determination, prove that

$$\lim_{n \to \infty} n(\sqrt[n]{z} - 1)$$

gives the principal determination of $\log z$.

**2.** For $|z - 1| \leq r < 1$ and the principal determination of $\log z$, prove that

$$|\log z| \leq \log (1 - r)^{-1}.$$

**3.** Prove (6.2.13).

**4.** Discuss the Riemann surface of $\log (1 + z^2)$.

**6.3. Arbitrary powers; the binomial series.** We can now define arbitrary powers of the two forms

$$z^\alpha \quad \text{and} \quad a^z.$$

Suppose that $z \neq 0, \infty$, and let $\alpha$ be an arbitrary fixed complex number. We set

(6.3.1) $$z^\alpha = e^{\alpha \log z}.$$

This function is single-valued if and only if $\alpha$ is an integer; it has $q$ distinct values if $\alpha = p/q$ where the integers $p$ and $q$ are relatively prime. In all other cases the function is infinitely many-valued. It is single-valued on the Riemann surface of the logarithm. We can also make $z^\alpha$ single-valued by introducing a cut in the $z$-plane, say along the negative real axis, from $z = 0$ to $z = -\infty$. Crossing the cut from the upper to the lower half-plane gives

(6.3.2) $$z^\alpha \to e^{2\pi i \alpha} z^\alpha.$$

Relations like

(6.3.3) $$z^\alpha z^\beta = z^{\alpha+\beta},$$

(6.3.4) $$(z_1 z_2)^\alpha = z_1{}^\alpha z_2{}^\alpha$$

have to be handled with care. The first is correct if the three powers involved are based on the same determination of $\log z$. The second relation presupposes that

$$\log(z_1 z_2) - \log z_1 - \log z_2 = 0$$

and not merely congruent to zero modulo $2\pi i$.

We have

(6.3.5) $$\frac{d}{dz} z^\alpha = \alpha z^\alpha z^{-1},$$

provided the same determination of $z^\alpha$ is used on both sides of the equality. If

$$z = re^{i\theta}, \quad \alpha = \alpha_1 + i\alpha_2,$$

then

(6.3.6) $$z^\alpha = r^{\alpha_1} e^{-\alpha_2 \theta} e^{i(\alpha_2 \log r + \alpha_1 \theta)}.$$

If $\alpha_2 \neq 0$, the behavior of $z^\alpha$ in the neighborhood of the origin is very complicated. If $z \to 0$ in such a manner that $\theta$ stays bounded, $r^{\alpha_1}$ is the decisive factor, but if $z$ approaches zero along a spiral the factor $e^{-\alpha_2 \theta}$ comes into play and may become dominant.

We shall also say a few words about the symbol $a^z$. Here we assume $a \neq 0$, $1, e, \infty$, and define

(6.3.7) $$a^z = e^{z \log a},$$

where the determination of $\log a$ is at our disposal. Once $\log a$ is chosen, $a^z$ is a single-valued holomorphic function of $z$. There is a great difference between the two symbols $z^\alpha$ and $a^z$. Both are infinitely many-valued. The first one represents the various branches of one and the same analytic function, and we can pass from one branch to another by a suitable circuit around the origin. The second symbol represents infinitely many distinct entire functions of $z$. We can pass from one of these functions to another by multiplying by

$$e^{2k\pi i z}$$

for a suitable choice of the integer $k$, but we cannot bring about this passage by letting $z$ move around.

Finally we shall consider the binomial series

$$(6.3.8) \qquad \sum_{n=0}^{\infty} \binom{\alpha}{n} z^n \equiv B(z, \alpha).$$

For a fixed complex $\alpha$ the series converges for any $z$ with $|z| < 1$ and defines a holomorphic function of $z$ in the unit circle. For $z$ fixed, $|z| = r < 1$, $B(z, \alpha)$ is an entire function of $\alpha$. To prove this, we first note that for $|\alpha| \leq N$, $N$ a positive integer, we have

$$\sum_{n=0}^{\infty} \left| \binom{\alpha}{n} \right| r^n \leq \sum_{n=0}^{\infty} N(N+1) \cdots (N+n-1) \frac{r^n}{n!} = (1-r)^{-N}.$$

Here the third member is obtained by differentiating the geometric series

$$\sum_{n=0}^{\infty} r^n = (1-r)^{-1}$$

$(N-1)$ times and dividing the result by $(N-1)!$. The estimate is the best possible of its kind; the upper bound is reached for $\alpha = -N$, $z = -r$. In the notation of (5.7.7),

$$B(z, \alpha) = 1 + \sum_{n=1}^{\infty} \sum_{m=1}^{n} (-1)^{n-m} C_{n-m,\, n} \alpha^m \frac{z^n}{n!}.$$

This is a double series, and according to the above estimate it converges absolutely and uniformly for $|z| \leq r$, $|\alpha| \leq N$. We can consequently sum by columns instead of by rows as in (6.3.8), obtaining

$$(6.3.9) \qquad B(z, \alpha) = 1 + \sum_{m=1}^{\infty} \left\{ \sum_{n=m}^{\infty} (-1)^{n-m} C_{n-m,\, n} \frac{z^n}{n!} \right\} \alpha^m.$$

For fixed $z$, $|z| \leq r < 1$, this is a power series in $\alpha$, convergent for $|\alpha| \leq N$, and, since $N$ is arbitrary, it follows that $B(z, \alpha)$ is an entire function of $\alpha$.

By a closer study of the factorial coefficients $C_{k,\, n}$ we could sum the power series in $z$ which act as coefficients of the powers of $\alpha$. The following procedure is more interesting; it goes back to Abel's paper on the binomial series which appeared in Volume 1 of *Crelle's Journal* (1826). We shall show that

$$(6.3.10) \qquad B(z, \alpha)B(z, \beta) = B(z, \alpha + \beta);$$

that is, as a function of $\alpha$ with $z$ fixed, $B(z, \alpha)$ satisfies the functional equation (6.1.6) of the exponential function. Now $B(z, \alpha)$ has been shown to be an entire function of $\alpha$, and the only analytic solutions of (6.1.6) are exponential functions by Theorem 6.1.1. Hence there exists a function $f(z)$, defined for $|z| < 1$, such that

$$B(z, \alpha) = e^{\alpha f(z)}.$$

We can get $f(z)$ by computing the coefficient of the first power of $\alpha$ in (6.3.9). Since $C_{n-1, n} = (n - 1)!$, we get

$$f(z) = \sum_{n=1}^{\infty} \frac{(-1)^{n-1}}{n} z^n = \log (1 + z)$$

by (6.2.11), where the logarithm has its principal value. Hence

$$(6.3.11) \qquad B(z, \alpha) = e^{\alpha \log (1+z)} = (1 + z)^{\alpha}$$

with the principal determination of the logarithm and, hence, also of the power.

It remains to prove (6.3.10). This relation is certainly true whenever $\alpha$ and $\beta$ are positive integers, $\alpha = j$, $\beta = k$, since by the binomial theorem

$$\sum_{n=0}^{k} \binom{k}{n} z^n = (1 + z)^k.$$

Replacing $k$ by $j$, multiplying the two polynomials, and comparing coefficients, we obtain an identity between binomial coefficients which will be useful in a moment, namely

$$(6.3.12) \qquad \binom{j + k}{n} = \sum_{m=0}^{n} \binom{j}{m} \binom{k}{n - m}.$$

In order to prove (6.3.10) it will be sufficient to show that for each $n$ the coefficient of $z^n$ on the right equals the coefficient of $z^n$ in the Cauchy product series on the left, that is, the validity of

$$(6.3.13) \qquad \binom{\alpha + \beta}{n} = \sum_{m=0}^{n} \binom{\alpha}{m} \binom{\beta}{n - m}$$

for all $n$ and all values of $\alpha$ and $\beta$. We note that (6.3.13) holds for $\alpha = j$, $\beta = k$ by (6.3.12). This fact we use as follows: Denote the difference between the left and the right members of (6.3.13) by $S_n(\alpha, \beta)$. Let $k$ be a positive integer. Then $S_n(\alpha, k)$ is a polynomial in $\alpha$ of degree $n$ at most, since it is the difference of two polynomials in $\alpha$ each of degree $n$. But (6.3.12) says that

$$S_n(j, k) = 0$$

for every pair of positive integers $j$, $k$, that is, $S_n(\alpha, k)$ must be identically zero. We then give $\alpha$ a fixed value and consider $S_n(\alpha, \beta)$ as a function of $\beta$. It is a polynomial in $\beta$ of degree at most $n$, and this polynomial is zero whenever $\beta$ is a positive integer. This requires $S_n(\alpha, \beta) \equiv 0$ for all $n$. Hence (6.3.13) holds and thus also (6.3.10) and (6.3.11).

### EXERCISE 6.3

**1.** The symbol $i^i$ is infinitely many-valued. Find these values. They are all real positive.

**2.** Find the supremum of $\mid (1 + z)^i \mid$ in the open unit disk if the power is given its principal determination.

**3.** Show that

$$n \left| \binom{i}{n} \right| \equiv b_n$$

is bounded. Actually

$$\lim b_n = \left( \frac{\sinh \pi}{\pi} \right)^{\frac{1}{2}}.$$

**4.** Find three paths on the Riemann surface of the logarithm leading from $z = 1$ to $z = 0$ along which $\mid z^{1-i} \mid$ tends to 0, 1, $\infty$ respectively.

**5.** Show that the principal determination of $(1 + z)^\alpha$ is bounded in the open unit disk if and only if $\Re(\alpha) \geq 0$.

**6.** Compute $C_{n-2,\,n}$ and verify that

$$\sum_{n=2}^{\infty} (-1)^n C_{n-2,\,n} \frac{z^n}{n!} = \tfrac{1}{2}[\log (1 + z)]^2.$$

**6.4. The trigonometric functions.** We can base the definition of $\sin z$ and of $\cos z$ for complex values of $z$ either on the Maclaurin series or on formula (6.1.9), which links the exponential function with the trigonometric functions. The first choice gives

(6.4.1)      $\sin z = \sum_{k=0}^{\infty} \frac{(-1)^k}{(2k + 1)!} z^{2k+1}, \quad \cos z = \sum_{k=0}^{\infty} \frac{(-1)^k}{(2k)!} z^{2k},$

the second gives

(6.4.2)      $\sin z = \dfrac{1}{2i} [e^{iz} - e^{-iz}], \quad \cos z = \tfrac{1}{2}[e^{iz} + e^{-iz}].$

It is a simple matter to verify that the two definitions agree.

We see that $\sin z$ and $\cos z$ are entire functions of $z$, the former odd, the latter even:

(6.4.3)      $\sin (-z) = -\sin z, \quad \cos (-z) = \cos z.$

Differentiation of (6.4.1) or (6.4.2) gives

(6.4.4) $$\frac{d}{dz}\sin z = \cos z, \quad \frac{d}{dz}\cos z = -\sin z,$$

and the Taylor series

$$\sin z = \sin a \sum_{k=0}^{\infty} \frac{(-1)^k}{(2k)!}(z-a)^{2k} + \cos a \sum_{k=0}^{\infty} \frac{(-1)^k}{(2k+1)!}(z-a)^{2k+1}$$

$$= \sin a \cos (z-a) + \cos a \sin (z-a).$$

Setting $a = z_1$, $z - a = z_2$, we get the addition theorem

(6.4.5) $$\sin (z_1 + z_2) = \sin z_1 \cos z_2 + \cos z_1 \sin z_2.$$

By the same method we get the addition theorem for the cosine

(6.4.6) $$\cos (z_1 + z_2) = \cos z_1 \cos z_2 - \sin z_1 \sin z_2.$$

The basic properties of $\sin z$ and $\cos z$ follow from these formulas. In particular, setting $z_2 = -z_1 = z$ in (6.4.6) we get

(6.4.7) $$\sin^2 z + \cos^2 z = 1.$$

We can also express $\sin (x + iy)$ and $\cos (x + iy)$ in terms of trigonometric and hyperbolic functions of $x$ and $y$ respectively. With the aid of the addition theorems we get

(6.4.8) $$\sin (x + iy) = \sin x \cosh y + i \cos x \sinh y,$$

(6.4.9) $$\cos (x + iy) = \cos x \cosh y - i \sin x \sinh y.$$

These relations show that $\sin z$ and $\cos z$ are real not merely on the real axis but also on a system of vertical lines in the plane. The student should verify the following formulas:

$$\sin [k\pi + iy] = i(-1)^k \sinh y,$$

$$\cos [k\pi + iy] = (-1)^k \cosh y,$$

(6.4.10) $$\sin \left[ (2k + 1)\frac{\pi}{2} + iy \right] = (-1)^k \cosh y,$$

$$\cos \left[ (2k + 1)\frac{\pi}{2} + iy \right] = i(-1)^{k-1} \sinh y,$$

and should show that they exhaust the possibilities of $\sin z$ and $\cos z$ being either real or purely imaginary for a non-real value of $z$.

The addition theorems show that $\sin z$ and $\cos z$ have no other zeros than the well-known real ones, $\{k\pi\}$ for $\sin z$ and $\left\{(2k+1)\dfrac{\pi}{2}\right\}$ for $\cos z$. These theorems also show that if $p$ is a period of $\sin z$ or of $\cos z$, then

$$\sin p = 0, \quad \cos p = 1, \quad \text{and} \quad p = 2k\pi.$$

The trigonometric functions have vertical period strips. A convenient choice is

$$(2k-1)\pi < x \le (2k+1)\pi, \quad k = 0, \pm1, \pm2, \cdots.$$

As we shall see later, every finite value is taken on twice in each strip.

In passing, we note the relations

$$(6.4.11) \qquad\qquad |\sin(x+iy)|^2 = \sin^2 x + \sinh^2 y,$$

$$(6.4.12) \qquad\qquad |\cos(x+iy)|^2 = \cos^2 x + \cosh^2 y - 1.$$

They show that $|\sin z|$ and $|\cos z|$ lie between $\sinh|y|$ and $\cosh y$ and hence are uniformly large when $|y|$ is large.

Formula (6.4.8) shows that $w = \sin z$ maps the vertical line $x = a$, where $a \not\equiv 0 \pmod{\pi}$, onto the curve

$$u = \sin a \cosh y, \atop v = \cos a \sinh y, \qquad -\infty < y < \infty.$$

The point $(u, v)$ obviously lies on the hyperbola

$$(6.4.13) \qquad\qquad H_a: \quad \left(\frac{u}{\sin a}\right)^2 - \left(\frac{v}{\cos a}\right)^2 = 1$$

and traces the left or the right branch of the hyperbola according as $\sin a < 0$ or $\sin a > 0$. The other branch is obtained by replacing $a$ by $-a$. Every semihyperbola corresponds to infinitely many straight lines obtained by replacing $a$ by $a + 2k\pi$. The horizontal line $y = b$, $b \ne 0$, is mapped onto the curve

$$u = \sin x \cosh b, \atop v = \cos x \sinh b, \qquad -\infty < x < \infty.$$

This is the ellipse

$$(6.4.14) \qquad\qquad E_b: \quad \left(\frac{u}{\cosh b}\right)^2 + \left(\frac{v}{\sinh b}\right)^2 = 1,$$

which is described infinitely often. These two conics have the same foci $w = \pm 1$. Thus the hyperbolas and ellipses arising under this mapping form a confocal family of conics. Through every point of the $w$-plane, not on the real axis, passes exactly one semihyperbola $H_a$ and one ellipse $E_b$ of this family; this implies that $\sin(a+ib) = u + iv$.

Consider the domain

$$(6.4.15) \qquad\qquad D_1: \quad 0 < x < \frac{\pi}{2}, \quad 0 < y$$

in the $z$-plane. It follows from the preceding discussion that $w = \sin z$ maps this half-strip conformally onto the interior of the first quadrant of the $w$-plane. The mapping is continuous on the boundary, even conformal, except at the one point $z = \dfrac{\pi}{2}$ , where angles are doubled. The positive imaginary axes of the two planes correspond, the $z$-interval $\left[0, \dfrac{\pi}{2}\right]$ is mapped onto $[0, 1]$, and the half-line $x = \dfrac{\pi}{2}$ , $y > 0$ maps onto the interval $[1, \infty]$ in the $w$-plane. See Figures 15 and 16. We shall return to these mappings in the next section.

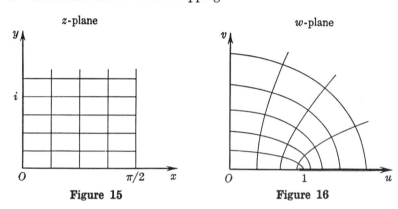

$z$-plane         $w$-plane

Figure 15       Figure 16

It remains to discuss briefly the function $\tan z$. We define

$$(6.4.16) \qquad\qquad \tan z = \frac{\sin z}{\cos z} \, .$$

This is the quotient of two entire functions. Such a function is said to be *meromorphic*. It is holomorphic in any domain not containing any zeros of the denominator. Such zeros give rise to singularities known as *poles*. We have

$$(6.4.17) \quad \lim_{z \to (2k+1)\pi/2} \tan z = \infty, \quad \lim_{z \to (2k+1)\pi/2} \left[ z - (2k+1)\frac{\pi}{2} \right] \tan z = -1,$$

no matter how $z$ approaches its limit. With a terminology that will be explained in Chapter 8, we say that the points $z = (2k+1)\dfrac{\pi}{2}$ are poles of order 1 of $\tan z$.

We get the addition theorem for $\tan z$ from those of $\sin z$ and $\cos z$ by division and simplification: Thus

$$(6.4.18) \qquad\qquad \tan (z_1 + z_2) = \frac{\tan z_1 + \tan z_2}{1 - \tan z_1 \tan z_2}$$

and in particular

$$(6.4.19) \quad \tan (x + iy) = \frac{\tan x \, (1 - \tanh^2 y)}{1 + \tan^2 x \, \tanh^2 y} + i \, \frac{(1 + \tan^2 x) \tanh y}{1 + \tan^2 x \, \tanh^2 y} \, .$$

This formula shows that $\tan z$ is real if and only if $z$ is real. It is purely imaginary whenever $x$ is a multiple of $\pi/2$. The only zeros are $z = k\pi$, and $\tan z$ is a periodic function of period $\pi$. Each of the strips

$$(2k - 1)\frac{\pi}{2} < x \le (2k + 1)\frac{\pi}{2}$$

is a period strip. In each such strip $\tan z$ takes on every value once and only once with two notable exceptions:

(6.4.20)                              $\tan z \ne \pm i$

for $z$ finite. On the other hand

(6.4.21)                              $\lim_{y \to +\infty} \tan (x \pm iy) = \pm i.$

## EXERCISE 6.4

**1.** Describe the curves $|\sin (x + iy)| = a$ for $0 < a < 1$, $a = 1$, and $1 < a$.

**2.** Find holomorphic solutions of the functional equation

$$|f(x + iy)|^2 = |f(x)|^2 + |f(iy)|^2.$$

**3.** How does the function $w = \cos z$ map the domain $D_1$ of (6.4.15)?

**4.** Prove (6.4.17).

**5.** Prove (6.4.18) and (6.4.19).

**6.** Prove (6.4.20) and (6.4.21).

**\*7.** If $w \ne \pm i$, prove that $\tan z = w$ has one root in each period strip. (*Hint:* Show that horizontal and vertical lines in the $z$-plane map into circles in the $w$-plane.)

**8.** Solve the transcendental equations
    **a.**  $\sin z = 2$.      **b.**  $\cos z = \sqrt{3}$.      **c.**  $\tan z = 2i$.

**9.** Show that the sign of the imaginary part of $\tan z$ is the same as that of $z$.

**10.** Show that $|\tan (x + iy)|^2$ satisfies the functional equation

$$S(x + iy) = \frac{S(x) + S(y)}{1 + S(x)S(y)}.$$

**11.** Show that

$$\frac{\sin \sqrt{z}}{\sqrt{z}} \quad \text{and} \quad \frac{1 - \cos \sqrt{z}}{z}$$

are entire functions of $z$ which tend to zero if $z \to \infty$ in such a manner that $\Im(z)$ stays bounded and $\Re(z) \to +\infty$.

**12.** We define $\cot z$ as the reciprocal of $\tan z$. Prove the analogues of (6.4.18) and (6.4.19).

**13.** Prove that $\lim\limits_{z \to 0} z \cot z = 1$.

**14.** Find the derivatives of $\tan z$ and of $\cot z$.

**15.** We define $\sec z$ as the reciprocal of $\cos z$. Where does $\sec z$ become infinite? Prove an analogue of formula (6.4.17) with $\tan z$ replaced by $\sec z$.

**6.5. Inverse trigonometric functions.** We discuss the inverse function of $w = \sin z$ with the aid of conformal mapping. We have seen that $w = \sin z$ maps the domain $D_1$ of (6.4.15) onto the interior, $Q_1$ say, of the first quadrant of the $w$-plane. The mapping is conformal because it is one-to-one, $\sin z$ is a holomorphic function, and its derivative $\cos z \neq 0$ in $D_1$. Since $\sin z$ takes on conjugate values for conjugate values of $z$ by (6.4.8), we conclude that the half-strip

$$D_4: \quad 0 < x < \frac{\pi}{2}, \quad y < 0,$$

is mapped conformally onto the interior $Q_4$ of the fourth quadrant. Further, $\sin(-z) = -\sin z$. Hence, with obvious notation, if $D_2 = -D_4$ and $D_3 = -D_1$, the half-strips $D_2$ and $D_3$ are mapped onto $Q_2$ and $Q_3$, the interiors of the second and third quadrants of the $w$-plane. The strip

$$D: \quad -\frac{\pi}{2} < x < \frac{\pi}{2}$$

is the union of $D_1$, $D_2$, $D_3$, $D_4$, with the imaginary axis and the line segment $\left(-\frac{\pi}{2}, \frac{\pi}{2}\right)$ added. Hence $D$ is mapped conformally onto the union of $Q_1, Q_2, Q_3,$ $Q_4$ plus the imaginary axis and the line segment $(-1, 1)$, conformally because the correspondence is still one-to-one, and $\cos z \neq 0$ in $D$. The image of $D$ then is the whole $w$-plane with the exception of the intervals $(-\infty, -1]$ and $[1, \infty)$ of the real axis. We see now that the equation

(6.5.1)                              $\sin z = w$

has a unique solution in $D$ for every value of $w$ not on the excluded intervals. In order to admit these excluded values, we have only to add part of its boundary to $D$. We take the line segments $\left[\frac{\pi}{2}, \frac{\pi}{2} + i\infty\right)$ and $\left(-\frac{\pi}{2} - i\infty, -\frac{\pi}{2}\right]$, and we denote by $R$ the region obtained in this way. The equation (6.5.1) now has a unique solution in $R$ for every finite $w$. This solution is, by definition, the principal determination of the infinitely many-valued function

(6.5.2)                          $z = \text{arc sin } w.$

Since $\sin z$ is periodic with period $2\pi$, other determinations are obtained by adding multiples of $2\pi$. Finally we have

$$\sin (\pi - z) = \sin z,$$

and this means that $\pi - \text{arc sin } w$ is also a solution. The totality of solutions is given by the two sets

(6.5.3)                  $\text{arc sin } w + 2k\pi,$

                                                              $k = 0, \pm 1, \pm 2, \cdots.$

(6.5.4)                  $-\text{arc sin } w + (2k + 1)\pi,$

These determinations have as their ranges $R + 2k\pi$ for the first set and $-R + (2k + 1)\pi$ for the second set. It is important to realize that the different ranges have no points in common and together cover the whole plane. The verification is left to the student.

The properties of the conformal mapping give us the clue to the structure of the Riemann surface of the arc sine. It consists of infinitely many copies $S_k$ of the $w$-plane slit along the line segments $(-\infty, -1]$ and $[1, \infty)$ of the real axis. These are the branch lines. Branch points lie over the points $w = -1$, $+1$, and $\infty$. The latter is of infinite order and does not belong to the surface, just as $w = 0, \infty$ were omitted from the surface of the logarithm. Over each of the points $w = \pm 1$ there are infinitely many branch points of order one, that is, each branch point belongs to exactly two sheets. The sheet $S_0$ is the image of $R, S_1$ corresponds to $\pi - R, S_2$ to $R + 2\pi$; in general, $S_{2k+1}$ corresponds to $-R + (2k + 1)\pi$, $S_{2k}$ to $R + 2k\pi$. These correspondences indicate what identifications have to be made along the branch lines. Since $R + 2k\pi$ and

$-R + (2k + 1)\pi$ have the line $x = (4k + 1)\dfrac{\pi}{2}$ as common boundary, the

sheets $S_{2k}$ and $S_{2k+1}$ have the line segment $[1, \infty)$ as common boundary. We can pass from one region to the other in the $z$-plane; hence we must be able to pass back and forth from $S_{2k}$ to $S_{2k+1}$. In fact, if $w$ describes a circle $\mid w - 1 \mid = r < 2$ in the negative sense, starting from a point $w_0$ in the upper half of $S_{2k}$, then after one circuit we are back at $w_0$ but in $S_{2k+1}$, and two full circuits are necessary before we are back at $w_0$ in $S_{2k}$. This double circuit on the $w$-surface

corresponds to a simple closed curve in the $z$-plane surrounding $z = (4k + 1)\dfrac{\pi}{2}$

once in the negative sense. Similarly a circuit on the $w$-surface along a circle $\mid w + 1 \mid = r < 2$ takes $w$ from a point $w_0$ in $S_{2k}$ to $w_0$ in $S_{2k-1}$, and a second circuit is needed in order to get back to $S_{2k}$. Thus, $S_{2k}$ is connected both with $S_{2k+1}$ and with $S_{2k-1}$, but along different branch lines. Similarly $S_{2k+1}$ is connected with $S_{2k}$ and $S_{2k+2}$ along different branch lines. Thus, the surface is connected, and it is possible to pass from any one sheet, $S_m$, to any other, $S_n$, by making circuits around the two finite branch points, alternating from one to the other.

There is another, more analytical but less instructive, way of introducing the arc sine. We can solve the equation

$$(6.5.5) \qquad w = \frac{1}{2i}(e^{iz} - e^{-iz})$$

for $z$, obtaining

$$(6.5.6) \qquad z = -i \log [iw + (1 - w^2)^{\frac{1}{2}}],$$

where now the square root and the logarithm are given all possible determinations. In particular, we get the principal determination of arc sin $w$ as follows: We draw the two cuts in the $w$-plane from $-\infty$ to $-1$ and from $+1$ to $+\infty$. In the cut plane both the square root and the logarithm are single-valued if their initial values have been chosen at one point in the domain. We take $w = 0$ to be that point and set $\sqrt{1} = +1$ and $\log 1 = 0$.

We already know that arc sin $w$, as the inverse of a holomorphic function, is locally holomorphic. Differentiating (6.5.6) and using the chain rule, we get

$$(6.5.7) \qquad \frac{d}{dw} \text{ arc sin } w = (1 - w^2)^{-\frac{1}{2}}.$$

This is a two-valued function, but it is single-valued in the cut plane once the determination has been chosen, say at $w = 0$. For the determinations (6.5.3) we take $\sqrt{1} = +1$, for those of type (6.5.4) $\sqrt{1} = -1$.

From (6.5.7) we also get the integral representation

$$(6.5.8) \qquad \text{arc sin } w = \int_0^w (1 - t^2)^{-\frac{1}{2}} dt$$

for the principal determination, where the integral is taken along the ray from $0$ to $w$ in the cut plane. This formula does not hold for $w$ on the cuts.

The discussion of arc cos $w$ can be based on the relation

$$\cos z = \sin \left( \frac{\pi}{2} - z \right),$$

whence

$$(6.5.9) \qquad \text{arc cos } w = \frac{\pi}{2} - \text{arc sin } w.$$

The discussion of arc tan $w$ may be based upon the geometrical results obtained by the student in solving Problem 7 of Exercise 6.4. We take as principal determination of arc tan $w$ that solution of

$$(6.5.10) \qquad \tan z = w$$

for which

$$(6.5.11) \qquad -\frac{\pi}{2} < x \leqq \frac{\pi}{2} \, .$$

The general solution of (6.5.10) is then of the form

(6.5.12)                                arc tan $w + k\pi$.

Since the strip (6.5.11) is mapped conformally by $w = \tan z$ on the $w$-plane slit along the imaginary axis from $-i\infty$ to $-i$ and from $i$ to $i\infty$, the Riemann surface of the arc tangent is easy to construct. It consists of infinitely many copies $S_k$ of the $w$-plane slit from $+i$ to $-i$ via $w = \infty$ (visualize on the sphere!). The only branch points are $w = \pm i$, which are of infinite order and do not belong to the surface. The surface has the same structure as the Riemann surface of the logarithm, only with a different location of the branch points. These lie over $w = 0$, $\infty$ in the case of log $w$, over $w = +i$, $-i$ in the case of arc tan $w$.

This similarity in structure of the two surfaces is no accident. We have

(6.5.13)                          $-i \, \dfrac{e^{iz} - e^{-iz}}{e^{iz} + e^{-iz}} = w,$

and solving for $z$ we get

(6.5.14)          arc tan $w = \dfrac{1}{2i} \log \dfrac{1 + iw}{1 - iw} = \dfrac{1}{2i} \log \dfrac{i - w}{i + w}.$

Here we get the principal determination of arc tan $w$ by taking log $1 = 0$ and restricting $w$ to the cut plane.

From (6.5.14) we get

(6.5.15)                    $\dfrac{d}{dw}$ arc tan $w = \dfrac{1}{1 + w^2},$

and conversely

(6.5.16)                    arc tan $w = \displaystyle\int_0^w \dfrac{dt}{1 + t^2},$

where the integral is taken along the ray from $0$ to $w$ in the cut plane.

## EXERCISE 6.5

**1.** Find the Maclaurin series of the principal branch of arc sin $w$.

**2.** Same problem for arc tan $w$.

**3.** Show that the imaginary part of arc tan $w$ has the same sign as that of $w$.

**4.** Give a geometrical construction of the real part of the principal determination of arc tan $w$.

**5.** If $\Re(w) \to +\infty$ $(-\infty)$, show that the principal determination of the function arc tan $w$ tends to $+\dfrac{\pi}{2} \left(-\dfrac{\pi}{2}\right)$.

**6.** Show that $| \operatorname{arc} \sin z | < \dfrac{\pi}{2}$ for $| z | < 1$ and the principal determination of the arc sine.

**7.** If for some choice of the square roots

$$w_3 = w_1 \sqrt{1 - w_2{}^2} + w_2 \sqrt{1 - w_1{}^2},$$

show that

$$\operatorname{arc} \sin w_3 = \operatorname{arc} \sin w_1 + \operatorname{arc} \sin w_2$$

for some determination of the arc sines.

## COLLATERAL READING

BAIRE, R.  *Leçons sur les Théories Générales d'Analyse,* Vol. II, Chap. 4. Gauthier-Villars, Paris, 1908.

HURWITZ, A., and COURANT, R.  *Vorlesungen über allgemeine Funktionentheorie und elliptische Funktionen,* Second Edition, Part I, Chap. 4, and Part III, Chap. 4.  Springer-Verlag, Berlin, 1925.

# 7

# COMPLEX INTEGRATION

**7.1. Integration in the complex plane.** We shall now turn to a systematic study of integration in the complex plane. Since the days of Cauchy, this device has been one of the most useful tools in the theory of analytic functions.

Let $C$ be a rectifiable curve in the complex plane

$$(7.1.1) \qquad C: \quad z = z(t), \quad 0 \leq t \leq 1,$$

where $z(t)$ is a continuous function of bounded variation in $[0, 1]$. $C$ is oriented by the parametrization, that is, the point $z_1 = z(t_1)$ precedes the point $z_2 = z(t_2)$ on $C$ if and only if $t_1 < t_2$. We write $-C$ for the same curve with opposite orientation

$$(7.1.2) \qquad -C: \quad z = z(1 - t).$$

According to Theorem 2.4.2 the length of $C$ is the total variation of $z(t)$

$$(7.1.3) \qquad l(C) = V_0^1[z].$$

Suppose now that a continuous function $f(z)$ is given on $C$. Then $f[z(t)]$ is a continuous function of $t$, $f[z(t)] \in C[0, 1]$. By Theorem C.3.1 the Riemann-Stieltjes integral of $f[z(t)]$ with respect to $z(t)$ exists.

DEFINITION 7.1.1.   *We define*

$$(7.1.4) \qquad \int_C f(z)\, dz \equiv \int_0^1 f[z(t)]\, dz(t).$$

We refer the reader to Appendix C for the properties of the Riemann-Stieltjes integral used in this chapter. It should be noted that if $C$ is a line segment and if $f(z)$ is holomorphic in a domain $D$ containing $C$, then the definition of the integral given in Section 4.2 coincides with the present one.

There are a number of properties of the integral (7.1.4) which follow from the definition by virtue of the corresponding properties of the Riemann-Stieltjes integral. We list some of these for future reference.

I. *The integral is linear with respect to the integrand:*

$$(7.1.5) \qquad \int_C [\alpha f_1(z) + \beta f_2(z)]\, dz = \alpha \int_C f_1(z)\, dz + \beta \int_C f_2(z)\, dz.$$

II. *The integral is additive with respect to the path.*

If $C_1$ and $C_2$ are two rectifiable curves such that $C_2$ begins where $C_1$ ends,

160

and if by $C_1 + C_2$ we understand the path consisting of $C_1$ followed by $C_2$, then

$$(7.1.6) \qquad \int_{C_1+C_2} f(z)\, dz = \int_{C_1} f(z)\, dz + \int_{C_2} f(z)\, dz.$$

III. *Reversing the orientation of the path replaces the integral by its negative:*

$$(7.1.7) \qquad \int_{-C} f(z)\, dz = - \int_C f(z)\, dz,$$

since the approximating sums are replaced by their negatives.

IV. We have also the following important *inequalities*:

$$(7.1.8) \qquad \left| \int_C f(z)\, dz \right| \leq \operatorname*{Max}_{z \in C} |f(z)| \cdot l(C)$$

and

$$(7.1.9) \qquad \left| \int_C f(z)\, dz \right| \leq \int_C |f(z)|\ |dz| = \int_0^1 |f[z(t)]|\ ds(t),$$

where $s(t)$ is the arc length on $C$ and $l(C)$ is the length. This estimate follows from (C.3.10) together with the observation that the total variation of $z(t)$ from $t = 0$ to $t = \alpha$ equals $s(\alpha)$, the length of the arc of $C$ from $z(0)$ to $z(\alpha)$.

V. In the applications we normally deal with *paths of integration consisting of a finite number of arcs along each of which $z(t)$ has a continuous derivative.* For such an arc $C$ formula (C.3.12) gives

$$(7.1.10) \qquad \int_C f(z)\, dz = \int_0^1 f[z(t)]\, z'(t)\, dt.$$

To conclude this discussion of the elementary properties of the integral, we work out three examples. The results will be needed in the next section.

EXAMPLE 1.    We start with the simplest of all integrals:

$$(7.1.11) \qquad \int_a^b dz = b - a$$

for any rectifiable path of integration leading from $z = a$ to $z = b$. Indeed, suppose that (7.1.1) is such a path, and $z(0) = a$, $z(1) = b$. By definition,

$$\int_C dz = \int_0^1 dz(t) = z(1) - z(0) = b - a.$$

EXAMPLE 2.    Similarly for all rectifiable paths we have

$$(7.1.12) \qquad \int_a^b z\, dz = \tfrac{1}{2}[b^2 - a^2],$$

since

$$\int_a^b z\, dz = \int_0^1 z(t)\, dz(t) = [z(1)]^2 - [z(0)]^2 - \int_0^1 z(t)\, dz(t)$$

by formula (C.3.13) for integration by parts in a Stieltjes integral. (This formula applies since $z(t)$ is both continuous and of bounded variation.) It follows that (7.1.12) holds.

EXAMPLE 3.    Finally, we take

$$(7.1.13) \qquad \int_C \frac{dz}{z - a} = 2\pi i, \quad C: \; z = a + re^{i\theta}, \;\; 0 \le \theta \le 2\pi.$$

Here $r$ is arbitrary, $r > 0$. Since $z = z(\theta)$ is a differentiable function of $\theta$, we can use (7.1.10), and the result is immediate.

Examples 1 and 2 suggest the following generalization: If $G(t)$, $0 \le t \le 1$, is a complex-valued function having a continuous derivative with respect to $t$, then

$$\int_0^1 G'(t) \, dt = G(1) - G(0),$$

as we can see by separating reals and imaginaries. Suppose now that $f(z)$ is holomorphic in a domain $D$ and that we know a function $F(z)$, also holomorphic in $D$, such that $F'(z) = f(z)$. Let $a$ and $b$ be two points in $D$ which are joined by a simple arc

$$C: \quad z = g(t), \quad 0 \le t \le 1, \quad g(0) = a, \, g(1) = b,$$

such that $g'(t)$ exists and is continuous. Then

$$\int_C f(z) \, dz = \int_C F'(z) \, dz = \int_0^1 F'(g(t)) \, g'(t) \, dt$$

$$= \int_0^1 \frac{d}{dt} [F(g(t))] \, dt = F(g(1)) - F(g(0)).$$

Thus, the classical formula of the integral calculus

$$(7.1.14) \qquad \int_a^b f(z) \, dz = F(b) - F(a) \quad \text{if} \quad F'(z) = f(z)$$

holds under suitable restrictions on the path of integration joining $a$ with $b$, provided $f(z)$ and $F(z)$ are holomorphic. It is clear that the same result is valid for a path $C: \; z = g(t)$ such that $g'(t)$ is piecewise continuous. In particular, we see that under the stated conditions the integral along a closed contour equals 0. This result will be proved under much more general conditions in the next section.

## EXERCISE 7.1

**1.** The following paths of integration are used below: $C_1$: $[-i, +i]$; $C_2$: $[-i, -i + 1, i + 1, i]$; $C_3$: $[-i, -i - 1, i - 1, i]$; $C_4$: $[i + 1, i - 1,$

$-i - 1, -i + 1, i + 1$]. These are polygonal lines joining the points indicated. Further, $C_5$ is $|z| = 1$. Compute the integrals:

**a.** $\displaystyle\int_{C_1} x \, dz.$    **c.** $\displaystyle\int_{C_2} |z| \, dz.$    **e.** $\displaystyle\int_{C_4} y \, dz.$    **g.** $\displaystyle\int_{C_2} \frac{dz}{z}.$

**b.** $\displaystyle\int_{C_2} x \, dz.$    **d.** $\displaystyle\int_{C_3} |z| \, dz.$    **f.** $\displaystyle\int_{C_5} y \, dz.$    **h.** $\displaystyle\int_{C_4} \frac{dz}{z}.$

**2.** Estimate $\left| \displaystyle\int_C (1 + z^2)^{-1} \, dz \right|$, where $C$ is the upper half of the circle $|z| = 2$.

**3.** Work out $\displaystyle\int_C z^m \, dz$, where $C$ is $z = re^{i\theta}$, $0 \leq \theta \leq 2\pi$, and $m$ is an integer $\neq -1$.

**4.** It is desired to integrate $(z + 1)^{-2}$ along the imaginary axis from $-\infty i$ to $+\infty i$. Show that this improper integral exists and find its value.

**7.2. Cauchy's theorem.**   This name is given to the following proposition, which is the fundamental theorem in the theory of analytic functions:[1]

THEOREM 7.2.1.   *Let $f(z)$ be holomorphic in a domain $D$. Let $C$ be a simple closed rectifiable curve in $D$ such that $f(z)$ is holomorphic inside and on $C$. Then*

$$(7.2.1) \qquad\qquad \int_C f(z) \, dz = 0.$$

---

[1] Cauchy published this result in his *Mémoire sur les Intégrales Définies Prises entre des Limites Imaginaires* (Bure Frères, Paris, 1825), but he had the basic idea as early as 1814. The early proofs proceeded by consideration of the two line integrals obtained by separating real and imaginary parts in the left side of (7.2.1). These line integrals are independent of the path of integration, in view of the Cauchy-Riemann equations, and hence they vanish. In order to apply this method, it is necessary to assume that $f'(z)$ is continuous, and that the contour $C$ satisfies fairly severe regularity conditions.

New ideas for the proof were introduced by E. Goursat in 1883 [*Acta Mathematica*, Vol. 14 (1884), and *Transactions of the American Mathematical Society*, Vol. 1 (1900)]. Goursat worked directly with the integral. He partitioned the interior of $C$ into a large number of small squares, plus some irregular residual portions near the contour $C$, and he expressed (7.2.1) as the sum of the integrals over the boundaries of these subregions. He also used the fact that the theorem holds for the particular functions $f(z) = 1$ and $f(z) = z$. In his second paper he assumed merely the existence of $f'(z)$. While avoiding continuity assumptions on $f'(z)$, he had to prove a delicate lemma concerning uniform approximation of the difference quotient by the derivative. Next, E. H. Moore, also in Vol. 1 of the *Transactions*, gave an indirect proof of the theorem, based on the nesting principle. Finally, in Vol. 2 of the *Transactions*, A. Pringsheim revised the proof of Goursat's lemma. He also launched the idea of first proving the desired results for triangles and of avoiding difficulties along the boundary by approximating $C$ by a polygon.

REMARK.    The proof given below is that of A. Pringsheim in the arrangement of K. Knopp.[1] The use of Goursat's lemma is avoided, and Pringsheim's argument is modified so as to give directly the validity of the theorem for triangles. It is an indirect proof using the nesting principle, the existence of $f'$ at the point to which the nested triangles converge, and again the special cases $f(z) = 1$ and $f(z) = z$. Since arbitrary polygons may be triangulated, the truth of (7.2.1) for polygons follows from the Cauchy theorem for triangles. Finally, we use the facts that a simple closed rectifiable curve can be approximated by a polygon, and that the integral over $C$ is the limit of Riemann-Stieltjes sums.

*Proof of Theorem 7.2.1.*    Following the program outlined above, we let $C = \triangle = [z_1, z_2, z_3, z_1]$ be a triangle, and we denote by $\triangle^*$ the union of $\triangle$ and its interior. The vertices of $\triangle$ are numbered, and the triangle is oriented, in such a way that if $z_4$ is the center of gravity of $\triangle$, then $\arg(z - z_4)$ increases by $2\pi$ as $z$ describes $\triangle$ once. Finally, let $L$ be the (length of the) perimeter of $\triangle$.

If now $\triangle^* \subset D$, we obtain a contradiction from the assumption that

$$\int_\triangle f(z)\, dz \neq 0.$$

We begin a nesting process, by joining the midpoints of the sides of $\triangle$ by straight-line segments. This gives four subtriangles, $\triangle_{01}$, $\triangle_{02}$, $\triangle_{03}$, and $\triangle_{04}$, where $\triangle_{04}$ contains $z_4$, while for $k = 1, 2, 3$, $z_k$ is a vertex of $\triangle_{0k}$. Each of these triangles is now oriented: $\triangle_{01}$, $\triangle_{02}$, and $\triangle_{03}$ with the orientation induced by that of $\triangle$, and $\triangle_{04}$ by requiring that $\arg(z - z_4)$ increase by $2\pi$ when $z$ describes $\triangle_{04}$. We then have

$$(7.2.2) \qquad \int_\triangle f(z)\, dz = \sum_{k=1}^{4} \int_{\triangle_{0k}} f(z)\, dz,$$

since the integral along $\triangle_{04}$ cancels the contributions from the other three triangles which do not correspond to segments from the sides of $\triangle$. See Figure 17.

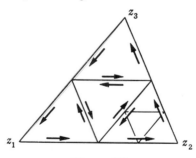

**Figure 17**

[1] *Funktionentheorie*, Second Edition, Vol. I (Sammlung Göschen, Berlin, 1918).

By assumption, the left side of (7.2.2) is different from zero. Hence, at least one of the integrals on the right side must have the same property. We select the integral of largest absolute value; if there is a tie, preference is given to the integral whose triangle has the lowest subscript. The triangle of the integral so chosen is relabeled $\triangle_1$, and we have the inequality

$$(7.2.3) \qquad \left| \int_{\triangle} f(z)\, dz \right| \leq 4 \left| \int_{\triangle_1} f(z)\, dz \right|.$$

$\triangle_1$ is now subdivided into triangles $\triangle_{11}, \triangle_{12}, \triangle_{13}, \triangle_{14}$ in the same manner as $\triangle$ was, so that

$$\int_{\triangle_1} f(z)\, dz = \sum_{k=1}^{4} \int_{\triangle_{1k}} f(z)\, dz.$$

Next, we select the triangle $\triangle_2$ to be the triangle $\triangle_{1k}$, which satisfies

$$\left| \int_{\triangle_1} f(z)\, dz \right| \leq 4 \left| \int_{\triangle_{1k}} f(z)\, dz \right|,$$

and for which $k$ is as small as possible.

In this manner we obtain a sequence of triangles $\{\triangle_n\}$, and, if $\triangle_n{}^*$ denotes the union of $\triangle_n$ with its interior, we have

$$\triangle_0{}^* \supset \triangle_1{}^* \supset \triangle_2{}^* \supset \cdots \supset \triangle_n{}^* \supset \cdots.$$

The length of the perimeter of $\triangle_n$ is clearly $2^{-n}L$, and the diameter of $\triangle_n{}^*$ is at most half of the perimeter of $\triangle_n$. It follows that there is one and only one point $z_0$ common to all the triangular regions $\triangle_n{}^*$. Further, we have

$$(7.2.4) \qquad \left| \int_{\triangle} f(z)\, dz \right| \leq 4^n \left| \int_{\triangle_n} f(z)\, dz \right|.$$

Now $z_0 \in D$, and $f(z)$ has a derivative at $z = z_0$; thus, given an $\varepsilon > 0$ we can find a $\delta = \delta(\varepsilon)$ such that

$$\left| \frac{f(z) - f(z_0)}{z - z_0} - f'(z_0) \right| < \varepsilon \quad \text{if} \quad 0 < |z - z_0| < \delta.$$

Hence

$$f(z) = f(z_0) + (z - z_0)f'(z_0) + \eta(z, z_0),$$

where

$$|\eta(z, z_0)| \leq \varepsilon |z - z_0| \quad \text{if} \quad |z - z_0| < \delta.$$

For this $\delta$ we can find an $n$ so large that $\triangle_n{}^*$ is interior to the circle $|z - z_0| = \delta$. Therefore

$$\int_{\triangle_n} f(z)\, dz = \int_{\triangle_n} [f(z_0) + (z - z_0)f'(z_0) + \eta(z, z_0)]\, dz$$

$$= [f(z_0) - z_0 f'(z_0)] \int_{\triangle_n} dz + f'(z_0) \int_{\triangle_n} z\, dz + \int_{\triangle_n} \eta(z, z_0)\, dz,$$

where we have used the linearity of the operation of integration. The first two integrals in the last member are zero by virtue of (7.1.11) and (7.1.12), and by (7.1.8),

$$\left| \int_{\triangle_n} \eta(z, z_0)\, dz \right| \leq \tfrac{1}{2}\varepsilon \{l[\triangle_n]\}^2 = \tfrac{1}{2}\varepsilon\, 4^{-n} L^2,$$

since $|\,\eta(z, z_0)\,| < \varepsilon\,|\,z - z_0\,| < \tfrac{1}{2}\varepsilon l[\triangle_n]$. Hence

$$\left| \int_{\triangle_n} f(z)\, dz \right| < \tfrac{1}{2}\varepsilon\, 4^{-n} L^2$$

and

$$\left| \int_{\triangle} f(z)\, dz \right| \leq 4^n \left| \int_{\triangle_n} f(z)\, dz \right| < \tfrac{1}{2}\, 4^n 4^{-n} L^2 \varepsilon = \tfrac{1}{2} L^2 \varepsilon.$$

Here $\varepsilon$ is arbitrary. It follows that

$$(7.2.5) \qquad\qquad \int_{\triangle} f(z)\, dz = 0$$

for any triangle $\triangle$ whose interior and boundary belong to $D$. This is the first step of the proof.

Let $\Pi$ be any simple closed polygon whose interior and boundary lie in $D$. By Theorem B.2.1 of Appendix B, any simple closed polygon may be *triangulated*; that is, we can find triangles

$$\triangle_1, \triangle_2, \cdots, \triangle_n$$

such that (i) the interiors of the triangles $\triangle_j$ and $\triangle_k$ have no point in common for $j \neq k$, and (ii) $\cup \triangle_k{}^* = \Pi^*$, where, as above, asterisks are used to denote the union of interior and boundary of the simple closed contour under consideration. It follows that

$$\int_{\Pi} f(z)\, dz = \sum_{k=1}^{n} \int_{\triangle_k} f(z)\, dz = 0,$$

and thus (7.2.1) holds for arbitrary polygons. See Figure 18. This is the second step.

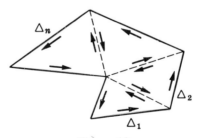

$\triangle_n$      $\triangle_2$      $\triangle_1$

**Figure 18**

We come now to the third and last step. Suppose that $C$ is an arbitrary simple closed rectifiable curve such that $C^* \subset D$, in the notation used above. Here we have to use the definition of the integral (7.2.1) as the limit of suitably chosen sums. First, we note that $C^*$ has a positive distance $d$ from the complement of $D$. Choose a quantity $\rho$, $0 < \rho < d$, and let $R$ be the closed set of points in the plane which either belong to $C^*$ or have a distance from $C^*$ not exceeding $\rho$. Thus $C^* \subset R \subset D$, where in both cases the inclusion is proper. Let $\omega(\delta, f)$ be the modulus of continuity of the restriction of $f(z)$ to $R$, that is,

$$\omega(\delta, f) = \sup |f(z_1) - f(z_2)| \quad \text{for} \quad |z_1 - z_2| \leq \delta, \quad z_1, z_2 \in R.$$

Let $\varepsilon$ be given, $\varepsilon > 0$. We can then find a finite set of points $z_1, z_2, \cdots, z_n$ on $C$ satisfying the following requirements:

　(i) $z_j$ precedes $z_k$ on $C$ if $j < k$;
　(ii) all line segments $[z_k, z_{k+1}]$ belong to $R$, where $z_{n+1} = z_1$;
　(iii) the polygon $\Pi = [z_1, z_2, \cdots, z_n, z_1]$ does not intersect itself;
　(iv) if $\delta$ is so chosen that $\omega(\delta, f) < \varepsilon/(2l)$, where $l$ is the length of $C$, then for each $k$, $|z_k - z_{k+1}| < \delta$; and

(v)
$$\left| \int_C f(z)\, dz - \sum_{k=1}^{n} f(z_k)(z_{k+1} - z_k) \right| < \frac{\varepsilon}{2}.$$

It is clear that these conditions, except possibly (iii), can be satisfied. The purpose of (iii) is to ensure that

(7.2.6)
$$\int_\Pi f(z)\, dz = 0,$$

which was proved above for simple closed polygons. But if (iii) is violated, this will still be true, for $\Pi$ is the union of a finite number of simple closed polygons plus, possibly, some double segments, which are traversed twice in opposite directions when $z$ describes $\Pi$. In any case (7.2.6) holds. Writing the finite sum in (v) as an integral along $\Pi$, we get

$$\left| \int_\Pi f(z)\, dz - \sum_{k=1}^{n} f(z_k)(z_{k+1} - z_k) \right| = \left| \sum_{k=1}^{n} \int_{z_k}^{z_{k+1}} [f(z) - f(z_k)]\, dz \right|$$
$$\leq \omega(\delta, f) \cdot l < \frac{\varepsilon}{2l} \cdot l = \frac{\varepsilon}{2},$$

where we have used the fact that the length of $\Pi$ does not exceed that of $C$. Combining this estimate with (v) and (7.2.6), we see that

$$\left| \int_C f(z)\, dz \right| < \varepsilon.$$

It follows that (7.2.1) holds, and the theorem is proved.

In order to be able to formulate some important corollaries of Cauchy's theorem we need

DEFINITION 7.2.1.    *The domain D in the finite plane is said to be simply-connected if the complementary set* $\mathbf{C}[D]$ *is connected with respect to the sphere.*

Note that we do not require that $\mathbf{C}[D]$ be arcwise connected; we insist merely that $\mathbf{C}[D] = S_1 \cup S_2$, where $S_1 \neq \emptyset \neq S_2$, imply $(\bar{S}_1 \cap S_2) \cup (S_1 \cap \bar{S}_2) \neq \emptyset$.

COROLLARY 1.    *If D is simply-connected, then* (7.2.1) *holds for every simple closed rectifiable curve C in D.*

COROLLARY 2.    *If D is simply-connected, and if a and b are any two points in D, then*

$$(7.2.7) \qquad\qquad \int_a^b f(z)\, dz$$

*is independent of the rectifiable path joining a and b in D.*

*Proof.*    If $C_1$ and $C_2$ are two rectifiable arcs joining $a$ with $b$ in $D$, and if $C_1$ and $C_2$ have only the points $a$ and $b$ in common, then $C_1 - C_2$ is a simple closed rectifiable curve to which Theorem 7.2.1 applies, so that

$$(7.2.8) \qquad\qquad \int_{C_1} f(z)\, dz = \int_{C_2} f(z)\, dz.$$

In the general case we can approximate the integrals along $C_1$ and $C_2$ with an error less than a preassigned $\varepsilon$ by means of integrals along polygonal lines $\Pi_1$ and $\Pi_2$. It may be shown that $\Pi_1 - \Pi_2$ is the sum of a finite number of simple closed polygons. Hence the integral along $\Pi_1 - \Pi_2$ equals the sum of a finite number of integrals along simple closed polygons, and each of these integrals is zero. Since this holds for every $\varepsilon > 0$, it follows that (7.2.8) must hold, and that the integral in (7.2.7) is independent of the path.

This result has an obvious bearing on formula (7.1.14), that is, the validity of

$$\int_a^b f(z)\, dz = F(b) - F(a) \quad \text{when} \quad F'(z) = f(z).$$

We see now that if $f(z)$ is holomorphic in a simply-connected domain $D$, then the formula holds for any rectifiable path in $D$ joining $a$ with $b$.

## EXERCISE 7.2

**1.** Suppose that $f(z)$ and $g(z)$ are holomorphic inside and on a simple closed rectifiable curve $C$ and find

$$\int_C f(z)\, g(z)\, dz.$$

**2.** Suppose that $C$ is the circle $z = a + re^{i\theta}$, $0 \leq \theta \leq 2\pi$, inside and on which $f(z)$ is holomorphic. Let $m$ be any positive integer and evaluate

$$\int_0^{2\pi} f(a + re^{i\theta})e^{mi\theta}\, d\theta.$$

**3.** Under the assumptions of Problem 2, $f(a + re^{i\theta})$ is a continuous differentiable function of $\theta$ with period $2\pi$. As such it has a Fourier series

$$f(a + re^{i\theta}) \sim \sum_{n=-\infty}^{\infty} f_n(r)e^{ni\theta}.$$

Interpret the result found in Problem 2 in terms of these Fourier coefficients.

**4.** Under the same assumptions, let $z_0$ be a fixed point interior to the circle, and let $z_0^*$ be obtained from $z_0$ by an inversion in $C$, that is,

$$(\bar{z}_0 - \bar{a})(z_0^* - a) = r^2.$$

Find the value of

$$\int_C \frac{f(z)}{z - z_0^*}\, dz.$$

**5.** Evaluate the integral $\int_C (z^2 + 4)^{-1}\, dz$, where $C$ is the unit circle.

**6.** Evaluate $\int_C \dfrac{e^z - e^a}{z - a}\, dz$, where $a$ is fixed and the simple closed rectifiable curve $C$ does not pass through $z = a$.

**7.** Suppose that the curve $C$ of Theorem 7.2.1 is the perimeter of a rectangle. Separate reals and imaginaries, and use the Cauchy-Riemann equations to evaluate the resulting line integrals.

**8.** A function $f(z)$ is holomorphic in $|z| < 1 + \delta$, $0 < \delta$. Evaluate the integral of $f(z)$ over the unit disk. (*Hint:* Use polar coordinates in the double integral.)

**9.** Given that $f(z)$ is holomorphic in $0 < |z| < 1$ and that (7.2.1) holds for every circle $|z| = r$, $0 < r < 1$. Show by an example that $f(z)$ need not be holomorphic at the origin.

**7.3. Extensions.** We give several extensions of Cauchy's theorem. It is necessary to introduce some geometrical concepts, but we shall try to hold the topological difficulties to a minimum.

DEFINITION 7.3.1.    *The region $R$ is starlike with respect to the point $z_0$, $z_0 \in \text{Int}(R)$, if with $z_1 \in R$ all the points $\alpha z_0 + (1 - \alpha)z_1$, $0 < \alpha < 1$, belong to Int $(R)$.*

THEOREM 7.3.1.    *Let $C$ be a simple closed rectifiable curve, let $C_i$ be the interior of $C$, and $C^*$ be the union of $C_i$ and $C$. If $C^*$ is starlike with respect to a point $z_0 \in C_i$, and if $f(z)$ is holomorphic in $C_i$ and continuous in $C^*$, then*

(7.3.1)                              $$\int_C f(z)\, dz = 0.$$

*Proof.*    Since $f(z) \in C[C^*]$, in the notation of Section 4.1, it follows that $f(z)$ is uniformly continuous in $C^*$. Let the curve $C$ be given by the equation $z = z(t)$, $0 \leq t \leq 1$, and define a family of curves $C_\alpha$ by

$$C_\alpha: \quad z = z_0 + \alpha[z(t) - z_0], \quad 0 < \alpha < 1.$$

By Definition 7.3.1 each curve $C_\alpha$ lies in $C_i$, the interior of $C$. Further, $C_\alpha$ is closed and rectifiable, since $z(t)$ is a continuous function of bounded variation. We have by (7.1.4)

$$\int_{C_\alpha} f(z)\, dz = \alpha \int_0^1 f\{z_0 + \alpha[z(t) - z_0]\}\, dz(t),$$

and this holds also for $\alpha = 1$ with $C_1 = C$. Here the left side is zero for $0 < \alpha < 1$, and owing to the uniform continuity of $f(z)$ in $C^*$, the integral is a continuous function of $\alpha$ in the closed interval $[0, 1]$. Then (7.3.1) holds.

The assumption that $C^*$ is starlike is used only to assure the existence of a family of curves $C_\alpha$ in $C_i$, such that $C_\alpha$ converges to $C$ and that

$$\int_{C_\alpha} f(z)\, dz \to \int_C f(z)\, dz.$$

Actually, one can always find such a family, so (7.3.1) holds without further restrictions on $C^*$.

DEFINITION 7.3.2.    *Let $C$: $z = z(t)$, $0 \leq t \leq 1$, be a simple closed rectifiable curve.  $C$ is said to have positive (negative) orientation if $z_0 \in C_i$ implies that $\arg[z(t) - z_0]$ increases (decreases) by $2\pi$ when $t$ goes from 0 to 1.*

Since, for fixed $t$, $\arg[z(t) - z_0]$ is a continuous function of $z_0$ as long as $z_0$ stays in $C_i$, it may be shown that the increase or decrease of the argument is independent of $z_0$ in $C_i$.

THEOREM 7.3.2.    *Suppose that $C_1$ and $C_2$ are two simple closed rectifiable curves having the same orientation and that $C_1$ lies in the interior of $C_2$. Suppose that $f(z)$ is holomorphic in a curvilinear annulus containing $(C_2)^* \ominus (C_1)_i$. Then*

(7.3.2)                       $$\int_{C_1} f(z)\, dz = \int_{C_2} f(z)\, dz.$$

*Proof.*    The assertion is trivially true if $f(z)$ is holomorphic in all of $(C_2)^*$ since then both integrals are zero. In the general case, pick a point $z_0$ interior to $C_1$ and, hence, a fortiori, interior to $C_2$, and draw a horizontal line through $z_0$. This line extended will intersect $C_1$ at least twice and $C_2$ at least twice.

But the intersection of closed sets is a closed set.  Therefore, in going from $z_0$ along the line to the right we encounter a last point $z_1$ on $C_1$, and a first subsequent point $z_2$ on $C_2$.  The line segment $(z_1, z_2)$ lies entirely in $(C_2)_i \cap (C_1)_e$.  Similarly, going to the left from $z_0$, we encounter a last point $z_3$ on $C_1$ and a first subsequent point $z_4$ on $C_2$, and the line segment $(z_4, z_3)$ also lies in $(C_2)_i \cap (C_1)_e$.  These two line segments sever the domain $(C_2)_i \cap (C_1)_e$ into two parts, an upper and a lower one, each of which is a domain.  See Figure 19.  To fix the ideas, suppose

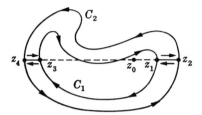

**Figure 19**

that $C_1$ and $C_2$ have positive orientation.  Let $\Gamma_1$ be the boundary of the upper domain, $\Gamma_2$ the boundary of the lower one, both contours having the positive orientation induced by $C_2$.  Here $\Gamma_1$ consists of an arc $C_{21}$ of $C_2$, an arc $-C_{11}$ of $-C_1$, and the two line segments $(z_1, z_2)$ and $(z_4, z_3)$.  Similarly $\Gamma_2$ consists of the remaining arc $C_{22}$ of $C_2$, the remaining arc $-C_{12}$ of $-C_1$, and the line segments $(z_2, z_1)$ and $(z_3, z_4)$.  We have

$$C_{21} + C_{22} = C_2, \quad -C_{11} - C_{12} = -C_1,$$

and the subarcs have only endpoints in common.  But

$$0 = \int_{\Gamma_1} + \int_{\Gamma_2} = \int_{C_{21}} - \int_{C_{11}} + \int_{C_{22}} - \int_{C_{12}}$$

since the four integrals along the line segments cancel in pairs.  Regrouping the terms we see that (7.3.2) results.

Theorem 7.3.2 is only a special case of a more general theorem which we shall state without proof.

**THEOREM 7.3.3.**    *Let* $C_0, C_1, \cdots, C_n$ *be* $(n + 1)$ *simple closed rectifiable curves in the plane having positive orientation.  Suppose that* $C_j \subset (C_k)_e \cap (C_0)_i$ *for* $j, k = 1, 2, \cdots, n, j \neq k$, *and that* $f(z)$ *is holomorphic in a domain* $D$ *containing the closure of the domain*

(7.3.3)              $$D_0 = (C_0)_i \cap (C_1)_e \cap (C_2)_e \cap \cdots \cap (C_n)_e.$$

*Then*

(7.3.4)              $$\int_{C_0} f(z)\, dz = \sum_{k=1}^{n} \int_{C_k} f(z)\, dz.$$

The proof depends upon the possibility of finding crosscuts severing $D_0$

into two or more simply-connected pieces. The integrals along the boundaries of these pieces add up to the difference of the two sides of (7.3.4), which consequently equals zero. See Figure 20.

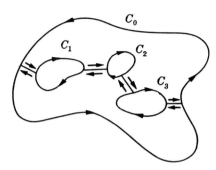

**Figure 20**

The last two theorems are standard tools for contour integration. We shall give three applications. First we consider

$$\int_C \frac{dz}{z - a},$$

where $C$ is a simple closed rectifiable positively oriented curve which does not pass through $z = a$. If $a$ lies in $C_e$, we see that the integrand is holomorphic in $C^*$, so the integral is zero by Cauchy's theorem. If $a \in C_i$, we can use Theorem 7.3.2, according to which the integral has the same value for $C$ as for a small circle with center at $z = a$ and described in the positive sense. The latter equals $2\pi i$ by formula (7.1.13).

To avoid verbiage we make the

CONVENTION.    *A simple closed rectifiable positively oriented curve will be called a "scroc" in the following.*

As the second example we take

$$\int_C \frac{dz}{z^2 + 4},$$

where $C$ is a "scroc" not passing through $z = \pm 2i$. Compare Problem 5 of Exercise 7.2, where $C$: $|z| = 1$. If neither of the points $\pm 2i$ lies in $C_i$, then the value of the integral is zero by Cauchy's theorem. In any other case, we write the integrand as the sum of partial fractions

$$\frac{1}{z^2 + 4} = \frac{1}{4i} \frac{1}{z - 2i} - \frac{1}{4i} \frac{1}{z + 2i}.$$

Since the integral of the sum is the sum of the integrals, the preceding example applies. The value of the integral is $\dfrac{\pi}{2}$ if only $2i$ is interior to $C$, $-\dfrac{\pi}{2}$ if only $-2i$ is interior, and $0$ if both points belong to $C_i$.

Our third example is

$$\int_C (z - a)^m \, dz,$$

where $m$ is an arbitrary integer and the "scroc" $C$ is allowed to pass through $z = a$ if $m \geq 0$, otherwise not. If $m \geq 0$, the value of the integral is $0$ by Cauchy's theorem. If $m = -1$, we are back to the first example. If $m < -1$ and $a \in C_i$, we use Theorem 7.3.2 to replace $C$ by a circle with center at $z = a$. This integral equals

$$r^{m+1} \int_0^{2\pi} e^{(m+1)i\theta} \, d\theta = 0$$

for every $r \neq 0$.

With the aid of the last example, we see that we can work out the integral of any rational function over any "scroc" not passing through a zero of the denominator. We need only expand the rational function in partial fractions and integrate each term separately. We note that only the terms of the form $c(z - a)^{-1}$ contribute to the value of the integral. This is a particular case of Cauchy's residue theorem. See Chapter 9.

The final result in this section is a theorem of a more special nature, but its proof illustrates the deformation principle, which is used all the time in contour integration.

THEOREM 7.3.4.   *Suppose that $f(z)$ is holomorphic in the sector*

$$\alpha_1 < \arg z < \alpha_2, \quad \alpha_2 - \alpha_1 < 2\pi, \quad 0 < |z| < \infty.$$

*Suppose also that positive quantities $\delta$, $\varepsilon$, $M_1$, $M_2$ exist such that*

(7.3.5)      $|f(z)| < \begin{cases} M_1 |z|^{-1+\delta}, & 0 < |z| \leq 1, \\[2mm] M_2 |z|^{-1-\varepsilon}, & 1 < |z|, \end{cases} \qquad \alpha_1 < \arg z < \alpha_2.$

*Then the integral*

(7.3.6)                      $J_\alpha = \displaystyle\int_0^\infty f(z) \, dz$

*converges when taken along the ray $\arg z = \alpha$, $\alpha_1 < \alpha < \alpha_2$, and its value is independent of $\alpha$.*

*Proof.*   We have

$$J_\alpha = e^{i\alpha} \int_0^\infty f(re^{i\alpha}) \, dr,$$

and this integral exists, since

$$\int_0^\infty |f(re^{i\alpha})| \, dr < M_1 \int_0^1 r^{-1+\delta} \, dr + M_2 \int_1^\infty r^{-1-\varepsilon} \, dr.$$

The main point of the theorem is the assertion that $J_\alpha$ is independent of $\alpha$. To prove this, we take $\alpha_1 < \alpha < \beta < \alpha_2$, and integrate along a "scroc" $C_{\rho,\,R}$ consisting of two rays and two circular arcs, namely,

$$\arg z = \alpha, \, 0 < \rho \leq |z| \leq R < \infty; \quad \arg z = \beta, \, \rho \leq |z| \leq R;$$

$$|z| = \rho, \, \alpha \leq \arg z \leq \beta; \quad |z| = R, \, \alpha \leq \arg z \leq \beta.$$

The value of this integral is zero. The absolute value of the integral along the small arc is dominated by

$$M_1 \rho^{-1+\delta}(\beta - \alpha)\rho = (\beta - \alpha)M_1\rho^\delta$$

and tends to zero with $\rho$. For the big arc we get the upper bound

$$M_2 R^{-1-\varepsilon}(\beta - \alpha)R = (\beta - \alpha)M_2 R^{-\varepsilon},$$

and this tends to zero when $R \to \infty$. The integral along $\arg z = \alpha$ tends to $J_\alpha$ when $\rho \to 0$, $R \to \infty$, and the integral along $\arg z = \beta$, which is taken toward the origin, tends to $-J_\beta$. It follows that $J_\alpha - J_\beta = 0$ as asserted.

## EXERCISE 7.3

**1.** Integrate the following functions along the circle $|z| = 3$ in the positive sense:

    **a.** $\dfrac{z^3 + 1}{z(z - 2)}$.     **b.** $\dfrac{z(z - 2)}{z^3 + 1}$.     **c.** $\dfrac{z^2 - 1}{z}$.     **d.** $\dfrac{z}{z^2 - 1}$.

**2.** The integral

$$J_\alpha = \int_0^\infty (1 + z^3)^{-1}\, dz, \quad \arg z = \alpha,$$

is well defined unless $\alpha$ is an odd multiple of $\pi/3$. Why? Show that $J_\alpha$ has only three distinct values, and find the relation holding between them. Determine the value of $J_0$ by contour integration, using the rays $\arg z = 0$ and $\arg z = 2\pi/3$ and the circle $|z| = R$.

**3.** Consider the principal branch of $(1 + z)^\alpha$, $\alpha$ real, $|z| \leq 1$. Show that for $\alpha > -1$

$$\int_{|z|=1} (1 + z)^\alpha\, dz = 0.$$

For $0 > \alpha > -1$, the integral can be thought of as the limit, as $\delta \to 0$, of the sum of the integrals over the two arcs $|z| = 1$, $|z + 1| \geq \delta$, and $|z + 1| = \delta$, $|z| \leq 1$. Both integrals have the limit zero.

**4.** If $\alpha = -1$ in the preceding problem, show that the two integrals mentioned have limits different from zero and find these limits.

**7.4. Cauchy's integral.** This term refers to the following fundamental consequence of Cauchy's theorem:

THEOREM 7.4.1.    *If $C$ is a "scroc," if $f(z)$ is holomorphic in a domain $D$ containing $C^*$, and if $z \in C_i$, then*

(7.4.1) $$f(z) = \frac{1}{2\pi i} \int_C \frac{f(t)}{t-z} dt.$$

*Proof.*    The integrand is a holomorphic function of $t$ in $C_i$, except for $t = z$. By Theorem 7.3.2 we have then

$$\frac{1}{2\pi i} \int_C \frac{f(t)}{t-z} dt = \frac{1}{2\pi i} \int_\gamma \frac{f(t)}{t-z} dt,$$

where $\gamma$ is the circle $| t - z | = \rho$, and $\rho$ is so small that $\gamma$ lies in $C_i$. We recall that

$$\frac{1}{2\pi i} \int_\gamma \frac{dt}{t-z} = 1.$$

Hence

$$\frac{1}{2\pi i} \int_\gamma \frac{f(t)}{t-z} dt - \frac{f(z)}{2\pi i} \int_\gamma \frac{1}{t-z} dt = \frac{1}{2\pi i} \int_\gamma \frac{f(t) - f(z)}{t-z} dt.$$

In the last integral, the integrand is of the form

$$f'(z) + \eta(t, z), \quad \text{and} \quad \eta(t, z) \to 0 \text{ as } t \to z.$$

Thus, the integrand is bounded, say by $M$, for $z$ fixed and $t$ in $C_i$. The integral, therefore, does not exceed $M\rho$ in absolute value, and it tends to zero with $\rho$. On the other hand, the right-hand side in (7.4.1) is independent of $\rho$. Its value must then be equal to $f(z)$, as asserted.

For the validity of (7.4.1) it is sufficient if $f(z)$ is holomorphic in $C_i$ and continuous in $C^*$. When $C^*$ is starlike, we can get this result by combining the techniques of Theorems 7.3.1 and 7.3.2. It suffices to observe that

$$\frac{1}{2\pi i} \int_C \frac{f(t)}{t-z} dt = \lim_{\alpha \to 1} \frac{1}{2\pi i} \int_{C_\alpha} \frac{f(t)}{t-z} dt,$$

and for $z \in (C_\alpha)_i$ the integral on the right equals $f(z)$.

There is an important complement to formula (7.4.1):

(7.4.2) $$\frac{1}{2\pi i} \int_C \frac{f(t)}{t-z} dt = 0 \quad \text{if} \quad z \in C_e.$$

The integral in (7.4.1) has a sense whenever $z$ is not on $C$. Since $C$ separates the plane into two domains, $C_i$ and $C_e$, the formula defines two distinct analytic functions: $f(z)$ if $z \in C_i$, 0 if $z \in C_e$. We shall elaborate on this observation in Section 7.6.

For $z$ on $C$, the integral is not well defined, but it may be interpreted as a limit in several different ways. Suppose that $z_0 \in C$. Then

$$(7.4.3) \qquad \lim_{z \to z_0} \frac{1}{2\pi i} \int_C \frac{f(t)}{t-z}\, dt = \begin{cases} f(z_0), & z \in C_i, \\ 0, & z \in C_e, \end{cases}$$

as follows from (7.4.1) and (7.4.2), since $f(z) \in H[C_i] \cap C[C^*]$. There is another interpretation as a limit, suggested by the analogy with functions of a real variable. The integral

$$\int_{-1}^{1} \frac{dt}{t}$$

does not make sense, but

$$\mathbf{PV}\left\{ \int_{-1}^{1} \frac{dt}{t} \right\} \equiv \lim_{\rho \to 0} \left\{ \int_{-1}^{-\rho} + \int_{\rho}^{1} \right\} \frac{dt}{t} = 0$$

exists and is known as Cauchy's *principal value* of the integral. This suggests defining

$$(7.4.4) \qquad \mathbf{PV}\left\{ \frac{1}{2\pi i} \int_C \frac{f(t)}{t-z_0}\, dt \right\} = \lim_{\rho \to 0} \left\{ \frac{1}{2\pi i} \int_{C(\rho)} \frac{f(t)}{t-z_0}\, dt \right\},$$

where $C(\rho)$ is that part of $C$ on which $|t - z_0| > \rho$. The existence and value of the limit will depend not merely upon the value of $f(t)$ at $t = z_0$ but also upon the geometrical configuration. To simplify, take $C$ to be a piecewise smooth curve having a continuously turning tangent, except for a finite number of vertices $z_1, z_2, \cdots, z_n$, where the inclinations of the positive semi-tangents have saltus of $\alpha_1, \alpha_2, \cdots, \alpha_n$, respectively. Let us define the *local interior angle*, $\omega(z_0)$, at $z = z_0$ on $C$ to be $\pi$, at any point $z_0$ where $C$ has a unique tangent, and to be $\pi - \alpha_k$ at $z = z_k$.

THEOREM 7.4.2.    *Under these assumptions, we have*

$$(7.4.5) \qquad \mathbf{PV}\left\{ \frac{1}{2\pi i} \int_C \frac{f(t)}{t-z_0}\, dt \right\} = \frac{1}{2\pi}\, \omega(z_0) f(z_0).$$

*Proof.*    If $\rho$ is sufficiently small, the circle $\gamma(\rho)$: $|t - z_0| = \rho$ intersects $C$ in exactly two points, $z_1(\rho)$ and $z_2(\rho)$, where $z_1(\rho)$ is the initial and $z_2(\rho)$ the terminal point of $C(\rho)$, whose orientation of course is that induced by $C$. These two points determine two arcs of $\gamma(\rho)$; let $\gamma_1(\rho)$ be the arc in $C_i$. By Cauchy's theorem the integral along $C(\rho)$ has the same value as the integral along $\gamma_1(\rho)$ from $z_1(\rho)$ to $z_2(\rho)$. But

$$\frac{1}{2\pi i} \int_{\gamma_1(\rho)} \frac{f(t)}{t-z_0}\, dt = \frac{f(z_0)}{2\pi i} \int_{\gamma_1(\rho)} \frac{dt}{t-z_0} + \frac{1}{2\pi i} \int_{\gamma_1(\rho)} \frac{f(t)-f(z_0)}{t-z_0}\, dt.$$

Since $f(t)$ is continuous at $t = z_0$, the second integral on the right tends to zero with $\rho$. In the first integral, we set

$$t = z_0 + \rho e^{i\theta}, \quad z_k(\rho) = z_0 + \rho e^{i\theta_k(\rho)}, \quad k = 1, 2$$

and obtain

$$\frac{1}{2\pi i} \int_{\gamma_1(\rho)} \frac{dt}{t - z_0} = \frac{1}{2\pi} \int_{\theta_1(\rho)}^{\theta_2(\rho)} d\theta = \frac{1}{2\pi} [\theta_2(\rho) - \theta_1(\rho)] \to \frac{1}{2\pi} \omega(z_0).$$

This proves the assertion.

It is desirable to give Theorem 7.4.1 a somewhat more general form using Theorem 7.3.3.

THEOREM 7.4.3.   *Suppose that contours $C_0, C_1, \cdots, C_n$ are defined as in Theorem 7.3.3 and that $f(z)$ is holomorphic for $z$ in a domain $D$ containing the domain $D_0$ of (7.3.3). Then for $z \in D_0$*

$$(7.4.6) \qquad f(z) = \frac{1}{2\pi i} \int_C \frac{f(t)}{t - z} \, dt, \quad C = C_0 - C_1 - C_2 - \cdots - C_n.$$

*Proof.*   The integrand is a holomorphic function of $t$ in $D_0$ save for the point $t = z$. Let $\gamma$ be the circle $|t - z| = \rho$ with positive orientation, and let $\rho$ be so small that $\gamma$ lies in $D$. Adding $\gamma$ to the other contours and using Theorem 7.3.3, we see that the right member of (7.4.6) equals

$$\frac{1}{2\pi i} \int_C \frac{f(t)}{t - z} \, dt = f(z)$$

by the argument used in the proof of Theorem 7.4.1.

## EXERCISE 7.4

**1.** Discuss the following integrals, where $C$ is a "scroc" which does not pass through any singularity of the integrand, and $f(t)$ is holomorphic in $C_i \cup C$:

**a.** $\dfrac{1}{2\pi i} \displaystyle\int_C \dfrac{e^{t-z}}{t - z} f(t) \, dt.$

**c.** $\dfrac{1}{2\pi i} \displaystyle\int_C \dfrac{t}{t^2 - z^2} f(t) \, dt.$

**b.** $\dfrac{1}{2\pi i} \displaystyle\int_C \dfrac{f(t) \, dt}{(t - 1)(t - z)}.$

**d.** $\dfrac{1}{2\pi i} \displaystyle\int_C \dfrac{t^2 + z^2}{t^2 - z^2} f(t) \, dt.$

**2.** Under what assumptions does the following formula hold:

$$f(z) = \frac{1}{2\pi i} \int_C \cot (t - z) f(t) \, dt?$$

(*Hint:*   The function $t \cot t$ is holomorphic in the neighborhood of $t = 0$ if it is defined to equal 1 at $t = 0$.)

**3.** What is the value of

$$\frac{1}{2\pi i} \int_C \frac{[f(t)]^n}{t - z} dt$$

where $n$ is a positive integer?

**4.** Find $\mathbf{PV} \left\{ \int_{|t|=1} \frac{dt}{t - 1} \right\}$.

**5.** Find $\mathbf{PV} \left\{ \int_{|t|=1} \frac{dt}{1 + t^2} \right\}$.

**6.** Let the "scroc" $C$ be made up of the two rays $\arg z = \pm \alpha, 0 \leq |z| \leq R$, and the arc of the circle $|z| = R$ joining the rays and containing $z = R$. Determine $\omega(z_0)$ for $z_0$ on $C$.

**7.5. Cauchy's formulas for the derivatives.** Formula (7.4.1) suggests that a representation of $f'(z)$ should be obtainable by differentiation with respect to $z$ under the sign of integration. This is indeed the case. We shall prove:

THEOREM 7.5.1. *Under the assumptions of Theorem 7.4.1 we have*

$$(7.5.1) \qquad f'(z) = \frac{1}{2\pi i} \int_C \frac{f(t)}{(t - z)^2} dt.$$

*Proof.* We shall not attempt a direct justification of the differentiation under the sign of integration, but we note that the right-hand side of (7.5.1) has a sense as long as $z$ does not lie on $C$. For $z \in C_i$, it defines a function of $z$ which for the time being we denote by $f_1(z)$, and we shall prove that

$$\frac{1}{h}[f(z + h) - f(z)] - f_1(z) \to 0$$

as $h \to 0$ as long as $z$ and $z + h$ are in $C_i$. Since

$$\frac{1}{h}\left[ \frac{1}{t - z - h} - \frac{1}{t - z} \right] - \frac{1}{(t - z)^2} = \frac{h}{(t - z)^2(t - z - h)},$$

we get

$$(7.5.2) \qquad \frac{1}{h}[f(z + h) - f(z)] - f_1(z) = \frac{h}{2\pi i} \int_C \frac{f(t)\, dt}{(t - z)^2(t - z - h)}.$$

Suppose that we restrict $z$ and $z + h$ to stay in that part of $C_i$ which has a distance $\geq \delta$ from $C$. We denote this set by $R(\delta)$. For $z, z + h \in R(\delta)$ the right-hand side of (7.5.2) is dominated by

$$(7.5.3) \qquad (2\pi)^{-1} \delta^{-3} M(f;\ C)\, l(C)\ |h|,$$

where $M(f;\ C)$ is the maximum of $|f(z)|$ on $C$, and $l(C)$ is the length of $C$. This estimate shows that the left-hand side of (7.5.2) tends to zero with $|h|$, uniformly with respect to $z$ in $R(\delta)$. But the difference quotient converges to the derivative $f'(z)$, since $f(z)$ is holomorphic in $C_i$. This proves that $f_1(z) = f'(z)$ and establishes (7.5.1). We have also proved the

COROLLARY.    *The difference quotient converges uniformly to the derivative in* $R(\delta)$.

Since $f'(z)$ is the uniform limit of a sequence of continuous functions in $R(\delta)$, it follows that $f'(z) \in C[R(\delta)]$. We shall prove a much stronger result below.

If $D$ is not simply-connected, it is often advantageous to use Cauchy's integral in the form furnished by Theorem 7.4.3. The proof of Theorem 7.5.1 applies also to this situation and gives

THEOREM 7.5.2.    *Suppose that* $f(z)$ *is holomorphic for* $z$ *in a domain* $D$ *containing the closure of the domain* $D_0$ *of* (7.3.3). *Then for* $z \in D_0$ *we have*

$$(7.5.4) \quad f'(z) = \frac{1}{2\pi i} \int_C \frac{f(t)}{(t-z)^2}\, dt, \quad C = C_0 - C_1 - C_2 - \cdots - C_n.$$

*If* $K$ *is any compact*[1] *set in* $D_0$, *then the difference quotient of* $f(z)$ *converges uniformly to* $f'(z)$ *in* $K$.

Thus we see that formal differentiation under the sign of integration leads to the correct result. But this is an operation which may be repeated as often as we please and which leads to cogent results as long as $z$ belongs to a domain $D_0$ whose closure lies in $D$. We may consequently make the bold conjecture that $f(z)$ has derivatives of all orders and that these are represented by the integrals obtained by formal differentiation. This is indeed so. In order to prove it, we shall need a lemma, which is analogous to Lemma 5.5.2.

LEMMA 7.5.1.    *For* $n \geq 1$ *we have the identity*

$$(7.5.5) \quad \frac{b^{-n} - a^{-n}}{b - a} + na^{-n-1}$$
$$= (b - a)[a^{-2}b^{-n} + 2a^{-3}b^{-n+1} + \cdots + na^{-n-1}b^{-1}].$$

*Proof.*    This is evidently true for $n = 1$. Suppose that the identity holds for $n = k$. Denoting the quantity inside the brackets by $T_n(a, b)$, we have

$$b^{-1}T_n(a, b) + (n + 1)a^{-n-2}b^{-1} = T_{n+1}(a, b).$$

---

[1] See Definition 2.3.1 (7).

It follows that

$$\frac{b^{-k-1} - a^{-k-1}}{b - a} = \frac{b^{-k-1} - b^{-1}a^{-k} + b^{-1}a^{-k} - a^{-k-1}}{b - a}$$

$$= b^{-1}\frac{b^{-k} - a^{-k}}{b - a} - a^{-k-1}b^{-1}$$

$$= b^{-1}[(b - a)T_k(a, b) - ka^{-k-1}] - a^{-k-1}b^{-1}$$

$$= (b - a)b^{-1}T_k(a, b) - (k + 1)a^{-k-1}b^{-1}$$

$$= (b - a)[T_{k+1}(a, b) - (k + 1)a^{-k-2}b^{-1}] - (k + 1)a^{-k-1}b^{-1}$$

$$= (b - a)T_{k+1}(a, b) - (k + 1)a^{-k-2},$$

and this is the desired identity for $n = k + 1$. Thus the relation holds for all $n$.

We are now ready to prove the basic

THEOREM 7.5.3.　*A function holomorphic in a domain $D$ has derivatives of all orders in $D$. If $D_0$ is the domain defined by (7.3.3) and if the closure of $D_0$ lies in $D$, then for $z \in D_0$ we have*

$$(7.5.6) \qquad f^{(m)}(z) = \frac{m!}{2\pi i} \int_C \frac{f(t)\,dt}{(t - z)^{m+1}}, \quad m = 1, 2, 3, \cdots.$$

*Proof.*　We use induction on $m$. Suppose that we have established the existence of derivatives of orders $m \leq k$ for all $z$ in $D$, and the validity of (7.5.6) for such values of $m$ and points $z$ in $D_0$. We denote the right-hand side of (7.5.6) by $f_m(z)$. This function is evidently defined for every integer $m$, when $z \in D_0$. Next we consider the difference

$$(7.5.7) \qquad \triangle_k(z, h; f) = \frac{1}{h}[f_k(z + h) - f_k(z)] - f_{k+1}(z).$$

Using Lemma 7.5.1, we see that

$$(7.5.8) \qquad \triangle_k(z, h; f) = h\frac{k!}{2\pi i} \int_C K_k(t, z, h) f(t)\,dt,$$

where

$$(7.5.9) \qquad K_k(t, z, h) \equiv \sum_{j=1}^{k+1} j(t - z)^{-j-1}(t - z - h)^{-k+j-2}$$

and

$$C = C_0 - C_1 - C_2 - \cdots - C_n.$$

In order to get the expression for $K_k(t, z, h)$ we set

$$n = k + 1, \quad a = t - z, \quad b = t - z - h, \quad b - a = -h$$

in (7.5.5), factor out $h$, and change the sign.

Suppose now that $K$ is a compact subset of $D_0$ whose distance from $C$ equals $\delta$. Then if $z, z + h \in K$ we have

$$|t - z| \geq \delta, \quad |t - z - h| \geq \delta,$$

and hence

$$|K_k(t, z, h)| \leq \tfrac{1}{2}(k + 1)(k + 2)\delta^{-k-3},$$

so that

(7.5.10)     $|\triangle_k(z, h; f)| \leq (4\pi)^{-1}(k + 2)! \, \delta^{-k-3} M(f; \, C) \, l(C) \, |h|.$

It follows that

$$\lim_{h \to 0} \triangle_k(z, h; f) = 0$$

uniformly with respect to $z$ in $K$. This proves the existence of the $(k + 1)$th derivative of $f(z)$ and the validity of (7.5.6) for $m = k + 1$. Thus derivatives of all orders exist, and formula (7.5.6) holds for all $n$.

It is important to realize the fundamental difference between functions of a real variable and functions of a complex variable, when it comes to differentiation. A continuous function $f(x)$, defined in the interval $(a, b)$ on the real line, normally does not have a derivative. If it has a first derivative, nothing can be said about the existence of second or higher derivatives. A continuous function $f(z)$, defined in a domain $D$ of the complex plane, normally does not have a derivative with respect to $z$. If, however, $f'(z)$ exists, then it is also a differentiable function, so that derivatives of all orders exist. Thus, if $f(z)$ is holomorphic in $D$, its derivatives are also holomorphic functions in $D$. In Section 5.5 we saw that this was the case for holomorphic functions defined by power series. Now we see that it holds for all holomorphic functions.

From the results of this section we conclude that some of the restrictions imposed in Chapter 4 are superfluous. Thus, we assumed the continuity of $f'(z)$ in (4.2.11) and in Theorems 4.5.1 and 4.6.1. We now see that continuity is a consequence of the existence of the derivative. Likewise, in the discussion of Laplace's equation in Section 4.4, we had to assume the existence of second-order partial derivatives. Again, this is a consequence of the assumption that $U + iV$ is holomorphic.

### EXERCISE 7.5

The following problems have no direct bearing on the theorems of Section 7.5, but they do involve the existence of first- and second-order derivatives.

**1.** If $f(z)$ is holomorphic in $D$ and its real (imaginary) part is constant in $D$, show that $f(z)$ is a constant.

**2.** Find the Laplacean of $|f(z)|^p$, where $f(z)$ is a function holomorphic in $D$, $f(z) \neq 0$ in $D$, and $p$ is a real number. Show that if $|f(z)|$ is constant in $D$, so is $f(z)$.

**3.** If $f(z)$ is holomorphic in $D$ and $f(z) \neq 0$, then $\log |f(z)|$ is harmonic in $D$. (*Hint:* Use the preceding problem and Problem 2b of Exercise 4.3.)

**4.** Let $f(z)$ be continuous in $D$ together with its first- and second-order partials with respect to $x$ and $y$, $z = x + iy$. Show that $f(z)$ is holomorphic in $D$ if and only if $f(z)$ and $zf(z)$ are harmonic in $D$.

**5.** What is the value of $\displaystyle\int_C \frac{f(t) - f(a)}{t - a}\, dt$ if $f(t)$ is holomorphic in $C \cup C_i$?

**7.6. Integrals of the Cauchy type.** Some sixty years ago, Émile Borel (1871–1956) made the observation that Cauchy's integral (7.4.1) defines a (locally) holomorphic function of $z$, because the Cauchy kernel

$$(7.6.1) \qquad\qquad\qquad \frac{1}{t - z}$$

has this property. Multiplication by $f(t)$ and integration along a rectifiable curve $C$ do not interfere with the holomorphy of the result as long as $z$ does not lie on the path of integration. We shall encounter several important consequences of this observation in the next chapter. At this point we direct attention to the subordinate roles played by the multiplier $f(t)$ and by the path $C$. In proving that the function

$$\frac{1}{2\pi i} \int_C \frac{f(t)}{t - z}\, dt, \quad z \text{ not on } C,$$

has a derivative given by

$$\frac{1}{2\pi i} \int_C \frac{f(t)}{(t - z)^2}\, dt,$$

no property of $f(t)$ on $C$ was used, beyond the mere fact that $f(t)$ was continuous. Likewise the only property of $C$ that mattered was that $C$ was rectifiable, so that the integral has a sense provided that $z$ does not lie on $C$. This observation leads to the consideration of *integrals of the Cauchy type*:

$$(7.6.2) \qquad\qquad\qquad \frac{1}{2\pi i} \int_\Gamma \frac{g(t)}{t - z}\, dt \equiv G(z).$$

Here $\Gamma$ is a rectifiable curve, not necessarily closed or simple. It is assumed, however, that $\Gamma$ intersects itself only a finite number of times. $\Gamma$ then divides the plane into a finite number of domains $D_0, D_1, \cdots, D_n$, each bounded by arcs of $\Gamma$. We let $D_0$ be the domain extending to infinity. Further, $g(t)$ is defined on $\Gamma$ as a continuous function of $t$, and $z$ does not lie on $\Gamma$. Then $G(z)$ is well defined. We shall prove that $G(z)$ is locally holomorphic, that is, it is holomorphic in each of the $(n + 1)$ domains $D_k$.

THEOREM 7.6.1.    $G(z)$ is a holomorphic function of z in each of the domains $D_k$, $k = 0, 1, 2, \cdots, n$. For z in $D_k$, we have

(7.6.3)
$$G^{(m)}(z) = \frac{m!}{2\pi i} \int_\Gamma \frac{g(t)}{(t-z)^{m+1}} \, dt.$$

*Proof.*    We denote the right-hand side of (7.6.3) by $G_m(z)$, with $G_0(z) = G(z)$, and we show that

$$\lim_{h \to 0} \frac{1}{h} [G_k(z+h) - G_k(z)] = G_{k+1}(z), \quad k = 0, 1, \cdots$$

by using the same argument as in the proof of Theorems 7.5.1 and 7.5.3. Except for trivial changes in the notation the same proof applies.

It should be observed once more that $G(z)$ is *locally*, or *piecewise, holomorphic.* If $D_k$ and $D_{k+1}$ are two adjacent domains separated by an arc $\Gamma_k$ of $\Gamma$, then normally there is no function holomorphic in $D_k \cup \Gamma_k \cup D_{k+1}$ which coincides with $G(z)$ in both $D_k$ and $D_{k+1}$. We have encountered this situation already in the case of the classical Cauchy integral, where

$$G(z) = \begin{cases} f(z), & z \in C_i, \\ 0, & z \in C_e. \end{cases}$$

Obviously $f(z)$ and $0$ are holomorphic functions of z in the domains indicated, but unless $f(z)$ is identically zero, one is not the analytic continuation of the other. The student will find further examples of analytic expressions defining piecewise holomorphic functions in Exercise 7.6.

THEOREM 7.6.2.    As $z \to \infty$ in $D_0$, $G(z)$ and all its derivatives tend to zero.

*Proof.*    Suppose that $\Gamma$ is located in the circle $|t| < R$ and that $|g(t)| \leq M$ on $\Gamma$. Denote the length of $\Gamma$ by $l(\Gamma)$. Then

(7.6.4)      $| G^{(n)}(z) | \leq (2\pi)^{-1} n! \, M \, l(\Gamma) [|z| - R]^{-n-1}, \quad |z| > R.$

We shall see later that $G(z)$ is actually holomorphic at $z = \infty$.

## EXERCISE 7.6

1. Let $\Gamma$ be the interval $[-1, 1]$ of the real axis, and let $g(t) \equiv 1$. Show that

$$\frac{1}{2\pi i} \int_{-1}^1 \frac{dt}{t-z} = \frac{1}{2\pi i} \log \frac{z-1}{z+1}$$

where z is restricted to the plane cut along $[-1, 1]$ and the logarithm is real if $z = x > 1$.

**2.** Let $\Gamma$ be the positively oriented unit circle and set $g(t) = t^{-1}$. Show that the corresponding function $G(z)$ is 0 or $-z^{-1}$ according as $z$ is inside $\Gamma$ or outside it.

**3.** Let $g(t) = t^{-2}$ and let $\Gamma$ be as in Problem 2. Investigate $G(z)$.

**4.** Let $g(t)$ be given on the real axis as a continuous function of $t$ such that

$$\int_{-\infty}^{\infty} |g(t)| \, dt < \infty.$$

Prove by an adaptation of the argument used above that

$$G(z) = \frac{1}{2\pi i} \int_{-\infty}^{\infty} \frac{g(t)}{t - z} \, dt$$

is holomorphic in the upper and lower half-planes. Prove also that $|\, yG(x + iy)\,|$, $y \neq 0$, is bounded and find an upper bound.

**5.** The function

$$G(z) = \lim_{n \to \infty} \frac{z^n - 1}{z^n + 1}$$

is defined for $|\, z\, | \neq 1$. Show that $G(z)$ is piecewise holomorphic. What are the two functions represented?

**6.** Show that the series

$$\sum \frac{z^n}{1 - z^{2n}},$$

where the summation extends over the powers of 2 beginning with the 0th power, converges to the sum $z(1 - z)^{-1}$ for $|\, z\, | < 1$ and to $(1 - z)^{-1}$ for $|\, z\, | > 1$.

$$\left( Hint: \qquad \frac{z}{1 - z^2} = \frac{z}{1 - z} - \frac{z^2}{1 - z^2}. \right)$$

**7.7. Analytic continuation: Schwarz's reflection principle.** The problems considered in the preceding section raise the question of when two functions defined in adjacent regions of the plane are analytic continuations of each other. We are now in a position to give a partial answer to this question.

THEOREM 7.7.1.   *Let $D_1$ and $D_2$ be simply-connected domains such that $D_1 \cap D_2 = \varnothing$, while $\bar{D}_1 \cap \bar{D}_2 \equiv \gamma$ is a simple rectifiable smooth arc and every point of Int $(\gamma)$ has a neighborhood in $D_1 \cup$ Int $(\gamma) \cup D_2 \equiv D$. Let $f_1(z)$ be holomorphic in $D_1$, continuous in $D_1 \cup \gamma$, and let $f_2(z)$ be holomorphic in $D_2$, continuous in $D_2 \cup \gamma$. Suppose that at every point $\zeta$ on*

$$\lim_{z \to \zeta} f_1(z) = \lim_{z \to \zeta} f_2(z) \equiv h(\zeta)$$

*for approach from $D_1$ in the first and from $D_2$ in the second limit. Then there exists a function $f(z)$, holomorphic in $D$, coinciding with $f_1(z)$ in $D_1$ and with $f_2(z)$ in $D_2$.*

*Proof.* By the statement that $\gamma$ is a smooth arc, we shall understand that $\gamma$ has a parametrization

$$t = g(s), \quad 0 \leq s \leq l(\gamma),$$

in terms of arc length $s$, and that $g(s)$, $g'(s)$, and $g''(s)$ are continuous. Then the "parallel curves" (see Figure 21)

$$\gamma_1(\varepsilon): \ t = g(s) + \varepsilon i\, g'(s),$$

and $$\varepsilon > 0,$$

$$\gamma_2(\varepsilon): \ t = g(s) - \varepsilon i\, g'(s),$$

will be rectifiable. For $\varepsilon$ sufficiently small, they will even be simple, since this is true for $\gamma$ itself. We suppose that the orientation of $\gamma$ is such that $\gamma_1(\varepsilon)$ lies in $D_1$ and $\gamma_2(\varepsilon)$ lies in $D_2$, except possibly for short pieces at the ends of the curves. Let $C$ be a "scroc" in $D$ such that $C$ intersects each of the curves $\gamma$, $\gamma_1(\varepsilon)$, and $\gamma_2(\varepsilon)$ in two points only. Let $C_1$ and $C_2$ be the subarcs of $C$ in $D_1$ and $D_2$ respectively. Since $C_1$ and $\gamma_1(\varepsilon)$ intersect in two points only, the two arcs of $C_1$ and of $\gamma_1(\varepsilon)$ between these points together form a "scroc" $C_1(\varepsilon)$ in $D_1$. For $z$ interior to this curve

$$f_1(z) = \frac{1}{2\pi i} \int_{C_1(\varepsilon)} \frac{f_1(t)}{t - z}\, dt.$$

Here we can let $\varepsilon \to 0$ and obtain

(7.7.1) $$f_1(z) = \frac{1}{2\pi i} \int_{C_1} \frac{f_1(t)}{t - z}\, dt + \frac{1}{2\pi i} \int_{\gamma_0} \frac{h(t)}{t - z}\, dt,$$

where $\gamma_0$ is the arc of $\gamma$ between the two intersections with $C$. This formula holds for any $z$ in $D_1 \cap C_i$, while for $z$ in $D_2 \cap C_i$ the sum of the integrals is 0 instead. Similarly we obtain

(7.7.2) $$f_2(z) = \frac{1}{2\pi i} \int_{C_2} \frac{f_2(t)}{t - z}\, dt - \frac{1}{2\pi i} \int_{\gamma_0} \frac{h(t)}{t - z}\, dt$$

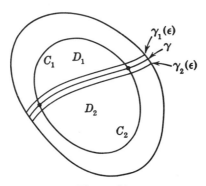

**Figure 21**

for $z$ in $D_2 \cap C_i$, while for $z$ in $D_1 \cap C_i$ the right-hand side is 0. We now define a function $g(t)$ on $C$ by

(7.7.3)
$$g(t) = \begin{cases} f_1(t), & t \text{ on } C_1, \\ f_2(t), & t \text{ on } C_2, \end{cases}$$

and set

(7.7.4)
$$f(z) = \frac{1}{2\pi i} \int_C \frac{g(t)}{t - z} dt.$$

By Theorem 7.6.1, $f(z)$ is holomorphic in $C_i$. Adding (7.7.1) and (7.7.2) gives

$$f(z) = f_1(z), \quad z \text{ in } D_1 \cap C_i, \quad f(z) = f_2(z), \quad z \text{ in } D_2 \cap C_i.$$

Thus, $f(z)$ has the desired property, and, since any point $z \in D$ can be made interior point of a suitably chosen contour $C$, the theorem is proved.

The first theorem dealing with questions of this nature was discovered by Schwarz[1] in 1869. It gives an actual method for analytic continuation.

THEOREM 7.7.2.   *Let $D_1$ be a simply-connected domain above (below) the real axis with an interval $(a, b)$ of the axis on its boundary such that every point of $(a, b)$ has a semi-circular neighborhood in $D_1$. Let $f(z)$ be holomorphic in $D_1$ and have real continuous boundary values on $(a, b)$ so that $f(z)$ is continuous in $D_1 \cup (a, b)$. Then $f(z)$ can be continued analytically across $(a, b)$ into the domain $D_2$, symmetric to $D_1$ with respect to the axis, by setting*

(7.7.5)
$$f(z) = \overline{f(\bar{z})}, \quad z \in D_2.$$

*Proof.*   (7.7.5) amounts to setting

(7.7.6)
$$f(x - iy) = U(x, y) - iV(x, y), \quad y > 0,$$

if

$$f(x + iy) = U(x, y) + iV(x, y).$$

If we give $x - iy$ increments $h$ and $ik$ and proceed as in Section 4.3, we see that the derivatives in the horizontal and the vertical directions are, respectively,

$$U_x(x, y) - iV_x(x, y) \quad \text{and} \quad V_y(x, y) + iU_y(x, y).$$

These expressions agree by virtue of the Cauchy-Riemann equations. We conclude that (7.7.6) defines a holomorphic function of $z$ in $D_2$. The boundary values on $(a, b)$ are $U(x, 0)$ in both cases. It follows that the conditions of Theorem 7.7.1 hold, and the theorem follows.

This result is usually referred to as Schwarz's *principle of symmetry* or as his *reflection principle*. It admits of considerable extensions by elementary

---

[1] Hermann Amandus Schwarz (1843–1921), a pupil of Weierstrass, and his successor in Berlin, contributed to the theory of minimal surfaces, conformal mapping, potential theory, hypergeometric functions, the Schwarzian derivative, interchange of order of differentiation, etc.

transformations. Thus linear fractional transformations on $z$ and $w$ give

COROLLARY.    *Let $D_1$ be a simply-connected domain interior (exterior) to the circle $c$: $|z - a| = r$ with an arc $\gamma$ of $c$ on its boundary such that every point of Int $(\gamma)$ has a "semi-circular" neighborhood in $D_1$. Let $f(z)$ be holomorphic in $D_1$ and continuous in $D_1 \cup \gamma$. Let the values of $f(z)$ on $\gamma$ fall on an arc $\Gamma$ of the circle $C$: $|w - b| = R$. Then $f(z)$ can be extended to the domain $D_2$ obtained by reflecting $D_1$ in $c$. If $z^*$ is the image of $z$ under the reflection:*

$$z^* - a = \frac{r^2}{\bar{z} - \bar{a}},$$

*we define*

(7.7.7)
$$f(z^*) - b = \frac{R^2}{\overline{f(z)} - \bar{b}}.$$

*The function so obtained is holomorphic in $D = D_1 \cup \gamma \cup D_2$.*

This case is reduced to that of Theorem 7.7.2 by fractional linear transformations mapping $c$ and $C$ onto the real axis. The corollary is exceedingly useful when it comes to conformal mapping problems in which the figures are bounded by straight lines and circles.

## EXERCISE 7.7

**1.** Why is the curve $t = g(s) + i\varepsilon\, g'(s)$ rectifiable when $t = g(s)$ is a smooth curve? Recall that $s$ is arc length, and that $g(s)$, $g'(s)$ and $g''(s)$ are continuous.

**2.** Show that the assumption that the equations

$$g(s_1) + i\varepsilon\, g'(s_1) = g(s_2) + i\varepsilon\, g'(s_2)$$

have solutions $s_1 \neq s_2$ for arbitrarily small values of $\varepsilon$ leads to a contradiction. $\left(\textit{Hint:} \text{ Note that } |\, g'(s)\,| = 1, \text{ and discuss the quotient } \dfrac{g'(s_1) - g'(s_2)}{g(s_1) - g(s_2)}\right).$

**3.** A function $f(z)$ is holomorphic in the strip $-\dfrac{\pi}{2} < x < \dfrac{\pi}{2}$ and continuous on the boundary, where it takes on real values only. It is real in the interval $\left(-\dfrac{\pi}{2}, \dfrac{\pi}{2}\right)$ and takes on conjugate imaginary values for conjugate imaginary values of $z$. Show that such a function can be continued analytically to the whole plane, and that it satisfies $f(\pi - z) = f(z)$, $f(z + 2\pi) = f(z)$. ($f(z)$ could be $\sin z$. Any other possibility?)

**4.** A function $f(z)$ is holomorphic in the square whose vertices are at $z = 0$, $1, 1 + i$, and $i$. It is real on the boundary and continuous, except at the vertices, where $f(z)$ becomes infinite. What functional equations will $f(z)$ satisfy, and how can they be used for analytic continuation? (This is an even, doubly periodic function.)

**7.8. The theorem of Morera.** Cauchy's theorem asserts that if $f(z)$ is holomorphic in a simply-connected domain $D$, then

(7.8.1)
$$\int_C f(z)\, dz = 0$$

for every "scroc" $C$ in $D$. This theorem has a converse, which was discovered by Giacinto Morera (1856–1909) in 1886.

THEOREM 7.8.1.   *Suppose that $f(z)$ is continuous in a simply-connected domain $D$ and suppose (7.8.1) holds whenever $C = \triangle$ is a triangle such that $\triangle^* \subset D$. Then $f(z)$ is holomorphic in $D$.*

*Proof.*   Suppose that $a \in D$ and let $D(a)$ be a domain starlike with respect to $a$ and contained in $D$. For $z \in D(a)$ we define

(7.8.2)
$$F(z, a) = \int_a^z f(t)\, dt,$$

where the integral is taken along the line segment $[a, z]$. If $\mid h \mid$ is small, then $z + h$ also belongs to $D(a)$ and

$$F(z + h, a) - F(z, a) = \int_a^{z+h} f(t)\, dt - \int_a^z f(t)\, dt$$
$$= \int_z^{z+h} f(t)\, dt,$$

since the triangle $[a, z, z + h, a]$ is in $D(a)$, and (7.8.1) holds for triangles. Hence

$$\frac{1}{h}[F(z + h, a) - F(z, a)] = \frac{1}{h}\int_z^{z+h} f(t)\, dt$$
$$= f(z) + \frac{1}{h}\int_z^{z+h}[f(t) - f(z)]\, dt.$$

By virtue of the continuity of $f(t)$, the second term in the last member tends to zero with $h$, that is,

$$\lim_{h \to 0} \frac{1}{h}[F(z + h, a) - F(z, a)] = f(z)$$

for every $z$ in $D(a)$. Hence, $F(z, a)$ is a holomorphic function of $z$ in $D(a)$. But by Theorem 7.5.2 this requires that its derivative, $f(z)$, is also holomorphic in $D(a)$. Here $a$ is arbitrary in $D$, whence it follows that $f(z)$ is holomorphic in the neighborhood of every point in $D$, and hence it is holomorphic in $D$.

COROLLARY.   *If $f(z)$ is holomorphic in a simply-connected domain $D$, then*

(7.8.3)
$$\int_a^z f(t)\, dt, \quad a, z \in D,$$

*is a holomorphic function of $z$ in $D$ whose derivative equals $f(z)$.*

It should be noted that if $D$ is not simply-connected, the integral is still locally holomorphic in $D$, but not necessarily single-valued. Examples are furnished by the integrals

(7.8.4) $$\int_1^z \frac{dt}{t} \quad \text{and} \quad \int_0^z \frac{dt}{t^2 + 1},$$

which represent the infinitely many-valued functions $\log z$ and arc tan $z$ respectively. In the first case, the integral is infinitely many-valued in the domain $D$ if $z$ can make circuits about the origin and still stay in $D$. The points $\pm i$ play a similar role in the second case.

## EXERCISE 7.8

**1.** $D$ is the complex plane punctured at $0$ and $\infty$. Give a complete discussion of the first integral under (7.8.4) for an arbitrary rectifiable path in $D$.

**2.** Same question for the second integral, $D$ now being the extended plane punctured at $z = \pm i$.

**7.9. The maximum principle.** This basic principle asserts that if $f(z)$ is holomorphic in a domain $D$ and continuous in $D \cup \partial D$, and if $M = M(f, \partial D)$ is the maximum of $|f(z)|$ on $\partial D$, then

(7.9.1) $$|f(z)| < M \quad \text{everywhere in } D,$$

unless $f(z)$ is a constant of absolute value $M$. At this juncture we shall prove only that $|f(z)| \leq M$ in $D$. The sharper result will be proved in Chapter 8. We give the proof in two steps.

THEOREM 7.9.1.    *If $f(z)$ is holomorphic in $D$, and if $C$ is a "scroc" in $D$ such that $C^* \subset D$, then for $z$ in $C_i$ we have*

(7.9.2) $$|f(z)| \leq M(f, C),$$

*where, as usual, $M(f, C)$ is the maximum of $|f(z)|$ on $C$.*

*Proof.*[1]    For $z \in C_i$ and any positive integer $k$, we have

(7.9.3) $$[f(z)]^k = \frac{1}{2\pi i} \int_C \frac{[f(t)]^k}{t - z} dt,$$

whence

$$|f(z)|^k \leq [2\pi d(z, C)]^{-1}[M(f, C)]^k,$$

where $d(z, C)$ is the distance of $z$ from $C$. Extracting the $k$th root and letting $k$ tend to infinity, we obtain (7.9.2).

---

[1] This argument is due to Edmund Landau (1877–1938), professor in Göttingen until 1933. Landau was a prolific writer and made profound contributions to function theory and analytic number theory. He introduced a telegram style in mathematics.

THEOREM 7.9.2.   [FIRST FORM OF THE MAXIMUM PRINCIPLE.]  *Suppose that $f(z)$ is holomorphic in a bounded domain $D$ and continuous in $D \cup \partial D$. Then for every $z$ in $D$*

$$(7.9.4) \qquad\qquad |f(z)| \leq M(f, \partial D),$$

*where $M(f, \partial D)$ is the maximum of $|f(z)|$ on $\partial D$.*

*Proof.*   If $\partial D$ is a simple smooth curve $C$, then (7.9.3) is still valid, and the conclusion follows from the proof of the preceding theorem.  For the general case we use an exhaustion argument.  We suppose that $D$ has the diameter $\delta$. Consider the subset of points of $D$ whose distance from $\partial D$ is $\geq 2^{-2}\delta$. This is a bounded closed non-void set, and by the Heine-Borel theorem we can cover it by a finite number of circles with centers in the set and radii equal to $2^{-3}\delta$. The union of the interiors of these circles is an open set $G_1$ where $G_1 \subset D$, $2^{-3}\delta \leq d(\partial D, G_1) \leq 2^{-2}\delta$, and $G_1$ contains every point of $D$ whose distance from $\partial D$ is $\geq 2^{-2}\delta$.  $G_1$ is bounded by a finite number of circular arcs.  We write $\partial G_1 = C_1$ and set

$$M(f, C_1) = M_1.$$

There exists at least one point $z_1$ on $C_1$ where $|f(z_1)| = M_1$.  $G_1$ need not be a domain, but it is the union of a finite number of disjoint domains, and applying Theorem 7.9.1 to each of the components, we see that $|f(z)| \leq M_1$ for $z \in G_1$.

This construction can now be repeated as often as we please;  it leads to a sequence of nested open sets $\{G_n\}$ with the following properties:  $G_n$ contains every point of $D$ whose distance from $\partial D$ is at least $2^{-n-1}\delta$ and, furthermore, $d(\partial D, G_n) \geq 2^{-n-2}\delta$.  Obviously,

$$G_1 \subset G_2 \subset G_3 \subset \cdots \subset G_n \subset D.$$

Every $G_n$ is bounded by a finite number of circular arcs, forming $\partial G_n \equiv C_n$.  If

$$M_n = M(f, C_n),$$

then $M_n \leq M_{n+1}$, since

$$(7.9.5) \qquad\qquad |f(z)| \leq M_n, \quad z \in G_n,$$

and there exists at least one point $z_n$ on $C_n$ where $|f(z_n)| = M_n$.  Since $|f(z)|$ is continuous in $D \cup \partial D$, the sequence $\{M_n\}$ is bounded and has a unique limit $M_0$.  The infinite sequence $\{z_n\}$ has at least one limit point $z_0$ which evidently lies on $\partial D$, and $|f(z_0)| = M_0$.  But this requires that $M_0 \leq M(f, \partial D)$ and implies (7.9.4).

If we examine the proof of this theorem, we realize that essentially the same argument can be used to prove a stronger form of the maximum principle.

**THEOREM 7.9.3.** [SECOND FORM OF THE MAXIMUM PRINCIPLE.] *Let $f(z)$ be holomorphic in a bounded domain $D$, and let $M$ be a positive number. Suppose that for every $\varepsilon > 0$ and every boundary point $z_0 \in \partial D$, there is a neighborhood $N(z_0, \varepsilon)$ of $z_0$ such that $|f(z)| < M + \varepsilon$ for all $z \in D \cap N(z_0, \varepsilon)$. Then, for every $z \in D$, we have that*

$$(7.9.6) \qquad\qquad |f(z)| \leqq M.$$

*Proof.* We can proceed as above. Here we do not know a priori that the sequence $\{M_n\}$ is bounded. But we still have a subsequence $\{z_{n_k}\}$ converging to a limit point $z_0$ on $\partial D$. For sufficiently large values of $k$, the point $z_{n_k}$ will belong to $N(z_0, \varepsilon)$, the neighborhood of $z_0$ in which $|f(z)| < M + \varepsilon$. It follows that $M_{n_k} < M + \varepsilon$, and, since the sequence $\{M_n\}$ is never decreasing, we must have

$$M_n < M + \varepsilon$$

for all $n$ and every $\varepsilon > 0$. This proves (7.9.6).

## EXERCISE 7.9

**1.** Find the maximum of the absolute value of the following functions for $|z| \leqq 1$:

    **a.** $z^2 - 1$;   **b.** $z(z-1)(z-2)$;   **c.** $\sin z$;   **d.** $\tan z$;   **e.** $(z^2 + 2)/(z^2 - 2)$.

**2.** Let $f(z) = U(x, y) + iV(x, y)$ be holomorphic in a domain $D$. Discuss maxima and minima of the function

$$[U(x, y)]^2 + [V(x, y)]^2,$$

and use the result to show that $|f(z)|$ can have no proper maximum in $D$ and no minimum other than 0.

**3.** Let $f(z)$ satisfy the assumptions of Theorem 7.9.1 and let there be a point $a \in C_i$ where $|f(a)| = M(f, C)$. Use the formula

$$k\,[f(a)]^{k-1}f'(a) = \frac{1}{2\pi i} \int_C \frac{[f(t)]^k}{(t-a)^2}\, dt$$

to prove that $f'(a) = 0$, and devise similar formulas to show that $f''(a)$ and all higher derivatives are 0. (This implies that $f(z)$ is a constant.)

**7.10. Uniformly convergent sequences of holomorphic functions.** In Section 4.2 we introduced the space $HB[D]$ of functions bounded and holomorphic in a given domain $D$. A norm was defined by

$$(7.10.1) \qquad\qquad \|f\| = \sup_{z \in D} |f(z)|.$$

A sequence of elements, $\{f_n\}$, of this space converges in the sense of the metric if

$$(7.10.2) \qquad \lim_{m,\,n\to\infty} \| f_m - f_n \| = 0.$$

Thus the sequence converges in $HB[D]$ if and only if the sequence of point functions $\{f_n(z)\}$ converges uniformly with respect to $z$ in $D$. The uniform convergence implies that $f(z) = \lim_{n\to\infty} f_n(z)$ is bounded and continuous in $D$. Does $f(z)$ have the property of being holomorphic in $D$, that is, does uniform convergence preserve holomorphy? Or, in other words, is the space $HB[D]$ complete in the metric defined by (7.10.1)? We are now able to answer these questions in the affirmative. We start by proving

**LEMMA 7.10.1.** *Let $C$ be a rectifiable arc, and let $\{f_n(t)\}$ be a sequence of functions, defined and continuous on $C$. Let $f_n(t)$ converge to a function $f(t)$ uniformly with respect to $t$ on $C$. Then*

$$(7.10.3) \qquad \lim_{n\to\infty} \int_C f_n(t)\,dt = \int_C f(t)\,dt.$$

*Proof.* Since $f(t)$ is continuous on $C$ by the uniform convergence, the right-hand side exists. Let $\varepsilon > 0$ be given, and let $N(\varepsilon)$ be such that for $n > N(\varepsilon)$

$$\sup_{t\in C} | f_n(t) - f(t) | \le \varepsilon.$$

Then

$$\left| \int_C f_n(t)\,dt - \int_C f(t)\,dt \right| = \left| \int_C [f_n(t) - f(t)]\,dt \right| \le \varepsilon l(C)$$

where $l(C)$ is the length of $C$. This implies (7.10.3).

We can now state and prove the basic

**THEOREM 7.10.1.** *Let the sequence $\{f_n(z)\}$ consist of functions holomorphic in a fixed domain $D$. Let $C$ be a "scroc" such that $C^* \subset D$, and suppose that the sequence $\{f_n(t)\}$ converges on $C$ uniformly with respect to $t$. Then there exists a function $f(z)$, holomorphic in $C_i$ and continuous in $C^*$, such that $f_n(z)$ converges to $f(z)$ uniformly with respect to $z$ in $C^*$. Moreover, if $S$ is any subset of $C_i$ having positive distance from $C$, and if $p$ is any positive integer, then the sequence $\{f_n^{(p)}(z)\}$ converges uniformly to $f^{(p)}(z)$ in $S$.*

*Proof.* Let $\varepsilon > 0$ be given and choose an integer $N(\varepsilon)$ such that

$$(7.10.4) \qquad \sup_{t\in C} | f_m(t) - f_n(t) | \le \varepsilon, \quad m,\, n > N(\varepsilon).$$

This we can do by virtue of the uniform convergence on $C$. But then the maximum principle shows that for such $m$ and $n$

$$\sup_{z\in C^*} | f_m(z) - f_n(z) | \le \varepsilon.$$

Since $\varepsilon$ is arbitrary, this implies that the sequence $\{f_n(z)\}$ converges uniformly in $C^*$ to a limit function $f(z)$ which is continuous in $C^*$. On the other hand, for any fixed $z$ in $C_i$ and $t$ on $C$ we have

$$\lim_{n \to \infty} \frac{f_n(t)}{t - z} = \frac{f(t)}{t - z}$$

uniformly in $t$. By Lemma 7.10.1 it follows that

$$f(z) = \lim_{n \to \infty} f_n(z) = \lim_{n \to \infty} \frac{1}{2\pi i} \int_C \frac{f_n(t)}{t - z}\, dt = \frac{1}{2\pi i} \int_C \frac{f(t)}{t - z}\, dt,$$

and by Theorem 7.6.1 the function in the last member is holomorphic in $C_i$. Thus $f(z)$ is holomorphic in $C_i$ as asserted.

We can now pass to the derivatives. By Theorem 7.6.1 we have

(7.10.5)
$$f^{(p)}(z) = \frac{p!}{2\pi i} \int_C \frac{f(t)}{(t - z)^{p+1}}\, dt,$$

and by Lemma 7.10.1 the right member equals

$$\lim_{n \to \infty} \frac{p!}{2\pi i} \int_C \frac{f_n(t)}{(t - z)^{p+1}}\, dt = \lim_{n \to \infty} f_n^{(p)}(z).$$

It remains to show that the convergence is uniform in any set $S$ whose distance from $C$ is positive. But if $m, n > N(\varepsilon)$ we have

$$|f_m^{(p)}(z) - f_n^{(p)}(z)| = \left| \frac{p!}{2\pi i} \int_C \frac{f_m(t) - f_n(t)}{(t - z)^{p+1}}\, dt \right| \leq p!(2\pi)^{-1}[d(z, C)]^{-p-1}l(C)\varepsilon,$$

where $d(z, C)$ is the distance of $z$ from $C$. Hence we have uniform convergence of the $p$th derivatives if $z$ is restricted to a subset of $C_i$ in which $d(z, C)$ has a positive lower bound. This completes the proof.

THEOREM 7.10.2. *$HB[D]$ is complete under the norm (7.10.1).*

*Proof.* Suppose that the sequence $\{f_n(z)\}$ belongs to $HB[D]$ and that (7.10.2) holds. The limit function $f(z)$ then belongs to $CB[D]$. But by Theorem 7.10.1, $f(z)$ is holomorphic in a neighborhood of any point of $D$ and, hence, holomorphic in $D$. In fact, if $a \in D$, then we can find a small circle $C$ with center at $a$ such that $C^* \subset D$, and Theorem 7.10.1 shows that $f(z)$ is holomorphic in $C_i$. Since $f(z)$ is bounded and holomorphic in $D$ it follows that $f(z) \in HB[D]$, and the theorem is proved.

We shall encounter many important applications of Theorem 7.10.1 in the next chapter. Here we observe only the following consequence:

THEOREM 7.10.3.    *Suppose that the functions*

$$\{w_n(z) \mid n = 0, 1, 2, \cdots\}$$

*are holomorphic in a fixed domain $D$, and suppose that the series*

(7.10.6)
$$\sum_{n=0}^{\infty} w_n(z) \equiv W(z)$$

*converges uniformly in any compact subset of $D$; then $W(z)$ is holomorphic in $D$, except possibly at $z = \infty$ if $\infty$ belongs to $D$. The series may be differentiated term by term as often as we please. The $p$-times-differentiated series converges to $W^{(p)}(z)$ uniformly on compact sets.*

*Proof.*    If $C$ is any "scroc" such that $C^* \subset D$, then the partial sums

$$W_n(z) = \sum_{k=0}^{n} w_k(z)$$

obey the conditions of Theorem 7.10.1, so that $W(z)$ is holomorphic in $C_i$. Thus $W(z)$ is holomorphic at every finite point of $D$. Term-by-term differentiation is also justified by Theorem 7.10.1.

The theorems of Section 5.5 are consequences of this theorem.

## EXERCISE 7.10

**1.** The sequence

$$f_n(z) = \sum_{k=1}^{\infty} \frac{nz^k}{nk^2 + 1}$$

converges uniformly for $|z| \leq 1$. What is the limit function? Show that the sequence $\{f_n{}'(z)\}$ does not converge at $z = 1$ and, hence, does not converge uniformly in any subset $S$ of the open unit disk having $z = 1$ as a limit point. Thus, the assertion concerning the derivatives in Theorem 7.10.1 cannot normally be sharpened.

**2.** Show that the Riemann zeta function $\zeta(z)$, defined by the Dirichlet series

$$\sum_{n=1}^{\infty} n^{-z}$$

when the latter converges, is holomorphic in the half-plane $\Re(z) > 1$.

**3.** Find a Dirichlet series representation of $\zeta^{(p)}(z)$ in $\Re(z) > 1$.

**4.** Show that the function defined by

$$\sum_{n=1}^{\infty} e^{-n} \sin nz$$

is holomorphic in the strip $-1 < \Im(z) < 1$. Find the sum of the series.

**5.** Show that the series $\sum_{n=1}^{\infty} \frac{1}{n} \arctan \frac{z}{n}$ defines a holomorphic function of $z$ in the plane cut along the imaginary axis from $z = i$ to $i\infty$ and from $-i$ to $-i\infty$. The arctangents are given their principal determinations. Find the first derived series.

**6.** A *factorial series* is an expansion of the form

$$\sum_{n=1}^{\infty} \frac{a_n n!}{z(z+1)(z+2)\cdots(z+n-1)}.$$

Show that if the series is absolutely convergent for a real positive value $a$, then the series converges absolutely and uniformly for $\Re(z) \geqq a$, and the sum is holomorphic for $\Re(z) > a$.

**7.** Take $a_n = 1$, and show that the corresponding series converges absolutely for $\Re(z) > 2$. (*Hint:* Use (5.1.18).)

**8.** In what region of the plane does the series

$$\sum_{n=1}^{\infty} 2^{-n}(z^n + z^{-n})$$

converge and represent a holomorphic function?

## COLLATERAL READING

As general references for this chapter consult

KNOPP, K. *Funktionentheorie*, Second Edition, Vol. I. Sammlung Göschen, Berlin, 1918. (An English translation of the fifth German edition, by Frederick Bagemihl, is published by Dover Publications, Inc., New York, under the title *Theory of Functions*, Part I.)

WATSON, G. N. *Complex Integration and Cauchy's Theorem*. Cambridge Tracts in Mathematics and Mathematical Physics, No. 15. Cambridge University Press, London, 1914.

In connection with Section 7.7 see

BEHNKE, H., and SOMMER, F. *Theorie der analytischen Funktionen einer komplexen Veränderlichen*, Chap. III, Section 2. Springer-Verlag, Berlin, 1955.

Further development of the subject matter of Section 7.10 is to be found in

LEAU, L. *Les Suites de Fonctions en Général* (*Domaine Complexe*). Mémorial des Sciences Mathématiques, Vol. 59. Gauthier-Villars, Paris, 1932.

# 8

# REPRESENTATION THEOREMS

**8.1. Taylor's series.** We recall Borel's remark to the effect that the analytical properties of Cauchy's integral reside in the Cauchy kernel. This kernel can be expanded in a Taylor's series about any point $z = a$. Using the geometric series, we obtain

$$\frac{1}{t-z} = \frac{1}{(t-a)-(z-a)} = \frac{1}{t-a} \cdot \frac{1}{1 - \dfrac{z-a}{t-a}}$$

(8.1.1)

$$= \frac{1}{t-a} + \frac{z-a}{(t-a)^2} + \frac{(z-a)^2}{(t-a)^3} + \cdots + \frac{(z-a)^n}{(t-a)^{n+1}} + \cdots.$$

This series converges for

$$|z-a| < |t-a|.$$

Moreover, if $|t-a| = R_1$ and $|z-a| \leqq \rho R_1$, where $\rho$ is a number between 0 and 1, the series converges uniformly with respect to both $z$ and $t$. This observation leads directly to Taylor's theorem in the form proved by Cauchy:

**THEOREM 8.1.1.** *Suppose that $f(z)$ is holomorphic in the circle $|z-a| < R$, then*

(8.1.2)
$$f(z) = \sum_{n=0}^{\infty} \frac{f^{(n)}(a)}{n!} (z-a)^n,$$

*where the series converges in the circle.*

*Proof.* Take any $R_1$ such that $0 < R_1 < R$. If $C_1$ is the circle $|t-a| = R_1$ with positive orientation, and if $|z-a| \leqq \rho R_1$, we have

$$f(z) = \frac{1}{2\pi i} \int_{C_1} \frac{f(t)}{t-z} \, dt.$$

Multiplication of the series (8.1.1) by $f(t)$, which is continuous and, hence, bounded on $C_1$, will not affect the uniform convergence. By Lemma 7.10.1, we may integrate term by term, and we obtain

(8.1.3)
$$f(z) = \sum_{n=0}^{\infty} (z-a)^n \frac{1}{2\pi i} \int_{C_1} \frac{f(t)}{(t-a)^{n+1}} \, dt.$$

By Theorem 7.5.3, then

$$\frac{1}{2\pi i} \int_{C_1} \frac{f(t)}{(t-a)^{n+1}} \, dt = \frac{f^{(n)}(a)}{n!},$$

so that formula (8.1.2) holds. The resulting series is absolutely and uniformly convergent for $|z-a| \leqq \rho R_1$. Now, $\rho$ and $R_1$ are arbitrary, subject to the conditions $0 < \rho < 1$, $0 < R_1 < R$. We conclude that the series converges for $|z-a| < R$, uniformly for $|z-a| \leqq \rho R$, and this is the desired result.

In Section 5.5 we saw that a power series defines a holomorphic function in its domain of convergence. We now observe that the converse is also true. This consequence of Theorem 8.1.1 is sufficiently important to merit restatement as

COROLLARY 1.    *A holomorphic function $f(z)$ may be represented by a power series in $(z-a)$ about any point $a$ in its domain of holomorphy. The series converges and represents the function in the largest circle with center $a$ whose interior belongs to the domain of holomorphy of $f(z)$.*

COROLLARY 2.    *On the circumference of the circle of convergence of a power series there is at least one point which does not belong to the domain of holomorphy of the function defined by the series.*

We saw in Section 5.7 that it may very well happen that no point of the circumference belongs to the domain of holomorphy. In this case the function defined by the series exists only in the circle.

THEOREM 8.1.2.    *If the point $a$ belongs to the domain of holomorphy of the function $f(z)$ and if the power series*

$$(8.1.4) \qquad\qquad \sum_{n=0}^{\infty} a_n(z-a)^n$$

*converges to $f(z)$ in some neighborhood of $a$, then*

$$(8.1.5) \qquad\qquad a_n = \frac{f^{(n)}(a)}{n!}.$$

*Proof.*    If (8.1.4) converges to $f(z)$ in some neighborhood of $z = a$, then Theorem 5.5.3, with $z$ replaced by $z - a$ and 0 by $a$, implies that the coefficients are given by (8.1.5).

Theorems 8.1.1 and 8.1.2 together assert that a holomorphic function has a unique power series development about any point in its domain of holomorphy.

COROLLARY.    *If*

$$\sum_{n=0}^{\infty} a_n(z-a)^n \equiv 0, \quad |z-a| < R,$$

*then all coefficients $a_n$ are zero.*

This is a great advantage of power series over some other expansions of the form

$$\sum_{n=0}^{\infty} a_n f_n(z),$$

where $\{f_n(z)\}$ is a given sequence of polynomials or entire functions. If such a series has a non-void region of uniform convergence, then it follows from Theorem 7.10.3 that it represents a holomorphic function there. If $a_n = 0$, $n = 0, 1, \cdots$, the series will represent the function 0; in general, however, there may exist representations of 0 for which not all $a_n$ vanish. An example is given by the class of *binomial series*; here the family $\{f_n(z)\}$ is given by

$$f_0(z) = 1, \quad f_n(z) = \frac{1}{n!} z(z-1) \cdots (z-n+1), \quad n > 0.$$

In this case, there are actually infinitely many ways in which 0 may be written as a convergent sum of the form $\Sigma a_n f_n(z)$. For example,

$$(8.1.6) \qquad 0 \equiv 1 + \sum_{n=1}^{\infty} (-1)^n \frac{1}{n!} z(z-1) \cdots (z-n+1)$$

is valid for $\Re(z) > 0$. On the other hand, such "null series" arise neither with power series, as we saw in the Corollary, nor in the related theory of Dirichlet series.

The Corollary to Theorem 8.1.2 asserts that a power series has zero coefficients if its sum is identically zero. The following theorem, due to Weierstrass, shows that the hypothesis may be weakened very much without changing the conclusion.

**THEOREM 8.1.3.** *If $f(z)$ is holomorphic in a domain $D$, if there exists a sequence of distinct points $\{z_n\}$, with at least one limit point in $D$, and if*

$$f(z_n) = 0, \quad n = 1, 2, 3, \cdots,$$

*then $f(z) \equiv 0$ everywhere in $D$.*

*Proof.* Without restricting the generality, we may assume that the sequence itself converges to a point $a$ in $D$, $z_n \to a$. Then there exists a circle $|z - a| < R$ in $D$ in which

$$(8.1.7) \qquad f(z) = \sum_{n=0}^{\infty} a_n (z-a)^n.$$

Here

$$a_0 = f(a) = \lim_{n \to \infty} f(z_n) = 0,$$

and the series has no constant term. Thus,

$$f(z) = (z-a) \sum_{n=0}^{\infty} a_{n+1}(z-a)^n \equiv (z-a) f_1(z).$$

But

$$a_1 = f'(a) = \lim_{n \to \infty} \frac{f(z_n) - f(a)}{z_n - a} = \lim_{n \to \infty} \frac{f(z_n)}{z_n - a} = 0,$$

since the numerator, $f(z_n)$, is 0 for all $n$. We may of course assume $z_n \neq a$ for all $n$. By complete induction one shows that $a_n = 0$ for each $n$. Thus, $f(z) \equiv 0$ in $|z - a| < R$. But this conclusion extends to all of $D$, for if $b$ is any point in $D$ outside the circle $|z - a| < R$, we can find a sequence of "intermediary" points $b_0 = a$, $b_1$, $b_2$, $\cdots$, $b_m = b$ with the following properties: Each point $b_k$ is center of a circle $|z - b_k| = R_k$, whose interior lies in $D$, and $|b_{k+1} - b_k| < R_k$. Form the Taylor series

$$\sum_{n=0}^{\infty} a_{n,k}(z - b_k)^n, \quad k = 0, 1, \cdots, m.$$

For $k = 0$, this is (8.1.7), so that $a_{n,0} = a_n = 0$ for all $n$. Now, the point $b_1$ lies in the circle of convergence of this series, whence it follows that

$$a_{n,1} = \frac{f^{(n)}(b_1)}{n!} = 0$$

for all $n$. But this implies that $f(z) \equiv 0$ also in the circle $|z - b_1| < R_1$, and the point $z = b_2$ belongs to this circle. After $m$ steps we arrive at $z = b_m = b$, and $f(b) = 0$. Thus $f(z) \equiv 0$ in $D$.

COROLLARY.  [THE IDENTITY THEOREM.] *Suppose that $f(z)$ and $g(z)$ are two functions holomorphic in the domains $D_1$ and $D_2$ respectively. Suppose that $D_1$ and $D_2$ intersect in a domain $D$ ($D = D_1 \cap D_2$) and there exists an infinite sequence of distinct points $\{z_n\}$ in $D$ having at least one limit point in $D$ such that*

(8.1.8)          $$f(z_n) = g(z_n), \quad n = 1, 2, 3, \cdots.$$

*Then $f(z) \equiv g(z)$ in $D$, and $g(z)$ is the analytic continuation of $f(z)$ in $D_2$, while $f(z)$ is the analytic continuation of $g(z)$ in $D_1$.*

   *Proof.*   The function

$$h(z) \equiv f(z) - g(z)$$

is holomorphic in $D$, and $h(z_n) = 0$ for all $n$. Thus, by Theorem 8.1.3, $h(z) \equiv 0$ in $D$, and $f(z) \equiv g(z)$ in $D$. We can then define a function

$$F(z) = \begin{cases} f(z), & z \in D_1, \\ g(z), & z \in D_2 \ominus D, \end{cases}$$

and $F(z)$ is clearly holomorphic in $D_1 \cup D_2$. This completes the proof.

Theorem 8.1.3 has a number of other interesting consequences. One of these is the following:

**THEOREM 8.1.4.** *Suppose that $f(z)$ is holomorphic in a circle $|z - a| < R$ and that $f(a) = 0$, while $f(z) \not\equiv 0$. Then there is an integer $n$ such that*

$$(8.1.9) \qquad f(a) = f'(a) = \cdots = f^{(n-1)}(a) = 0, \quad f^{(n)}(a) \neq 0.$$

*Moreover, there exists an $r$, $0 < r \leq R$, such that*

$$(8.1.10) \qquad f(z) \neq 0, \quad 0 < |z - a| < r.$$

*Proof.* We have $f(z) = \sum_{k=0}^{\infty} a_k(z - a)^k$, $|z - a| < R$. Now $a_0 = 0$, but not all $a_n = 0$. Hence there is a smallest integer $n$ such that $a_n \neq 0$. Since, by Theorem 8.1.3, $a$ cannot be a limit point of zeros, there is a disk of the form $|z - a| < r \leq R$ in which $z = a$ is the only zero of $f(z)$.

**DEFINITION 8.1.1.** *The function $f(z)$ has a zero of order (multiplicity) $n$ at $z = a$, if $f(z)$ is holomorphic in some neighborhood of $a$, and if there is a function $g(z)$ also holomorphic at $a$ such that $g(a) \neq 0$ and*

$$(8.1.11) \qquad f(z) = (z - a)^n g(z).$$

Theorem 8.1.4 asserts that a zero of $f(z)$ always has a certain finite order $n$.

Of course, a function may have infinitely many zeros in a domain $D$ in which it is holomorphic. However, unless $f(z) \equiv 0$, the limit points of these zeros are singularities of $f(z)$, and they must lie on the boundary of $D$. This is also true of the roots of the equation $f(z) = c$ for any $c$. Actually the frequency of the roots of such an equation in a neighborhood of a limit point is subject to severe limitations.

There is a certain principle of moderation in the theory of analytic functions which it is easy to sense but very hard to pin down and to express in exact terms. The reader may have noticed that orderly behavior has a way of propagating itself in this theory; thus, the existence of one derivative gave the existence of derivatives of all orders. A bound for the absolute value of $f(z)$ on the frontier of a domain of holomorphism is also a bound in the interior, and uniform convergence on the frontier implies uniform convergence in the interior. Theorem 8.1.3 can be regarded as another instance of the same phenomenon: equality in a set $\{z_n\}$ with a limit point in $D$ spreads to all of $D$. Later, we shall see that convergence of a sequence of holomorphic functions in such a point set implies convergence everywhere in the common domain of holomorphy and also the holomorphy of the limit, provided some condition of uniform boundedness is satisfied. In the next section we prove that an entire function either is a constant or is unbounded. Complex function theory has produced a large number of theorems of the following general type: A certain specified local behavior of $f(z)$ in a partial neighborhood of a possible singular point $a$ (such as boundedness or existence of a limit along a path or abundance of zeros) *either* has global implications (boundedness in the partial neighborhood, existence of the same limit for all paths, or the identical vanishing of the function), *or* the rate of growth of the maximum modulus of the function, for approach to $z = a$, must exceed certain specified bounds. All these theorems can be regarded as reflections of the principle of moderation: A function of modest rate of growth is allowed just so much freedom in its behavior *or else* $\cdots$

Let us return to subject matter of a more prosaic and more precise nature. The following theorem is frequently useful in the computation of Taylor's series:

THEOREM 8.1.5.    [WEIERSTRASS'S DOUBLE SERIES THEOREM.] *Suppose that*

$$(8.1.12) \qquad f(z) = \sum_{n=0}^{\infty} f_n(z),$$

*where each of the functions $f_n(z)$ is holomorphic in a circle $\mid z - a \mid < R$, and that the series converges uniformly for $\mid z - a \mid \leq \rho R$ for each $\rho$ with $0 < \rho < 1$. Moreover, suppose that*

$$(8.1.13) \qquad f_n(z) = \sum_{k=0}^{\infty} a_{k, n}(z - a)^k.$$

*Then each of the series*

$$(8.1.14) \qquad \sum_{n=0}^{\infty} a_{k, n} \equiv A_k, \quad k = 0, 1, 2, \cdots$$

*converges, and, for $\mid z - a \mid < R$,*

$$(8.1.15) \qquad f(z) = \sum_{k=0}^{\infty} A_k(z - a)^k.$$

*Proof.*    According to Theorem 7.10.3, the series (8.1.12) may be differentiated term by term $p$ times ($p = 1, 2, 3, \cdots$) and the resulting series converges uniformly to the $p$th derivative for $\mid z - a \mid \leq \rho R$. In particular we have

$$f^{(p)}(a) = \sum_{n=0}^{\infty} f_n^{(p)}(a) = p! \sum_{n=0}^{\infty} a_{p, n} = p! A_p.$$

This proves both (8.1.14) and (8.1.15). That the radius of convergence of the series (8.1.15) is at least $R$ follows from the fact that $f(z)$ is holomorphic at least in the circle $\mid z - a \mid < R$. This completes the proof. The name "Double Series Theorem" refers to the fact that the proof shows that

$$\sum_{n=0}^{\infty} \sum_{k=0}^{\infty} a_{k, n}(z - a)^k = \sum_{k=0}^{\infty} \left( \sum_{n=0}^{\infty} a_{k, n} \right)(z - a)^k$$

for $\mid z - a \mid < R$; that is, the double series on the left, which is convergent by rows, may also be summed by columns to the same sum.

## EXERCISE 8.1

**1.** Prove that the series (8.1.6) converges for $\Re(z) > 0$ with the aid of the remarks after Theorem 5.1.4, and justify that the sum is zero with the aid of Theorem 5.4.4 and formula (6.3.11).

**2.** Find the power series (Maclaurin series) of $(1 - z)^{\alpha}$.

**3.** Find the power series of $\cos \sqrt{z}$. Is this a two-valued function?

**4.** Find the power series of arc sin $z$. Note that the derivative is elementary.

**5.** Find the power series of arc tan $z$. Note that the derivative is elementary.

**6.** Find the power series of $[\text{arc tan } z]^2$.

**7.** Find the power series of $(1 + z) \log (1 + z) - z$. Note that the second derivative is elementary.

**8.** The function $\log (z + \sqrt{1 + z^2})$ is uniquely defined for $|z| < 1$ by the conventions $\sqrt{1} = 1$, $\log 1 = 0$. Find its power series. Note that the derivative is elementary.

**9.** Find the first five coefficients of the power series of $\exp \dfrac{1}{1 - z}$.

**10.** Same question for arc sin $(2 \sin z)$.

**11.** Expand the Riemann zeta function in powers of $(z - 2)$.

**8.2. The maximum modulus.** This section centers around the interplay of Taylor's series and the maximum modulus. We start with the so-called Cauchy's inequalities.

THEOREM 8.2.1.    *Suppose that $f(z)$ is holomorphic in $|z - a| < R$ and set*

$$(8.2.1) \qquad \max_{\theta} |f(a + re^{i\theta})| = M(r, a; f), \quad 0 < r < R.$$

*Then*

$$(8.2.2) \qquad |f^{(n)}(a)| \leq n! \, r^{-n} M(r, a; f).$$

*Proof.*    Apply the usual estimates of complex integrals to

$$(8.2.3) \qquad f^{(n)}(a) = \frac{n!}{2\pi i} \int_C \frac{f(t) \, dt}{(t - a)^{n+1}},$$

where the integral is taken along the circle $C$: $|t - a| = r$.

The estimates (8.2.2) show that the sequence $\{|f^{(n)}(a)|\}$ grows as $n!$, if we neglect terms of lower order. That this estimate cannot be essentially lowered follows from the elementary formula

$$\frac{d^n}{dz^n} \frac{1}{1-z} = \frac{n!}{(1-z)^{n+1}}.$$

This does not exclude, of course, that in special cases a much slower rate of growth can occur, as, for example, if $f(z)$ is an entire function.

This result has a bearing on functions of a real variable having derivatives of all orders. Such a function may or may not be holomorphic in some domain of the complex plane containing the interval of the real axis where it is originally defined. Formula (8.2.2) provides a crucial test: if at some point $x = a$ the derivatives grow faster than

$$n!A^n,$$

no matter how large $A$ is, then $f(x)$ certainly cannot be holomorphic at $x = a$. If the function passes this test, an examination of the remainder term in Taylor's formula is still necessary, for a function of a real variable may very well vanish, together with all its derivatives, at a point $x = a$ without vanishing identically. An example is given by the function $f(x) = \exp(-x^{-2})$, $x \neq 0$, $f(0) = 0$. Here $f^{(n)}(0) = 0$ for all $n$. As we have seen, such a behavior is not possible for holomorphic functions. Such a function must vanish identically if it vanishes together with all its derivatives at a single point. We note that the analytic function $\exp(-z^{-2})$ is not holomorphic at $z = 0$, and that it is not possible to assign any value to this function at $z = 0$.

The estimates (8.2.2) involve the arbitrary quantity $r$, where $0 < r < R$. In order to get as accurate an estimate as possible, one can consider the problem of finding an optimal value of $r$. If $f(z)$ is also bounded on $|z - a| = R$, we can of course choose $r = R$, and this will be the best possible choice. On the other hand, if $M(r, a; f) \to \infty$ as $r \to R$, one usually gets a fairly good estimate by taking

$$(8.2.4) \qquad\qquad r = \left(1 - \frac{1}{n}\right) R$$

in (8.2.2). It is not to be expected that the resulting estimates are very accurate. If $f(z)$ is an entire function, $R = \infty$, and (8.2.4) becomes meaningless, but the problem of minimizing the right member of (8.2.2) for a fixed $n$ still makes sense.

A striking and interesting application of Cauchy's inequalities is furnished by Theorem 8.2.2, generally known as the theorem of Liouville.[1]

---

[1] Joseph Liouville (1809–1882) was professor at the Collège de France in Paris, and the founder of the *Journal de Mathématiques Pures et Appliquées*, which started in 1835 and is still known as the *Journal de Liouville*. He founded the theory of boundary value problems for second-order linear differential equations and also the theory of transcendental numbers.

THEOREM 8.2.2.    *A bounded entire function is a constant.*

*Proof.*    Let

$$(8.2.5) \qquad\qquad f(z) = \sum_{n=0}^{\infty} a_n z^n$$

be an entire function, so that the series converges for all finite $z$. Set $M(r, 0; f)$ $= M(r; f)$. By assumption, there exists a positive number $M$ such that $M(r; f) \le M$ for all $r$. Formula (8.2.2) then gives

$$| a_n | \le r^{-n} M(r; f) \le M r^{-n}.$$

Here $r$ is arbitrary, and if $n > 0$, the last member tends to zero when $r \to \infty$. We conclude that

$$a_n = 0, \quad n \ge 1;$$

that is, $f(z) \equiv a_0$, a constant.

COROLLARY.    *Only constants are holomorphic in the extended plane.*

Compare the remarks after Theorem 4.2.1.

This theorem shows that the maximum modulus of an entire function, which is never decreasing by the maximum principle, must tend to infinity with $r$ unless $f(z)$ is a constant. We can go farther, as is shown by

THEOREM 8.2.3.    *If $f(z)$ is an entire function, and if*

$$(8.2.6) \qquad\qquad M(r; f) \le M r^{\alpha},$$

*for fixed positive values of $M$ and $\alpha$ and for a sequence of values of $r$ tending to infinity, then $f(z)$ is a polynomial in $z$ of degree not exceeding $\alpha$.*

*Proof.*    For the values of $r$ in question we have

$$| a_n | \le M r^{\alpha-n},$$

and this tends to zero with $1/r$, provided $n > \alpha$. Hence

$$a_n = 0 \quad \text{for} \quad n > \alpha,$$

and the result follows. Note that $\alpha$ need not be an integer.

We conclude from this theorem that *the maximum modulus $M(r; f)$ of a transcendental entire function tends to infinity faster than any fixed power of $r$.*

We return for a moment to formula (8.2.3), where we set $n = 0$ and $t = a + re^{i\theta}$, obtaining

THEOREM 8.2.4.    *The value of a holomorphic function at a point $z = a$ is the mean value, in the sense of the integral calculus, of its values on any circle with center at $a$ and radius less than the distance from $a$ to the boundary of the domain of holomorphy:*

$$(8.2.7) \qquad\qquad f(a) = \frac{1}{2\pi} \int_0^{2\pi} f(a + re^{i\theta}) \, d\theta.$$

Setting $f(z) = U(z) + iV(z)$, we see that

(8.2.8)
$$U(a) = \frac{1}{2\pi} \int_0^{2\pi} U(a + re^{i\theta}) \, d\theta,$$

and we can also replace $U$ by $V$. Here $U(z)$ is a real-valued harmonic function in the domain $D$ where $f(z)$ is holomorphic. Conversely, if $U(z)$ is harmonic in the circle $|z - a| < R$, then it follows from the discussion in Section 4.4 that there exists a conjugate function $V(z)$, also harmonic in the circle, such that $U(z) + iV(z)$ is holomorphic in the circle. From this we conclude that the mean-value property (8.2.8) *holds for arbitrary harmonic functions.* Actually it is characteristic for such functions.

COROLLARY.    *The absolute value of $f(z)$ satisfies the inequality*

(8.2.9)
$$|f(a)| \leqq \frac{1}{2\pi} \int_0^{2\pi} |f(a + re^{i\theta})| \, d\theta.$$

DEFINITION 8.2.1.    *A continuous real-valued function $F(z)$ of the complex variable $z$ is said to be subharmonic in the domain $D$ if for each $a$ in $D$ and for all small values of $r$*

(8.2.10)
$$F(a) \leqq \frac{1}{2\pi} \int_0^{2\pi} F(a + re^{i\theta}) \, d\theta.$$

In this sense, *the absolute value of a holomorphic function is subharmonic in $D$.* The theory of subharmonic functions, introduced by F. Riesz in 1922, has become an important adjunct to the theory of analytic functions.

We note that (8.2.9) implies that

(8.2.11)
$$|f(a)| \leqq M(r, a; f),$$

as required by the maximum principle. The time has now come to elaborate this principle. The following result is due to Weierstrass:

THEOREM 8.2.5.    [THIRD FORM OF THE MAXIMUM PRINCIPLE.]    *Let $f(z)$ be holomorphic in $D$ and not a constant. Let $a \in D$. Then in any neighborhood of $z = a$ there exist points $z_1$ where*

(8.2.12)
$$|f(z_1)| > |f(a)|,$$

*and, if $f(a) \neq 0$, also points $z_2$ where*

(8.2.13)
$$|f(z_2)| < |f(a)|.$$

*Proof.*    We assume $f(a) \neq 0$ and use the Taylor series

$$f(z) = a_0 + a_n(z - a)^n + \sum_{k=n+1}^{\infty} a_k(z - a)^k.$$

Here $a_0 = f(a) \neq 0$ by assumption and, since $f(z)$ is not a constant, there is at least one other coefficient which is different from zero. We assume that $a_n$ is the first such coefficient. We choose arg $(z - a)$ subject to the condition

(8.2.14)          $\arg a_n + n \arg (z - a) \equiv \arg a_0 \pmod{2\pi}$.

This gives $n$ possible directions, forming equal angles of opening $2\pi/n$. For each such choice we have

$$| a_0 + a_n(z - a)^n | = | a_0 | + | a_n(z - a)^n | > | a_0 |.$$

In order to prove the possibility of satisfying (8.2.12), it suffices to show that $| z - a |$ can be made so small that

$$\left| \sum_{k=n+1}^{\infty} a_k(z - a)^k \right| < \tfrac{1}{2} | a_n | | z - a |^n,$$

or, what amounts to the same thing, that

(8.2.15)          $$\left| \sum_{m=1}^{\infty} a_{n+m}(z - a)^m \right| < \tfrac{1}{2} | a_n |.$$

The left side of (8.2.15) tends to zero with $z - a$, and $a_n \neq 0$, by assumption. We conclude that there exists an $r > 0$ such that, for all $z$ with $| z - a | < r$, the inequality (8.2.15) holds. For any such $z$, provided that (8.2.14) holds, we have consequently

$$\begin{aligned}
|f(z)| &\geq | a_0 + a_n(z - a)^n | - \left| \sum_{k=n+1}^{\infty} a_k(z - a)^k \right| \\
&> | a_0 | + | a_n(z - a)^n | - \tfrac{1}{2} | a_n(z - a)^n | \\
&= | a_0 | + \tfrac{1}{2} | a_n(z - a)^n | > | a_0 | = |f(a)|.
\end{aligned}$$

Hence any such $z$ will satisfy (8.2.12).

To obtain points satisfying (8.2.13), we change the choice of arg $(z - a)$; we replace (8.2.14) by

(8.2.16)          $\arg a_n + n \arg (z - a) \equiv \arg a_0 + \pi \pmod{2\pi}$.

Again we have $n$ directions at our disposal; the new directions bisect the angles between the old ones. We have now

$$| a_0 + a_n(z - a)^n | = | a_0 | - | a_n(z - a)^n |$$

provided the right member is $\geq 0$. We choose $| z - a |$ subject to (8.2.15) and arg $(z - a)$ subject to (8.2.16), and we find that

$$\begin{aligned}
|f(z)| &\leq | a_0 + a_n(z - a)^n | + \left| \sum_{k=n+1}^{\infty} a_k(z - a)^k \right| \\
&< | a_0 | - | a_n(z - a)^n | + \tfrac{1}{2} | a_n(z - a)^n | \\
&= | a_0 | - \tfrac{1}{2} | a_n(z - a)^n | < | a_0 | = |f(a)|.
\end{aligned}$$

This completes the proof.

Corollary to the First Form of the Maximum Principle. *If $f(z)$ is holomorphic in the domain $D$ and continuous in $D \cup \partial D$, and if $M$ is the maximum of $|f(z)|$ on $\partial D$, then either*

$$(8.2.17) \qquad\qquad |f(z)| < M$$

*everywhere in $D$ or $f(z)$ is a constant of absolute value $M$.*

It is instructive to consider the surface

$$(8.2.18) \qquad\qquad S: \; u = |f(x + iy)|^2$$

in three-dimensional Euclidean space.[1] It is defined over any domain $D$ in which the function $f(z)$ is holomorphic. Theorem 8.2.5 shows that $S$ has no relative maxima in $D$, and minima occur only at the zeros of $f(z)$. At each point of $S$ where $f(a) \neq 0$, there is always a direction leading to a higher elevation. In general, the opposite direction will be that of steepest descent. At a place where $f'(a) = 0$ but $f''(a) \neq 0$, the surface has a saddle-point, a mountain pass in Jensen's language. Here there are two opposite directions of steepest descent and the perpendicular directions are those of steepest ascent. At a point where $f'(z)$ has a zero of multiplicity $(n - 1)$, there is a complicated mountain pass with $n$ valleys and $n$ separating ridges.

As an application of the maximum principle or, more precisely in this case, of the minimum principle, we shall prove the so-called fundamental theorem of algebra, which asserts that every algebraic equation over the complex field has a root in this field. In our terminology this becomes

Theorem 8.2.6. *Every polynomial of degree $\geq 1$ in $z$ has a zero.*

*Proof.* Suppose we are given the polynomial

$$P(z) = z^n + c_1 z^{n-1} + c_2 z^{n-2} + \cdots + c_n.$$

It is required to show the existence of a point $z = z_0$ where $P(z_0) = 0$. In other words, we want to show that the corresponding surface

$$u = |P(x + iy)|^2$$

has a lowest point where then, according to what we have just seen, $P(z)$ must assume the value zero. Now, on a closed disk $|z| \leq R$ the continuous function $|P(z)|^2$ assumes its minimum value, but this could, possibly, always take place on the boundary, in which case the minimum would not have to be zero. In order to disprove this possibility, it suffices to show that if $R$ is big enough, the minimum value on $|z| = R$ exceeds some value taken on in the interior, say the value at the origin. To show this, let us set

$$c = \max_k |c_k|.$$

---

[1] The Danish telephone engineer and mathematician J. L. W. V. Jensen (1859–1925) called attention to these surfaces, which he called "analytical landscapes." When it rains in such a landscape, the rain water collects in little lakes around the zeros of the function.

Then for $|z| > 1$,

$$|P(z)| \geq |z|^n - \left|\sum_{k=1}^{n} c_k z^{n-k}\right| \geq |z|^n - c\sum_{k=1}^{n} |z|^{n-k}$$

$$= |z|^n - c\frac{|z|^n - 1}{|z| - 1} > |z|^n \left\{1 - \frac{c}{|z| - 1}\right\}.$$

This shows that

(8.2.19)                    $P(z) \neq 0 \quad \text{for} \quad |z| > 1 + c.$

Further,

$$|P(z)| > \tfrac{1}{2}|z|^n \quad \text{if} \quad |z| > 1 + 2c.$$

Now

$$\tfrac{1}{2}(1 + 2c)^n > nc \geq c \geq |c_n| = |P(0)|.$$

Thus, we see that the minimum value of $|P(z)|$ on the disk $|z| \leq 1 + 2c$ is actually taken on in the interior of the disk. The minimum value must then be zero, and the theorem is proved.

Suppose that the minimum value is reached at $z = z_1$. Since

$$P(z) = (z - z_1)Q(z) + P(z_1) = (z - z_1)Q(z),$$

where $Q(z)$ is a polynomial of degree $n - 1$, as shown by long division, we can apply the theorem also to $Q(z)$. Thus, if $n > 1$, $Q(z)$ also has a zero, say $z = z_2$, where it is possible that $z_2 = z_1$. Proceeding in this manner, we obtain the factorization

(8.2.20)                    $$P(z) = \prod_{k=1}^{n} (z - z_k).$$

The factorization is obviously unique, except for the order of the factors.

### EXERCISE 8.2

**1.** What estimates may be obtained for the coefficients of the Maclaurin series of $(1 - z)^{-\alpha}$, $\alpha > 0$, by using (8.2.2) and (8.2.4)?

**2.** Same question for $\exp[-(1 - z)^{-1}]$.

**3.** Same question for $\sin z$, minimizing the right member of (8.2.2).

**4.** Let the entire function $f(z)$ satisfy $M(r; f) < M \exp(r^\alpha)$ where $M$ and $\alpha$ are positive constants. What estimates can be obtained for the Maclaurin coefficients of $f(z)$?

**5.** Prove that $U(z)$ cannot have a proper maximum or minimum in a domain where it is harmonic. (*Hint:* Use (8.2.8) or the methods of the calculus.)

**6.** Verify the formula

$$M_2(r; f) \equiv \left\{\frac{1}{2\pi}\int_0^{2\pi} |f(re^{i\theta})|^2 \, d\theta\right\}^{\frac{1}{2}} = \left\{\sum_{n=0}^{\infty} |a_n|^2 r^{2n}\right\}^{\frac{1}{2}}$$

for the quadratic mean value of a holomorphic function. Here $f(z) = \sum_0^\infty a_n z^n$ and $r < R$, the radius of convergence of the series.

**7.** Show that if $f(z)$ is an entire transcendental function, then $M_2(r; f)$ grows faster than any power of $r$. State and prove the analogues of Theorems 8.2.2 and 8.2.3 replacing $M(r; f)$ by $M_2(r; f)$.

**8.** Find the angle which the normal to the surface (8.2.18) makes with the vertical.

**9.** How would you derive the maximum principle from the interior mapping theorem? See the Corollary of Theorem 4.6.1.

**8.3. The Laurent expansion.** Taylor's series is an excellent local representation of a function which is holomorphic in a simply-connected domain $D$. If $D$ is not simply-connected, we need other expansions for the study of the local behavior of the function. If $D$ contains an *annulus*, that is, a ring-shaped region bounded by two concentric circles, the Laurent expansion meets the needs.[1]

THEOREM 8.3.1.    *If $f(z)$ is holomorphic in the annulus*

$$0 \leqq R_1 < |z - a| < R_2 \leqq \infty,$$

*then*

(8.3.1)      $$f(z) = \sum_{n=-\infty}^{\infty} a_n (z - a)^n, \quad a_n = \frac{1}{2\pi i} \int_C \frac{f(t)}{(t - a)^{n+1}} \, dt,$$

*where $C$: $|t - a| = r$, $R_1 < r < R_2$, and the series is absolutely convergent in the annulus.*

*Proof.*    Again Cauchy's integral forms a suitable point of attack. Let us introduce numbers $R_k$ such that

$$R_1 < R_3 < R_5 < R_6 < R_4 < R_2.$$

Then, for $R_5 \leqq |z - a| \leqq R_6$, and $C_1$ and $C_2$ defined by

$$C_1 : |t - a| = R_3, \quad C_2 : |t - a| = R_4,$$

we have, by Theorem 7.4.3,

(8.3.2)      $$f(z) = \frac{1}{2\pi i} \int_{C_2} \frac{f(t)}{t - z} \, dt - \frac{1}{2\pi i} \int_{C_1} \frac{f(t)}{t - z} \, dt.$$

The first of these integrals defines two functions, one, $f_2(z)$, for $z$ inside $C_2$, and

---

[1] Pierre Alphonse Laurent (1813–1854), a French military engineer (*chef de bataillon de génie* at the time of his death), discovered this theorem in 1843. It was known to Weierstrass in 1841.

another one for $z$ outside $C_2$. We are not concerned with this latter function. Similarly for the second integral; here we are interested in the function $f_1(z)$ defined for $z$ exterior to $C_1$.

We can expand $f_2(z)$ in ascending powers of $(z - a)$, using the series (8.1.1). Thus,

$$(8.3.3) \qquad f_2(z) = \sum_{n=0}^{\infty} (z - a)^n \frac{1}{2\pi i} \int_{C_2} \frac{f(t)}{(t - a)^{n+1}} \, dt.$$

In order to represent $f_1(z)$, we use the fact that the Cauchy kernel also admits of the expansion

$$(8.3.4) \qquad -\frac{1}{t - z} = \frac{1}{z - a - (t - a)} = \frac{1}{(z - a)\left(1 - \dfrac{t - a}{z - a}\right)}$$

$$= \frac{1}{z - a} + \frac{t - a}{(z - a)^2} + \frac{(t - a)^2}{(z - a)^3} + \cdots + \frac{(t - a)^n}{(z - a)^{n+1}} + \cdots,$$

which converges absolutely and uniformly with respect to $z$ and $t$ for

$$|t - a| = R_3, \quad |z - a| \geqq R_5.$$

We multiply this series by $f(t)$ and integrate termwise with respect to $t$ along $C_1$, obtaining

$$(8.3.5) \qquad -f_1(z) = \sum_{n=1}^{\infty} (z - a)^{-n} \frac{1}{2\pi i} \int_{C_1} (t - a)^{n-1} f(t) \, dt.$$

We now observe that the integrals in (8.3.3) and (8.3.5) are independent of the path of integration, as long as it lies in the annulus $R_1 < |t - a| < R_2$, and provided arg $(t - z)$ increases by $2\pi$ when the path is described once. But by (8.3.2), $f(z) = f_2(z) - f_1(z)$ so that (8.3.1) follows from addition of (8.3.3) and (8.3.5).

The proof shows that the resulting Laurent expansion converges absolutely and uniformly with respect to $z$ in $R_5 \leqq |z - a| \leqq R_6$. Here $R_5$ and $R_6$ are arbitrary, subject to the condition $R_1 < R_5 < R_6 < R_2$. It follows that the series converges absolutely in $R_1 < |z - a| < R_2$. This completes the proof.

The Laurent expansion associated with a given function and a given annulus is unique: *there is one and only one series* (8.3.1) *which converges and represents the function* $f(z)$ *in* $R_1 < |z - a| < R_2$. This does not imply, however, that there is at most one expansion in powers of $(z - a)$ associated with a given function. As a matter of fact, there may be infinitely many such expansions, because the point $z = a$ may be the center of infinitely many distinct annuli in each of which $f(z)$ is holomorphic and admits a Laurent expansion.  A case in point is given by the function cot $z$, which is holomorphic in each of the annuli

$$k\pi < |z| < (k+1)\pi, \quad k = 0, 1, 2, \cdots.$$

One of the main applications of the Laurent expansion is to the study of the singular points of $f(z)$.  In the hypothesis of Theorem 8.3.1 we had that $0 \leq R_1 < R_2 \leq \infty$, but the case where strict inequality holds throughout is of little interest for the study of singularities.  True, there are singular points on each of the circles $|z| = R_1$ and $|z| = R_2$, but the information about these singularities is usually well hidden in the sequence $\{a_n\}$.  We have a different situation if either $R_1 = 0$ or $R_2 = \infty$.  In the first case the series gives information about the character of the point $z = a$ as a singularity of $f(z)$; in the second case the information refers to the point at infinity.  We shall discuss these cases in some detail in the next section.

## EXERCISE 8.3

**1.** Expand the function

$$\frac{1}{z(z-1)(z-2)}$$

in powers of $z$ valid for  **(a)** $0 < |z| < 1$;  **(b)** $1 < |z| < 2$;  **(c)** $2 < |z|$.  It may help to use partial fractions.

**2.** Same question for

$$\frac{z^2 + 1}{z^3 - 3z^2 + 2z}.$$

**3.** Expand the preceding function in powers of $z + 1$ valid for $|z + 1| > 3$.

**\*4.** Prove the expansion

$$\exp\left[\frac{a}{2}\left(z - \frac{1}{z}\right)\right] = J_0(a) + \sum_{n=1}^{\infty} [z^n + (-z)^{-n}]J_n(a)$$

where $J_n(a)$ is the Bessel function

$$J_n(a) = \sum_{k=0}^{\infty} \frac{(-1)^k a^{n+2k}}{2^{n+2k}k!(n+k)!}.$$

**8.4. Isolated singularities.**  If $R_1 = 0$, the point $z = a$ is an isolated singularity of $f(z)$.  Similarly, if $R_2 = \infty$, then $z = \infty$ is an isolated singularity. We consider the two cases separately and start with:

I. $R_1 = 0$.  In this case the series converges in a neighborhood of $z = a$, normally with the exception of this point.  The negative powers of $(z - a)$

present in the expansion determine the character of the singularity. There are three distinct cases.

Case Ia. *No negative powers.* Here

$$f(z) = f_2(z) = \sum_{n=0}^{\infty} a_n(z-a)^n$$

for $0 < |z - a| < R_2$. But as $z \to a$, $f(z) \to a_0$, and we need merely define $f(a) = a_0$ to obtain a function holomorphic also at $z = a$. We say that $f(z)$ has a *removable singularity* at $z = a$. The point $z = a$ is not actually singular, it was only suspected to be a singular point on what turned out to be insufficient evidence.

Case Ib. *A finite number of negative powers.* Suppose that

$$a_{-n} = 0, \quad n > m, \quad \text{but} \quad a_{-m} \neq 0$$

so that

$$(8.4.1) \quad f(z) = \frac{a_{-m}}{(z-a)^m} + \frac{a_{-m+1}}{(z-a)^{m-1}} + \cdots + \frac{a_{-1}}{z-a} + \sum_{n=0}^{\infty} a_n(z-a)^n.$$

The point $z = a$ is now a *pole of order* (or *multiplicity*) $m$, and the polynomial of degree $m$ in $(z-a)^{-1}$ which precedes the infinite series is called the *principal part* of $f(z)$ at $z = a$. It is the part which becomes infinite when $z \to a$.

We note that

$$(8.4.2) \qquad (z-a)^m f(z) = \sum_{k=0}^{\infty} a_{-m+k}(z-a)^k, \quad 0 < |z-a| < R_2,$$

has a removable singularity at $z = a$, and

$$(8.4.3) \qquad \lim_{z \to a} (z-a)^m f(z) = a_{-m} \neq 0.$$

This gives an alternate method of defining a pole and its order.

DEFINITION 8.4.1. *A function $f(z)$, holomorphic in $0 < |z - a| < R_2$, has a pole at $z = a$ if there exists an integer $p$ such that $(z - a)^p f(z)$ has a removable singularity at $z = a$. The least admissible value of $p$ is the order of the pole.*

Set

$$M(r, a; f) = \text{Max}_{\theta} |f(a + re^{i\theta})|, \quad 0 < r < R_2.$$

Formula (8.4.3) shows that if $f(z)$ has a pole of order $m$ at $z = a$, then

$$(8.4.4) \qquad \lim_{r \to 0} r^m M(r, a; f) = |a_{-m}| \neq 0.$$

The converse is also true in the following sense:

THEOREM 8.4.1.   *If $f(z)$ is holomorphic in $0 < |z - a| < R_2$, and if there exist positive numbers $M$ and $\alpha$ such that*

$$M(r, a; f) \leq Mr^{-\alpha}$$

*for all small values of $r$, then $z = a$ is either a removable singularity of $f(z)$ or a pole of order not exceeding $\alpha$.*

*Proof.*   We proceed as in the proof of Theorem 8.2.3. In the Laurent series (8.3.1) we have

$$|a_{-n}| = \left| \frac{1}{2\pi i} \int_C f(t)(t - a)^{n-1} \, dt \right|$$

$$\leq M(r, a; f)\, r^n \leq Mr^{n-\alpha} \to 0, \quad n > \alpha,$$

as $r \to 0$. Thus, the singularity is at most a pole of order $\leq \alpha$.

But formula (8.4.3) really shows much more: it is not merely the maximum modulus which grows as $r^{-m}$, but all the values of the function are large. In view of this fact we add $z = a$ to the domain of definition of our function, and we set $f(a) = \infty$. This is analogous to the conventions made in Section 2.5, where we introduced the point at infinity and assigned the value $\infty$ to the function

$$\frac{1}{z}$$

at $z = 0$. In our present terminology, this function has a simple pole at $z = 0$.

Case Ic.   *Infinitely many negative powers.*   The principal part of $f(z)$ at $z = a$ is still given by

$$(8.4.5) \qquad\qquad -f_1(z) = \sum_{n=1}^{\infty} a_{-n}(z - a)^{-n},$$

but this is now a transcendental entire function of $(z - a)^{-1}$ and no longer a polynomial as in the case of a pole. We now say that $f(z)$ has an *essential singularity* at $z = a$. Theorem 8.4.1 shows that

$$(8.4.6) \qquad\qquad \lim_{r \to 0} r^n M(r, a; f) = \infty$$

for every $n$. But it is not merely the maximum modulus that has a different behavior; the character of the function in the neighborhood of an essential singular point is completely different from that associated with a pole. At a pole, $f(z)$ assumes a definite value, $\infty$, and in a small neighborhood of the pole the function is uniformly large.

To see what may happen at an essential singularity, let us consider the function (see Problem 4, Exercise 6.1)

$$e^{1/z},$$

which has an essential singularity at $z = 0$. We cannot assign any value to the function at $z = 0$; in fact, it takes on every value $c$, $c \neq 0$, in any neighborhood of $z = 0$. It has no limit for approach along the imaginary axis; it becomes infinite if $z \to 0$ through positive real values, but it tends to zero if $z \to 0$ through negative real values.

Actually, this strange behavior is typical for an essential singularity. In 1880, Émile Picard proved that in *the neighborhood of an essential singular point $z = a$ of $f(z)$, supposed holomorphic in $0 < |z - a| < R_2$, each equation*

$$(8.4.7) \qquad\qquad\qquad f(z) = c$$

*has infinitely many roots with at most one exceptional value $c$.* We are not in a position to prove this deep theorem here, but we shall prove the much weaker but still suggestive

**THEOREM 8.4.2.**  *If $f(z)$ is holomorphic in $0 < |z - a| < R_2$ and has an essential singularity at $z = a$, and if $c$ is any complex number, then the function*

$$(8.4.8) \qquad\qquad\qquad g_c(z) \equiv \frac{1}{f(z) - c}$$

*cannot stay bounded in any neighborhood of $z = a$, no matter how small.*

*Proof.*   If the equation (8.4.7) has infinitely many roots clustering at $z = a$, there is nothing to prove. We may therefore assume that the equation has no roots in $0 < |z - a| < \rho \leq R_2$. Thus, $g_c(z)$ is holomorphic in this domain. If $M(r, a; g_c)$ were bounded, Theorem 8.4.1 would show that $z = a$ must be a removable singularity of $g_c(z)$. Since

$$f(z) = c + \frac{1}{g_c(z)} \, ,$$

the point $z = a$ would then have to be a removable singularity of $f(z)$ or a pole, according as $g_c(a) \neq 0$ or $= 0$. This contradicts the assumption that $z = a$ is an essential singular point of $f(z)$. We conclude that either the equation (8.4.7) has infinitely many roots in any neighborhood of $z = a$ or else $M(r, a; g_c)$ cannot remain bounded as $r \to 0$.

II. $R_2 = \infty$.   The series converges outside the circle $|z - a| = R_1$, but normally not at infinity. Here it is the positive powers of $(z - a)$ which decide the character of the singular point at infinity. There are again three subcases.

Case IIa.  *No positive powers.*   In this case

$$f(z) = -f_1(z) = \sum_{n=0}^{\infty} a_{-n}(z - a)^{-n}$$

for $R_1 < |z - a| < \infty$. Here $z = \infty$ is a removable singularity, and we define $f(\infty) = a_0$ in order to obtain a function holomorphic at $z = \infty$.

Case IIb. *A finite number of positive powers.* If

$$a_n = 0, \quad n > m, \quad \text{but } a_m \neq 0,$$

we have

$$(8.4.9) \qquad f(z) = a_m(z - a)^m + \cdots + a_1(z - a) + \sum_{n=0}^{\infty} a_{-n}(z - a)^{-n}.$$

Now the point $z = \infty$ is a pole of order $m$, and the polynomial in $(z - a)$ is the principal part of $f(z)$ at the pole. Strictly speaking, we should, however, subtract the value of the polynomial at $z = 0$ in order to get the principal part. See further remarks below.

We note that

$$(z - a)^{-m} f(z)$$

has a removable singularity at $z = \infty$ and tends to a limit different from 0 when $z \rightarrow \infty$. The maximum modulus satisfies

$$(8.4.10) \qquad \lim_{r \to \infty} r^{-m} M(r, a; f) = |a_m|,$$

and if, conversely,

$$(8.4.11) \qquad M(r, a; f) \leq Mr^{\alpha},$$

then $z = \infty$ is either a removable singularity or a pole of order not exceeding $\alpha$. This is proved by the same argument as that used in proving Theorem 8.4.1.

Case IIc. *Infinitely many positive powers.* The principal part of $f(z)$ is now a transcendental entire function, namely $f_1(0) - f_1(z)$, where

$$- f_1(z) = \sum_{n=1}^{\infty} a_n(z - a)^n.$$

The point at infinity is an essential singularity of $f(z)$. We have

$$(8.4.12) \qquad \lim_{r \to \infty} r^{-n} M(r, a; f) = + \infty$$

for every $n$. The distribution of the values of $f(z)$ for large values of $z$ is governed by the theorem of Picard and, of course, by the analogue of Theorem 8.4.2. Thus, the function $g_c(z)$ of (8.4.8) either has infinitely many poles outside of any fixed circle or else is unbounded in such a manner that (8.4.12) holds with $f$ replaced by $g_c$. For the case that $f(z)$ is an entire function, this result is known as the Casorati-Weierstrass theorem.[1]

If $f(z)$ is holomorphic in $R_1 < |z - a| < \infty$, then $f(z)$ is also holomorphic in $R_1 + |a| < |z| < \infty$ (Why?), and it admits a Laurent expansion

$$(8.4.13) \qquad f(z) = \sum_{n=-\infty}^{\infty} c_n z^n,$$

---

[1] Felice Casorati (1835–1890) was a professor at Pavia and one of the founders of the school of analysts in Italy. The revival of Italian mathematics is conventionally dated from the journey which Enrico Betti (1823–1892), Francesco Brioschi (1824–1897), and Casorati made to France and Germany in 1858.

absolutely convergent in this domain. The preceding discussion applies also to this case, and we have the three subcases according as there are no, a finite number of, or infinitely many positive powers of $z$. It should be observed, however, that (8.4.13) belongs to the same subcase as the original expansion (8.3.1). The nature of the singularity of $f(z)$ at $z = \infty$ does not depend upon the particular representation of $f(z)$ but is an intrinsic property of the function. This is intuitively obvious. The following double series argument gives a formal proof and justifies the remarks made earlier concerning the principal part at infinity.

For $|z| > |a|$ we have the series

$$(8.4.14) \qquad (z - a)^{-n} = z^{-n} \sum_{k=0}^{\infty} \frac{1}{k!} n(n + 1) \cdots (n + k - 1)\left(\frac{a}{z}\right)^k.$$

Therefore

$$\sum_{n=1}^{\infty} a_{-n}(z - a)^{-n} = \sum_{n=1}^{\infty} a_{-n} z^{-n} \sum_{k=0}^{\infty} \frac{1}{k!} n(n + 1) \cdots (n + k - 1)\left(\frac{a}{z}\right)^k.$$

This double series is absolutely convergent if

$$|z| > R_1 + |a|,$$

and for such values of $z$ it may be rearranged according to descending powers of $z$. Thus, with obvious notation,

$$(8.4.15) \qquad \sum_{n=1}^{\infty} a_{-n}(z - a)^{-n} = \sum_{k=1}^{\infty} \gamma_{-k} z^{-k},$$

where the equality is valid in the common domain of convergence, and at least for $|z| > R_1 + |a|$. Similarly we have

$$(8.4.16) \qquad \sum_{n=1}^{\infty} a_n(z - a)^n = \sum_{n=1}^{\infty} a_n \sum_{k=0}^{n} \binom{n}{k}(-a)^{n-k} z^k \equiv \sum_{k=0}^{\infty} \gamma_k z^k,$$

where

$$\gamma_0 = \sum_{n=0}^{\infty} a_n(-a)^n.$$

Combining these two expansions, we see that

$$f(z) = a_0 + \sum_{k=-\infty}^{\infty} \gamma_k z^k,$$

convergent for $R_1 + |a| < |z| < \infty$. This series must coincide with the Laurent series (8.4.13) which is valid in the same annulus. Thus,

$$c_k = \gamma_k, \quad k \neq 0, \quad c_0 = a_0 + \gamma_0.$$

The series (8.4.16) shows that

$$a_n = 0, \quad n > m \geqq 0, \quad a_m \neq 0,$$

implies that

$$c_n = 0, \quad n > m \geqq 0, \quad c_m = a_m \neq 0,$$

and vice versa. Thus, the principal parts of the two expansions (8.3.1) and (8.4.13) correspond and are identical entire functions, provided we subtract the term $\gamma_0 = -f_1(0)$ from the left side of (8.4.16).

The function $f(z)$ may also be expanded in powers of $(z - b)$, where $b$ is any given fixed complex number, and the resulting expansion converges at least for $R_1 + |b - a| < |z - b| < \infty$.

## EXERCISE 8.4

**1.** Determine the poles and their principal parts for the function in Problem 1 of Exercise 8.3. Use partial fractions.

**2.** Same question for the function in Problem 2 of Exercise 8.3.

**3.** Determine the poles of $[e^z - 1]^{-1}$ and find the principal part at $z = 0$.

**4.** $P(z)$ and $Q(z)$ are two given polynomials, and $z = a$ is a zero of $Q(z)$. Find the principal part of $P(z)/Q(z)$ at $z = a$, if the zero is (**a**) simple; (**b**) double. It is assumed that $P(z)$ and $Q(z)$ have no zeros in common.

**5.** Where are the poles of $\cot z$? Determine multiplicities and principal parts.

**6.** Verify that the double series used in deriving (8.4.15) is absolutely convergent for $|z| > R_1 + |a|$.

**8.5. Meromorphic functions.** We start with

DEFINITION 8.5.1.    *A function $f(z)$ is said to be meromorphic in a domain D if it has no singularities other than poles in D.*

If $f(z)$ is called *meromorphic* (without specification of the domain) it is understood that $f(z)$ is meromorphic in the whole (finite or extended) plane.

Any rational function is meromorphic in the extended plane. The functions $\tan z$, $\cot z$, and $\sec z$ are meromorphic in the finite plane, and the point at infinity is a limit point of poles.

THEOREM 8.5.1.    *A function which is meromorphic in the extended plane is necessarily a rational function.*

*Proof.*    Since $f(z)$ has no singularities other than poles, it can have only a finite number of poles. Suppose that these are located at the points $z_1, z_2, \cdots, z_n$, and $\infty$. Let the principal part of $f(z)$ at $z = z_k$ be

$$R_k(z) = \sum_{j=1}^{\mu_k} a_{j,k}(z - z_k)^{-j}, \quad k = 1, 2, \cdots, n,$$

while the principal part at $\infty$ is given by

$$R_\infty(z) = \sum_{j=1}^{\mu} a_j z^j.$$

We form the rational function

$$R(z) = \sum_{k=1}^{n} R_k(z) + R_\infty(z),$$

and the difference

$$D(z) = f(z) - R(z).$$

The only possible singularities of $D(z)$ are poles located at the poles of $f(z)$. But in the neighborhood of $z = z_k$ the principal parts of $f(z)$ and of $R(z)$ are the same, namely $R_k(z)$. It follows that $D(z)$ is holomorphic at each of the points $z_k$, and the same argument applies at $z = \infty$. Thus, $D(z)$ is holomorphic in the extended plane, and, by the Corollary to Liouville's theorem 8.2.2, $D(z) = a_0$, a constant. Thus

(8.5.1) $$f(z) = \sum_{k=1}^{n} R_k(z) + a_0 + R_\infty(z).$$

This completes the proof.

Several comments are in order. Formula (8.5.1) is the representation of a rational function by partial fractions, which is familiar to the student from the calculus. The terms $a_0 + R_\infty(z)$ are present if and only if in the representation of $f(z)$ as the quotient of two polynomials, $f(z) = P(z)/Q(z)$, the degree of $P(z)$ at least equals that of $Q(z)$. In this case one uses long division and obtains the quotient $a_0 + R_\infty(z)$ plus a remainder. The $R_k(z)$ correspond to the zeros of $Q(z)$: the order of the pole of $f(z)$ at $z = z_k$ equals the order of the zero of $Q(z)$ at this point.

Further, we notice that in the present case the sum of the principal parts at the finite poles differs from the function $f(z)$ at most by a polynomial in $z$. If we want to get the most general meromorphic function having the same poles and principal parts, we need only add an entire function, which may be completely arbitrary. This raises the question whether a similar procedure works for meromorphic functions having infinitely many poles. If the poles and the corresponding principal parts are known, how can we represent the function with the aid of the principal parts? The expansion

(8.5.2) $$\frac{\pi^2}{\sin^2 \pi z} = \sum_{-\infty}^{\infty} \frac{1}{(z-n)^2},$$

which will be proved in Chapter 9, gives an example of a convergent series of principal parts. But normally, the series of principal parts of a given meromorphic function does not converge; it is then necessary to modify this series so that it becomes convergent, but this must be accomplished without introducing new, undesirable singularities. This problem was solved by the Swedish mathematician Mittag-Leffler in 1877.[1] The somewhat simpler solution given below is due to Weierstrass.

THEOREM 8.5.2.    *Given a sequence of distinct complex numbers $\{z_n\}$ having no limit point in the finite plane, and given a sequence of polynomials $\{P_n(z)\}$ which may be distinct or equal. Given $P_k(0) = 0$. Then there exists a meromorphic function $f(z)$ having the principal parts*

$$P_n\left(\frac{1}{z - z_n}\right) \quad at \quad z = z_n, \quad n = 0, 1, 2, \cdots.$$

*Assuming $z_0 = 0$, such a function may be given the form*

$$(8.5.3) \qquad f(z) = P_0\left(\frac{1}{z}\right) + \sum_{n=1}^{\infty} \left\{ P_n\left(\frac{1}{z - z_n}\right) - \sum_{k=0}^{k_n} A_{n,k} z^k \right\},$$

*where*

$$(8.5.4) \qquad P_n\left(\frac{1}{z - z_n}\right) \equiv \sum_{k=0}^{\infty} A_{n,k} z^k \quad for \quad |z| < |z_n|.$$

*It is possible to choose the sequence of integers $\{k_n\}$ in such a manner that the series (8.5.3) converges absolutely and uniformly on compact sets not containing any of the poles. The most general meromorphic function with these poles and principal parts is obtained by adding an arbitrary entire function to $f(z)$.*

*Proof.*    We may assume $|z_n| \leq |z_{n+1}|$ for all $n$. The expansion (8.5.4) is easily obtained by using the following analogue of formula (8.4.14), namely

$$(8.5.5) \quad (z - a)^{-m} = (-a)^{-m} \sum_{k=0}^{\infty} \frac{1}{k!} m(m + 1) \cdots (m + k - 1)\left(\frac{z}{a}\right)^k,$$

which converges for $|z| < |a|$. If now

$$P_n(z) = \sum_{m=1}^{m_n} B_{n,m} z^m,$$

then, for $|z| < |z_n|$,

$$P_n\left(\frac{1}{z - z_n}\right) = \sum_{m=1}^{m_n} B_{n,m}(-z_n)^{-m} \sum_{k=0}^{\infty} \frac{1}{k!} m(m + 1) \cdots (m + k - 1)\left(\frac{z}{z_n}\right)^k,$$

so that

$$(8.5.6) \qquad A_{n,k} = \frac{1}{k!} \sum_{m=1}^{m_n} (-1)^m B_{n,m} m(m + 1) \cdots (m + k - 1)(z_n)^{-k-m}.$$

---

[1] Gösta Mittag-Leffler (1846–1927), a pupil of Weierstrass, held professorships in Helsingfors (Helsinki) and, from 1881, in Stockholm, where he founded the international periodical *Acta Mathematica*. He was a colorful person and a great organizer. In addition to the theorem which carries his name, Mittag-Leffler introduced starlike regions of holomorphy and obtained numerous representations of the function in such a region. He started the Swedish school of analysts and had many prominent pupils.

We now choose arbitrarily a convergent series with positive terms; for instance,

$$\sum_{n=1}^{\infty} 2^{-n} = 1.$$

Since the series (8.5.4) is uniformly convergent for $|z| \leq \frac{1}{2}|z_n|$, we may choose an integer $k_n$ so large that, if

$$(8.5.7) \qquad\qquad f_n(z) \equiv P_n\left(\frac{1}{z - z_n}\right) - \sum_{k=0}^{k_n} A_{n,k} z^k,$$

then

$$(8.5.8) \qquad\qquad |f_n(z)| \leq 2^{-n} \quad \text{for} \quad |z| \leq \frac{1}{2}|z_n|.$$

With this choice of $k_n$, (8.5.3) has become the series

$$(8.5.9) \qquad\qquad f(z) \equiv \sum_{n=0}^{\infty} f_n(z), \quad f_0(z) = P_0\left(\frac{1}{z}\right).$$

Suppose now that $K$ is a compact set containing none of the poles $\{z_n\}$. We can then choose two positive numbers $\varepsilon$ and $R$ in such a manner that $K$ is contained in the set $S$ defined by the inequalities

$$|z| \leq R, \quad |z - z_n| \geq \varepsilon, \quad n = 0, 1, 2, \cdots.$$

We shall prove uniform convergence of (8.5.9) on $S$.

For this purpose we choose an integer $N$ such that

$$|z_n| > 2R, \quad n > N.$$

In the series

$$\sum_{n=N+1}^{\infty} f_n(z),$$

the terms satisfy (8.5.8), since $|z| \leq R < \frac{1}{2}|z_n|$ for $n > N$. By the Weierstrass M-test, the series is uniformly convergent for $|z| \leq R$. By Theorem 7.10.3, therefore, the sum of the series is a holomorphic function of $z$ in $|z| < R$. On the other hand, the finite sum

$$\sum_{n=0}^{N} f_n(z)$$

is a rational function, which is bounded in $S$, since $S$ is bounded away from every point $z_n$. Thus, the series (8.5.9) converges uniformly on $S$, and its sum is holomorphic there. It is clear that $f(z)$ has no singularities other than poles in $|z| < R$, and if $|z_n| < R$, then the principal part of $f(z)$ at $z = z_n$ is

$$P_n\left(\frac{1}{z - z_n}\right).$$

Thus $f(z)$ has all the desired properties. If $F(z)$ is any other meromorphic function having the same poles and principal parts, then $F(z) - f(z)$ is holomorphic in the finite plane and is consequently an entire function. This completes the proof.

It is clear that there is much arbitrariness in this construction and that in any given case a simpler result may be obtainable. There is one case which occurs sufficiently often to justify separate formulation.

THEOREM 8.5.3.    *If $P_n(z) = z$ for every $n$, and if there exists an integer $k$ such that*

$$(8.5.10) \qquad \sum_{n=1}^{\infty} |\, z_n\,|^{-k} = \infty, \quad \sum_{n=1}^{\infty} |\, z_n\,|^{-k-1} < \infty, \quad k \geqq 0,$$

*then we can take*

$$(8.5.11) \quad f(z) = \frac{1}{z} + \sum_{n=1}^{\infty} \left\{ \frac{1}{z - z_n} + \frac{1}{z_n} + \frac{z}{z_n^{\,2}} + \cdots + \frac{z^{k-1}}{z_n^{\,k}} \right\}, \quad if \; k > 0,$$

*but*

$$(8.5.12) \qquad\qquad f(z) = \sum_{n=0}^{\infty} \frac{1}{z - z_n}, \quad if \; k = 0.$$

*Proof.*    Here the last formula is immediate. In this case the sum of the principal parts is convergent and defines the desired meromorphic function. If $k > 0$, we observe that

$$\frac{1}{z - a} + \frac{1}{a} + \frac{z}{a^2} + \cdots + \frac{z^{k-1}}{a^k} = \frac{z^k}{a^k(z - a)}.$$

Hence, if the $n$th term of (8.5.11) is denoted by $f_n(z)$, and if $|\, z\,| \leqq R < \frac{1}{2} |\, z_n\,|$, we have for $n \geqq 1$

$$|\, f_n(z)\,| \leqq 2 \frac{|\, z\,|^k}{|\, z_n\,|^{k+1}}.$$

It follows that (8.5.11) converges uniformly on any compact set not containing any of the poles. Further, $f(z)$ is a meromorphic function having the principal part $(z - z_n)^{-1}$ at the pole $z = z_n$, $n = 0, 1, 2, \cdots$. This completes the proof.

## EXERCISE 8.5

**1.** Find the simplest meromorphic function having poles at the integers, if $P_n(z) = z$. Same question for $P_n(z) = nz$.

**2.** Determine the integer $k$ of Theorem 8.5.3, if $z_n$ equals

   **a.** $n^p$ ($p > 0$, fixed);   **b.** $\exp\,[\log n]^2$;   **c.** $\exp\, n$.

**3.** A meromorphic function is to have poles at all the Gaussian integers, that is, all numbers of the form $m + ni$, where $m$, $n$ are ordinary integers. Construct such a function, if $P_{m,n}(z)$ is (a) $z$; (b) $z^2$; (c) $z^3$.

**4.** What is the analogue of Theorem 8.5.3, if $P_n(z) = z^2$ for all $n$?

**5.** Suppose that $P_n(z) = z$ for all $n$, but $z_n = \log(n+1)$. Here there is no *fixed* integer $k$ which satisfies (8.5.10). What would be a reasonable choice of $k_n$ in Theorem 8.5.2 in this case?

**8.6. Infinite products.** Since we shall need infinite products in the discussion of entire functions, we give a brief account of the properties used later. An infinite product is an expression of the form

$$(8.6.1) \qquad \Pi \equiv \prod_{k=1}^{\infty} (1 + w_k),$$

where $\{w_k\}$ is a sequence of complex numbers. If

$$(8.6.2) \qquad \Pi_n = \prod_{k=1}^{n} (1 + w_k),$$

we might be tempted to say that the product (8.6.1) is convergent, provided $\lim \Pi_n$ exists as a finite quantity. This turns out to be too wide a definition. For one thing, if one of the factors is zero, $\Pi_n$ has the limit zero, regardless of the other factors. It is not desirable to exclude zero factors, since this case arises in the applications that we have in mind, but we shall be justified in requiring that only a finite number of factors are zero, and that, after these factors have been omitted, the remaining product converges. Further, it is desirable to require that a convergent product shall have the value zero if and only if one of its factors is zero. These considerations lead to

DEFINITION 8.6.1.   *Let*

$$(8.6.3) \qquad \Pi_{m,n} = \prod_{k=m}^{n} (1 + w_k), \quad 1 \leqq m < n.$$

*The product (8.6.1) is said to be convergent if (1) there exists an $m_0 \geqq 1$ such that $w_k \neq -1$ for $k \geqq m_0$, and (2) for $m \geqq m_0$*

$$\lim_{n \to \infty} \Pi_{m,n}$$

*exists as a finite quantity $\neq 0$.*

Cauchy's convergence principle shows that $\Pi$ will converge if and only if given any $\varepsilon > 0$ there exists an $N = N(\varepsilon)$ such that

$$(8.6.4) \qquad |\, \Pi_{n+1,\, n+p} - 1 \,| < \varepsilon, \quad N < n, \quad 1 \leqq p.$$

We observe in passing that the definition of convergence of the product is such that the product converges if and only if the series

$$\sum_{n=m_0}^{\infty} \log (1 + w_n)$$

converges. We shall not prove this assertion and it will not be used in the following.

THEOREM 8.6.1. *If $p_k \geqq 0$ for all $k$, then the infinite product*

$$\prod_{k=1}^{\infty}(1 + p_k)$$

*converges if and only if the series $\Sigma\, p_k$ converges.*

*Proof.* Let us write

$$P_n = \sum_{k=1}^{n} p_k, \quad Q_n = \prod_{k=1}^{n} (1 + p_k).$$

Then

(8.6.5) $$1 + P_n \leqq Q_n \leqq e^{P_n}.$$

The first half of this inequality is obvious; the second one follows from the well-known inequality

$$1 + x \leqq e^x,$$

valid for all real $x$. Since $\{P_n\}$ and $\{Q_n\}$ are increasing sequences, their equi-convergence follows from formula (8.6.5).

DEFINITION 8.6.2. *An infinite product is absolutely convergent if the product $\Pi(1 + |\, w_k\,|)$ converges.*

THEOREM 8.6.2. *An absolutely convergent product is convergent.*

*Proof.* Let $\Pi_{m,n}$ be defined by (8.6.3), and write $Q_{m,n}$ for the corresponding expression with $w_k$ replaced by $|\, w_k\,|$. Since $\Pi$ is absolutely convergent, Theorem 8.6.1 shows that $w_k \to 0$. Hence, we can find an $m_0$ such that $w_k \neq -1$ for $k \geqq m_0$. A simple calculation gives the inequality

(8.6.6) $$|\, \Pi_{n+1,\, n+p} - 1\,| \leqq Q_{n+1,\, n+p} - 1, \quad 1 \leqq p.$$

The absolute convergence of the product implies that the right member of this inequality can be made $< \varepsilon$ for $N < n$. Hence the left member is also $< \varepsilon$, and by (8.6.4) the product converges.

For the applications we must consider products with variable factors. This brings up the question of uniform convergence and its possible implications.

**DEFINITION 8.6.3.**    *Let $\{w_k(z)\}$ be a sequence of functions defined and continuous in a domain $D$ of the complex plane.  Let $S$ be a compact subset of $D$.  The infinite product*

$$(8.6.7) \qquad \prod_{k=1}^{\infty} [1 + w_k(z)] \equiv \Pi(z)$$

*is said to be uniformly convergent on $S$ if (1) there exists a fixed integer $m_0$ such that $w_k(z) \neq -1$ for $k \geq m_0$ and every $z$ in $S$, and (2) for any $\varepsilon > 0$ there exists a fixed $N = N(\varepsilon)$ such that for $n > N$, and every $z$ in $S$ we have*

$$(8.6.8) \qquad |\Pi_{n+1,\, n+p}(z) - 1| < \varepsilon, \quad 1 \leq p.$$

It is perhaps desirable to comment on the roles of these conditions.  If (1) is satisfied, then the bounded continuous function $\Pi_{m_0}(z)$ may possibly vanish somewhere in $S$, but no product $\Pi_{m,\, n}(z)$ with $m > m_0$ can vanish, and condition (2) ensures that the sequence $\{\Pi_{m,\, n}(z) \mid n > m\}$ converges uniformly on $S$.  It follows that the sequence $\{\Pi_n(z)\}$ also converges uniformly on $S$ since it differs from $\{\Pi_{m,\, n}(z)\}$ by the bounded continuous factor $\Pi_{m-1}(z)$.

From this definition and Theorems 4.1.2 and 7.10.1 we obtain

**THEOREM 8.6.3.**    *Let $w_n(z)$ be continuous (holomorphic) in $D$ for each $n$, and let the product (8.6.7) converge uniformly on every compact subset of $D$; then $\Pi(z)$ is continuous (holomorphic) in $D$.*

**COROLLARY.**    *If $w_n(z)$ is holomorphic in $D$ for each $n$, and if the series*

$$(8.6.9) \qquad \sum_{n=1}^{\infty} |w_n(z)|$$

*converges uniformly on compact subsets of $D$, then $\Pi(z)$ is holomorphic in $D$.*

## EXERCISE 8.6

**1.**  For what values of $p$, $p > 0$, does the product with $w_n = n^{-p}$ converge?

**2.**  Evaluate the product $\prod_{n=2}^{\infty} (1 - n^{-2})$.

**3.**  If $w_n = i/n$, show that the product $\Pi(1 + w_n)$ diverges while the product $\Pi \mid 1 + w_n \mid$ converges.

**4.**  Discuss the convergence of the product with $w_n = (-1)^{n-1}/n$.

**5.**  Show that the product

$$\pi z \prod_{n=1}^{\infty} \left(1 - \frac{z^2}{n^2}\right)$$

defines an entire function.  (Its value is $\sin(\pi z)$.)

**6.** Show that the product with

$$w_n(z) = \frac{1}{n} \sin \frac{z}{n}$$

defines an entire function.

**7.** Show that the product with

$$w_n(z) = \frac{1}{n} \tan \frac{z}{n}$$

defines a meromorphic function. Where are the poles?

**8.** If $p$ runs through the primes, show that $\Pi(1 - p^{-z})$ is absolutely convergent for $\Re(z) > 1$. Show that its value is $[\zeta(z)]^{-1}$.

**9.** Show that the infinite product

$$B(z) = \prod_{n=1}^{\infty} \frac{n^2 z - 1}{n^2 z + 1}$$

is absolutely convergent for $\Re(z) \geqq 0$, $z \neq 0$. Consider also $\Re(z) < 0$.

**10.** Show that $B(z)$ does not converge uniformly on $0 < |z| < 1$, $\Re(z) > 0$.

**11.** Show that $|B(z)| \leqq 1$ for $\Re(z) > 0$.

**12.** We have $B(n^{-2}) = 0, n = 1, 2, 3, \cdots$. Any contradiction with Theorem 8.1.3?

**8.7. Entire functions.** We recall that a function holomorphic in the finite plane is called an *entire function*. Such a function may be *rational*, that is, a polynomial, or it may be *transcendental*. In the latter case, the function is represented by a power series involving infinitely many powers of $z$ and converging for every finite value of $z$. The functions

$$e^z, \quad \sin z, \quad \text{and} \quad \cos z$$

are examples of elementary transcendental entire functions. Using power series, we can extend our collection of entire functions ad lib. Thus,

(8.7.1) $$\sum_{n=0}^{\infty} \frac{z^n}{(n!)^p} \, (p > 0), \quad \sum_{n=2}^{\infty} \frac{z^n}{(\log n)^n}, \quad \sum_{n=0}^{\infty} e^{-n^2} z^n$$

are examples of entire functions. In the preceding section we found that certain infinite products also define entire functions. Examples are

(8.7.2) $$\prod_{n=1}^{\infty} \left(1 - \frac{z}{n^p}\right), \quad p > 1, \quad \text{and} \quad \prod_{n=1}^{\infty} (1 - e^{-n}z),$$

and also the functions of Problems 5 and 6, Exercise 8.6.

A polynomial is determined, up to a constant factor, by its zeros:

$$(8.7.3) \qquad\qquad P(z) = A \prod_{k=1}^{n} (z - z_k).$$

We note that a constant is the only rational entire function having no zeros. These observations raise two questions: (1) What is the most general transcendental entire function having no zeros? (2) Is it possible to construct an entire function having a given infinite set of zeros? These questions will be answered by Theorems 8.7.1 and 8.7.3.

THEOREM 8.7.1.    *If $f(z)$ is an entire function without zeros, then there exists an entire function $g(z)$ such that*

$$(8.7.4) \qquad\qquad f(z) = e^{g(z)}.$$

*Proof.*    Without restricting the generality, we may assume that $f(0) = 1$. Then for each $\theta$, $0 < \theta < 1$, there exists a $\rho > 0$ such that $|f(z) - 1| < \theta$ for $|z| < \rho$. For $z$ in this disk, we can define the principal branch of $\log f(z) \equiv g(z)$ by the series

$$(8.7.5) \qquad\qquad g(z) = \log f(z) = - \sum_{n=1}^{\infty} \frac{1}{n} [1 - f(z)]^n,$$

where we have used (6.2.11). This series of holomorphic functions is uniformly convergent for $|z| < \rho$ and defines a holomorphic function in this circle. Using either the chain rule or term-by-term differentiation of (8.7.5), we get

$$(8.7.6) \qquad\qquad g'(z) = \frac{f'(z)}{f(z)}.$$

Here the right-hand side, as the quotient of two entire functions, can have no singularities other than poles in the finite plane. But poles could present themselves only at points where $f(z) = 0$, and, by assumption, there are no such points. It follows that $g'(z)$ is actually an entire function, and this implies that $g(z)$ is also entire. This completes the proof.

We come now to the second problem, that of constructing an entire function with given zeros $\{z_n\}$. It is clear that the set $\{z_n\}$ can have no limit point in the finite plane. The problem suggests the construction of an infinite product

$$\prod_{n=1}^{\infty} F_n(z, z_n),$$

in which each factor $F_n(z, z_n)$ is an entire function of $z$, having a single zero at $z = z_n$. Absolute convergence of the product requires absolute convergence of the series

$$\sum_{n=1}^{\infty} [F_n(z, z_n) - 1].$$

What we need is a "*prime function*" which can be adjusted to these requirements. Such a function was found by Weierstrass in 1876. We define

$$E(z, 0) = 1 - z,$$

(8.7.7)
$$E(z, p) = (1 - z) \exp \left[ \frac{z}{1} + \frac{z^2}{2} + \cdots + \frac{z^p}{p} \right].$$

The polynomial in the exponent is the $p$th partial sum of the Maclaurin series of

$$\log \frac{1}{1-z} = \sum_{k=1}^{\infty} \frac{z^k}{k}.$$

The basic property of $E(z, p)$ is given by

THEOREM 8.7.2.    *For* $|z| \leq 1$ *we have*

(8.7.8)               $|E(z, p) - 1| \leq |z|^{p+1}.$

*Proof.*[1]    The case $p = 0$ is trivial. If $p > 0$, we note that

$$E(z, p) = 1 + \sum_{k=1}^{\infty} A_{k, p} z^k$$

is an entire function. We shall determine some of the properties of the coefficients $A_{k, p}$. A simple calculation gives

$$E'(z, p) = -z^p \exp \left[ \frac{z}{1} + \cdots + \frac{z^p}{p} \right],$$

and expanding the exponential function in powers of $z$, we see that all its coefficients must be positive. This means that the power series for $E'(z, p)$ starts with the term $-z^p$, and all following terms have negative coefficients. But

$$E'(z, p) = \sum_{k=1}^{\infty} k A_{k, p} z^{k-1}.$$

Comparing the two expressions for the derivative, we see that

$$A_{1, p} = A_{2, p} = \cdots = A_{p, p} = 0, \quad A_{k, p} < 0 \text{ for } k > p.$$

On the other hand, by definition $E(1, p) = 0$, so that

$$\sum_{k=p+1}^{\infty} |A_{k, p}| = 1.$$

Hence, for $|z| \leq 1$ we have, as asserted,

$$|E(z, p) - 1| \leq \sum_{k=p+1}^{\infty} |A_{k, p}| \, |z|^k = |z|^{p+1} \sum_{k=p+1}^{\infty} |A_{k, p}| \, |z|^{k-p-1}$$

$$\leq |z|^{p+1} \sum_{k=p+1}^{\infty} |A_{k, p}| = |z|^{p+1}.$$

---

[1] This proof was communicated to me some forty years ago by my teacher Marcel Riesz. If I remember correctly, he ascribed it to Fejér. The proof does not seem to have been published.

We are now ready to state and prove Weierstrass's factorization theorem.

THEOREM 8.7.3.    *Given a sequence $\{z_n\}$ of distinct complex numbers, having no limit point in the finite plane, and given a sequence of positive integers $\{\mu_n\}$, there exists an entire function having a zero of multiplicity $\mu_n$ at $z = z_n$ for each n, and having no other zeros. If for all n, $z_n \neq 0$, then the product*

$$(8.7.9) \qquad f(z) = \prod_{n=1}^{\infty} \left\{ E\left(\frac{z}{z_n}, p_n\right) \right\}^{\mu_n}$$

*satisfies the above conditions, provided the integers $p_n$ are chosen so that the product converges uniformly on compact sets. The most general entire function with zeros of multiplicity $\mu_n$ at $z_n$ and multiplicity $\mu_0$ at $z_0 = 0$ is given by*

$$z^{\mu_0} f(z) \exp\left[g(z)\right],$$

*where $f(z)$ is the product (8.7.9), and $g(z)$ is an entire function.*

*Proof.*    We shall assume that $z_n \neq 0$, and that the enumeration is such that $|z_n| \leq |z_{n+1}|$ for all $n$. It is clear that the $n$th factor in the product has a zero of multiplicity $\mu_n$ at $z = z_n$, and it vanishes nowhere else. To justify the construction, we have merely to show that it is always possible to choose the integers $p_n$ in such a manner that the product converges absolutely and uniformly on any given disk.

Let $R > 0$ be given, and let $N$ be so large that for $n > N$ we have $|z_n| > R$. Theorem 8.7.2 then shows that

$$\left| E\left(\frac{z}{z_n}, p_n\right) - 1 \right| \leq \left| \frac{z}{z_n} \right|^{p_n+1}, \quad |z| \leq R, \quad n \geq N.$$

To prove absolute and uniform convergence of the product on $|z| \leq R$, it is thus sufficient to show that it is possible to choose the integers $p_n$ so that the series

$$(8.7.10) \qquad \sum_{n=1}^{\infty} \mu_n \left| \frac{z}{z_n} \right|^{p_n+1}$$

converges uniformly on the disk $|z| \leq R$, no matter how large $R$ is taken.

Suppose first that we can find a $k$ such that

$$(8.7.11) \qquad \sum_{n=1}^{\infty} \mu_n |z_n|^{-k-1}$$

converges. It is then clearly sufficient to take

$$(8.7.12) \qquad p_n = k \quad \text{for all } n.$$

In this case, (8.7.9) is called the *canonical product*, provided the least value of $k$ is chosen. But if the sequence $\{|z_n|\}$ grows very slowly, or the sequence $\{\mu_n\}$

grows very fast, no such integer $k$ may be available. We shall prove that in any event it is sufficient to choose the $p_n$'s subject to the condition

$$(8.7.13) \qquad\qquad p_n > \log (n^2 \mu_n).$$

Indeed, if $R > 0$ is given, we can choose an integer $N(R)$ so large that we have $|z_n| > eR$ for $n > N(R)$. We have then

$$\mu_n \left| \frac{R}{z_n} \right|^{p_n+1} < \mu_n e^{-p_n} < \mu_n \exp\left[-\log (n^2 \mu_n)\right] = \frac{1}{n^2},$$

and this is the general term of a convergent series. It follows that the series (8.7.10) converges uniformly on $|z| \leq R$ for any choice of $R$. This, in turn, implies that the infinite product (8.7.9) converges absolutely and uniformly on any such disk. Hence the product represents an entire function having the desired properties. The final observation follows from Theorem 8.7.1. For if $F(z)$ is another entire function having the same zeros with the same multiplicities, then $F(z)/f(z)$ has neither zeros nor poles. Thus, the quotient is an entire function without zeros, and Theorem 8.7.1 applies.

### EXERCISE 8.7

**1.** Determine the least possible value of the integer $k$ in (8.7.11), given that for $n = 1, 2, 3, \cdots$ we have **(a)** $z_n = n$, $\mu_n = 1$; **(b)** $z_n = n$, $\mu_n = n$; **(c)** $z_n = n^p$, $\mu_n = [n^q]$ where $p, q > 0$ and $[x]$ is the largest integer not exceeding $x$.

**2.** Same question for $z_n = e^n$ if **(a)** $\mu_1 = 1$; **(b)** $\mu_n = [e^n]$.

**3.** Construct an entire function having simple zeros at the Gaussian integers $\pm m \pm in$.

**4.** If $z_n = n$, $\mu_n = n!$, show that $p_n = [\beta n]$, $1 < \beta$ fixed, gives uniform convergence of (8.7.10).

**5.** Construct a canonical product having the same zeros as $(\pi z)^{-1} \sin \pi z$ with the same multiplicities.

**6.** Same question for $\cos \pi z$.

**8.8. The Gamma function.** We end this chapter with a discussion of the Gamma function of Euler and Gauss. This is a meromorphic function, and its reciprocal is entire, so that it can serve as an illustration of both classes of functions. We define the Gamma function by

$$(8.8.1) \qquad\qquad \Gamma(z) = \int_0^\infty t^{z-1} e^{-t}\, dt, \quad \Re(z) > 0.$$

This integral is Euler's definition of the Gamma function, familiar to the student from advanced calculus for real positive values of $z$. Since

$$(8.8.2) \qquad \Gamma(n+1) = n!, \quad n = 0, 1, 2, \cdots,$$

we see that $\Gamma(z)$ is a solution of the *interpolation problem* of finding a function $F(z)$ such that

$$(8.8.3) \qquad F(n+1) = n!.$$

Euler, Gauss, and Weierstrass were among the many mathematicians who have considered this important problem.

In accordance with (6.3.7), we set

$$t^z = e^{z \log t}, \quad t > 0, \quad \log t \text{ real.}$$

The integral in (8.8.1) is absolutely convergent for $\Re(z) = x > 0$, and since

$$|\, t^z \,| = t^x$$

we have

$$(8.8.4) \qquad |\, \Gamma(x+iy) \,| \leq \Gamma(x), \quad x > 0.$$

Integration by parts gives the fundamental relation

$$(8.8.5) \qquad \Gamma(z+1) = z\Gamma(z).$$

We first use Morera's theorem of Section 7.8 to show that $\Gamma(z)$ is holomorphic in the half-plane $\Re(z) > 1$ where, for fixed $z$, the integrand is continuous as a function of $t$ in $[0, \infty]$. Let $\triangle$ be any triangle in this half-plane. Then

$$\int_{\triangle} \Gamma(z)\, dz = \int_{\triangle} dz \int_0^{\infty} t^{z-1} e^{-t}\, dt = \int_0^{\infty} e^{-t} \left\{ \int_{\triangle} t^{z-1}\, dz \right\} dt = 0.$$

This proves the assertion. Here we have used the facts that the order of integration may be interchanged in an absolutely convergent iterated integral, and that, for each fixed $t$, the function $t^{z-1}$ is holomorphic in $z$, so that the integral along $\triangle$ in the third member is zero by Cauchy's theorem. Using Morera's theorem, we conclude that $\Gamma(z)$ is holomorphic.

Once we have established the fact that $\Gamma(z)$ is holomorphic in $\Re(z) > 1$, we can use (8.8.5) to obtain the analytic continuation of $\Gamma(z)$ in the whole plane. Since

$$(8.8.6) \qquad \Gamma(z) = \frac{\Gamma(z+1)}{z}$$

for $\Re(z) > 0$, we see first that $\Gamma(z)$ is actually holomorphic in $\Re(z) > 0$, where it is defined by (8.8.1). But we can then continue and define $\Gamma(z)$ in the strip $-1 < \Re(z) \leq 0$ with the aid of (8.8.6). The extension is holomorphic in

$\Re(z) > -1$, save for a simple pole at $z = 0$ due to the zero of the denominator in (8.8.6). By repeated use of (8.8.5) we get

$$(8.8.7) \qquad \Gamma(z) = \frac{\Gamma(z+m)}{z(z+1)\cdots(z+m-1)}.$$

This enables us to define $\Gamma(z)$ in $\Re(z) > -m$ as a holomorphic function save for poles at $z = 0, -1, \cdots, -m+1$. It follows that $\Gamma(z)$ is a meromorphic function of $z$ having simple poles at $0$ and at the negative integers.

The Gamma function, as we have seen, is a solution of the important *difference equation*

$$(8.8.8) \qquad f(z+1) = zf(z).$$

This equation has infinitely many meromorphic solutions. If $\omega(z)$ is a meromorphic function of period 1, then

$$(8.8.9) \qquad f(z) = \omega(z)\Gamma(z), \quad \omega(z+1) = \omega(z),$$

is also a meromorphic solution. If we choose for $\omega(z)$ an entire periodic function having zeros at the integers, then $f(z)$ will be an entire function. In particular we find that

$$(8.8.10) \qquad \Gamma_-(z) \equiv \frac{1}{\pi}\sin \pi z\, \Gamma(z)$$

is such a solution. It has simple zeros at the positive integers. We shall encounter this solution again in the following.

There are other methods available for obtaining the analytic continuation of $\Gamma(z)$. The following procedure gives the extension to the whole plane in a single operation and has an interesting bearing on the Mittag-Leffler theorem. The two integrals

$$(8.8.11) \qquad P(z) = \int_0^1 t^{z-1}e^{-t}\, dt, \quad Q(z) = \int_1^\infty t^{z-1}e^{-t}\, dt$$

are known as *incomplete Gamma functions*. Here $Q(z)$ is readily seen to be an entire function. In fact, the integral converges for all $z$, and the Morera theorem shows that $Q(z)$ is holomorphic in the finite plane. $P(z)$, on the other hand, is defined only for $\Re(z) > 0$. If, however, we substitute the exponential series for $e^{-t}$ under the sign of integration and assume $\Re(z) > 1$, we obtain a uniformly convergent series which may be integrated term by term. The result is

$$(8.8.12) \qquad P(z) = \sum_{n=0}^\infty \frac{(-1)^n}{n!(z+n)}.$$

This series converges absolutely and uniformly on any region of the $z$-plane having positive distance from 0 and the negative integers. Thus we have

$$(8.8.13) \qquad \Gamma(z) = \sum_{n=0}^\infty \frac{(-1)^n}{n!(z+n)} + Q(z).$$

This is a representation of the Gamma function as the sum of an entire function and the series of principal parts of $\Gamma(z)$.

There is still another method of analytic continuation, which leads to the important product representation of $\Gamma(z)$. Here we start with the observation that on any fixed interval $[0, a]$ the sequence of polynomials

$$\left(1 - \frac{t}{n}\right)^n$$

is ultimately monotone increasing and converges uniformly to the limit $e^{-t}$. Define

(8.8.14) $$\Gamma_n(z) = \int_0^n t^{z-1}\left(1 - \frac{t}{n}\right)^n dt, \quad \Re(z) > 0.$$

The substitution $t = ns$ gives

$$\Gamma_n(z) = n^z \int_0^1 s^{z-1}(1 - s)^n \, ds.$$

The integral is a rational function of $z$, having simple poles at $z = 0, -1, -2, \cdots$, $-n$. Using an induction argument or the calculus of finite differences, one verifies that

(8.8.15) $$\Gamma_n(z) = \frac{n^z n!}{z(z + 1) \cdots (z + n)}.$$

We shall now prove (1) that $\lim \Gamma_n(z)$ exists for every $z$ not equal to 0 or a negative integer, and (2) that the limit equals $\Gamma(z)$.

For this purpose we consider

(8.8.16) $$\frac{1}{\Gamma_n(z)} = n^{-z} z \prod_{k=1}^n \left(1 + \frac{z}{k}\right).$$

It is fairly obvious from this expression that the limit, if it exists, is going to be closely related to the Weierstrass product representation of an entire function having simple zeros at 0 and the negative integers. The canonical product for this case is

(8.8.17) $$z \prod_{k=1}^\infty \left\{\left(1 + \frac{z}{k}\right) e^{-\frac{z}{k}}\right\} \equiv G(z),$$

and we have

$$z \prod_{k=1}^n \left\{\left(1 + \frac{z}{k}\right) e^{-\frac{z}{k}}\right\} = \exp\left\{-z\left[\frac{1}{1} + \frac{1}{2} + \cdots + \frac{1}{n} - \log n\right]\right\} \frac{1}{\Gamma_n(z)}.$$

Now it is a familiar fact that

(8.8.18) $$\lim_{n \to \infty} \left\{\frac{1}{1} + \frac{1}{2} + \cdots + \frac{1}{n} - \log n\right\} \equiv C$$

exists. Here $C = .57721 \cdots$ is Euler's constant. It is known to several hundred decimal places, but nobody has been able to show that it is a transcendental number or even that it is irrational. It follows that

$$(8.8.19) \qquad \lim_{n \to \infty} \frac{1}{\Gamma_n(z)} = e^{Cz} G(z),$$

where the limit exists uniformly in any finite disk. This of course also implies the existence of $\lim \Gamma_n(z)$, when $z$ is neither 0 nor a negative integer.

It remains to prove that $\lim \Gamma_n(z) = \Gamma(z)$. Since the limit is holomorphic in the right half-plane, and $\Gamma(z)$ has the same property, it suffices to show that the equality holds on an interval of the real axis, for two holomorphic functions which coincide in an interval are identical, by the identity theorem. But if $z = x$ is real, then, by the monotoneity properties mentioned above, we have

$$\Gamma_n(x) = \int_0^n t^{x-1}\left(1 - \frac{t}{n}\right)^n dt < \int_0^n t^{x-1}\left(1 - \frac{t}{n+1}\right)^{n+1} dt < \Gamma_{n+1}(x)$$

$$= \int_0^{n+1} t^{x-1}\left(1 - \frac{t}{n+1}\right)^{n+1} dt < \int_0^{n+1} t^{x-1} e^{-t}\, dt < \Gamma(x).$$

Hence $\{\Gamma_n(x)\}$ is a monotone increasing sequence bounded by $\Gamma(x)$. On the other hand, for $0 < a < n$,

$$\Gamma_n(x) > \int_0^a t^{x-1}\left(1 - \frac{t}{n}\right)^n dt,$$

so that

$$\lim_{n \to \infty} \Gamma_n(x) \geq \int_0^a t^{x-1} e^{-t}\, dt$$

by the uniform convergence. Since this holds for every $a$, we must have

$$\lim \Gamma_n(x) \geq \Gamma(x),$$

or, finally,

$$\lim_{n \to \infty} \Gamma_n(x) = \Gamma(x).$$

We have consequently proved the product representation

$$(8.8.20) \qquad \frac{1}{\Gamma(z)} = e^{Cz} z \prod_{n=1}^{\infty} \left\{\left(1 + \frac{z}{n}\right) e^{-\frac{z}{n}}\right\}$$

valid for all $z$. This shows that the reciprocal of the Gamma function is an entire function.

Since

$$\Gamma(1 - z) = -z\Gamma(-z),$$

we see that

$$(8.8.21) \qquad \frac{1}{\Gamma(z)\Gamma(1 - z)} = z \prod_{n=-\infty}^{\infty}{}' \left\{\left(1 - \frac{z}{n}\right) e^{\frac{z}{n}}\right\},$$

where the prime indicates that the value $n = 0$ is omitted. It may be shown that this infinite product equals $\pi^{-1} \sin \pi z$, so that we finally have (cf. formula (8.8.10) and page 262)

$$(8.8.22) \qquad \Gamma(z)\Gamma(1-z) = \frac{\pi}{\sin \pi z}.$$

We give next a representation of the reciprocal of the Gamma function by a contour integral due to Hermann Hankel[1] (1864). It is

$$(8.8.23) \qquad \frac{1}{\Gamma(z)} = \frac{1}{2\pi i} \int_C e^t t^{-z}\, dt.$$

Here we cut the $t$-plane along the negative real axis and define

$$t^{-z} = e^{-z \log t}$$

with the imaginary part of $\log t$ between $-\pi$ and $\pi$. The contour of integration $C$ follows the lower edge of the cut from $-\infty$ to $-\delta$, $\delta > 0$, goes around the origin in the positive sense on the circle $|t| = \delta$, and then follows the upper edge of the cut from $-\delta$ to $-\infty$. The integral clearly has a sense for every finite value of $z$. Using Morera's theorem as above in this section, we can show that the integral represents a function holomorphic in the finite plane, that is, an entire function.

In order to identify this function with $1/\Gamma(z)$, we may restrict ourselves to the half-plane $\Re(z) < 1$. Along the edges of the cut we see that $e^t t^{-z}$ equals

$$e^{-\pi i z} e^{-s} s^{-z} \quad \text{on the upper edge and}$$

$$e^{\pi i z} e^{-s} s^{-z} \quad \text{on the lower,}$$

where $s = -t$ is real and positive. On the circle $\gamma : |t| = \delta$, we have

$$\left| \int_\gamma e^t t^{-z}\, dt \right| < 2\pi e^\delta \delta^{1-\Re(z)} e^{\pi |z|},$$

and this tends to zero with $\delta$ since $\Re(z) < 1$, uniformly with respect to $z$ in any bounded domain. It follows that the limit of the integral as $\delta \to 0$ equals

$$\frac{1}{2\pi i}(e^{\pi i z} - e^{-\pi i z}) \int_0^\infty e^{-s} s^{-z}\, ds = \frac{\sin \pi z}{\pi} \Gamma(1-z) = \frac{1}{\Gamma(z)}.$$

Since the value of the integral is independent of $\delta$, formula (8.8.23) must hold for $\Re(z) < 1$. But an entire function which coincides with $1/\Gamma(z)$ in a half-plane

---

[1] 1839–1873. Certain classes of determinants and of Bessel functions are also named after Hankel. He codified the formal laws of arithmetic (the field postulates of Section 1.1) and wrote a history of early mathematics. Most of his work was published after his untimely death.

clearly is identical with this function for all $z$. This completes the proof of (8.8.23).

Finally, we prove a version of Stirling's[1] formula, namely,

$$(8.8.24) \qquad \log \Gamma(z) = (z - \tfrac{1}{2}) \log z - z + \tfrac{1}{2} \log 2\pi + \omega(z).$$

Here $-\pi < \arg z < \pi$, and the logarithms are real when $z$ is real positive. Further, $\omega(z)$ is holomorphic in the $z$-plane cut along the negative real axis and tends to zero as $z \to \infty$ in such a manner that its distance from the cut becomes infinite. Several representations of $\omega(z)$ will be given.

To simplify, let us write

$$\gamma(z) = \log \Gamma(z), \quad L(z) = (z - \tfrac{1}{2}) \log z - z, \quad \alpha = \tfrac{1}{2} \log 2\pi,$$

so that (8.8.24) is equivalent to the identity

$$\gamma(z) = L(z) + \alpha + \omega(z).$$

But

$$L(z) = \int_2^z \log t \, dt - \tfrac{1}{2} \log z + \beta, \quad \beta = 2 \log 2 - 2,$$

and

$$\log t = \gamma(t + 1) - \gamma(t)$$

by (8.8.5), so that

$$(8.8.25) \qquad L(z) = \int_2^z [\gamma(t + 1) - \gamma(t)] \, dt - \tfrac{1}{2}[\gamma(z + 1) - \gamma(z)] + \beta.$$

Here the path of integration is a straight line segment from $t = 2$ to $t = z$ in the cut plane. If $z$ is temporarily restricted to the sector

$$S: \quad -\tfrac{5}{6}\pi < \arg (z - 2) < \tfrac{5}{6}\pi,$$

then every point of the path of integration is the center of a circle of radius $>1$ in which $\gamma(t)$ is holomorphic. By Taylor's theorem it follows that

$$\gamma(t + 1) - \gamma(t) = \sum_{n=1}^{\infty} \frac{1}{n!} \gamma^{(n)}(t),$$

where the series is absolutely and uniformly convergent with respect to $t$ on any bounded part of $S$. We can then substitute the series in (8.8.25) and integrate term by term. Combining terms involving derivatives of equal orders, we obtain

$$L(z) = \gamma(z) - \tfrac{1}{2} \sum_{n=2}^{\infty} \frac{n - 1}{n + 1} \frac{1}{n!} \gamma^{(n)}(z) + \delta$$

and

$$\omega(z) = \tfrac{1}{2} \sum_{n=2}^{\infty} \frac{n - 1}{n + 1} \frac{1}{n!} \gamma^{(n)}(z) - \alpha - \delta,$$

---

[1] James Stirling, 1692–1770.

where $\delta$ is a constant. We shall prove below that the sum of the infinite series converges to zero as $z \to \infty$ in $S$, whence it follows that $\omega(z)$ tends to the real limit $-\alpha - \delta$. Our next task is to show that this limit is actually zero.

For this purpose we use formula (8.8.22) with $z = \frac{1}{2} + iy$, $y > 0$. Combining

$$\log \Gamma(\tfrac{1}{2} + iy) + \log \Gamma(\tfrac{1}{2} - iy) = \log \pi - \log \cosh(\pi y)$$

with formula (8.8.24), we get

$$iy[\log(\tfrac{1}{2} + iy) - \log(\tfrac{1}{2} - iy)] - 1 + \log 2\pi + 2\Re[\omega(\tfrac{1}{2} + iy)]$$
$$= \log 2\pi - \pi y + \log[1 + e^{-2\pi y}].$$

Here the first term on the left reduces to

$$-2y \arctan 2y = -2y\left\{ \frac{\pi}{2} - \int_{2y}^{\infty} \frac{du}{1 + u^2} \right\}$$
$$= -2y\left[ \frac{\pi}{2} - \frac{1}{2y} + O\left(\frac{1}{y^3}\right) \right] = -\pi y + 1 + O\left(\frac{1}{y^2}\right).$$

It follows that

$$\Re[\omega(\tfrac{1}{2} + iy)] = O\left(\frac{1}{y^2}\right) \quad \text{as} \quad y \to \infty,$$

and hence, since the limit exists and is real, that

$$\lim_{z \to \infty} \omega(z) = 0 \quad \text{and} \quad \delta = -\alpha.$$

Thus, the following representation is valid at least in the sector $S$:

$$(8.8.26) \qquad \omega(z) = \tfrac{1}{2} \sum_{n=2}^{\infty} \frac{n-1}{n+1} \frac{1}{n!} \frac{d^n}{dz^n} \log \Gamma(z).$$

By logarithmic differentiation we obtain from formula (8.8.20) the series

$$(8.8.27) \qquad \frac{d}{dz} \log \Gamma(z) \equiv \Psi(z) = -C - \frac{1}{z} + \sum_{k=1}^{\infty} \left\{ \frac{1}{k} - \frac{1}{z+k} \right\},$$

whence

$$(8.8.28) \qquad \frac{d^n}{dz^n} \log \Gamma(z) = \Psi^{(n-1)}(z) = (-1)^{n-1}(n-1)! \sum_{k=0}^{\infty} \frac{1}{(z+k)^n}, \quad n \geq 2.$$

Substituting in (8.8.26), we get the double series

$$(8.8.29) \qquad \omega(z) = \tfrac{1}{2} \sum_{n=2}^{\infty} (-1)^{n-1} \frac{n-1}{n(n+1)} \sum_{k=0}^{\infty} \frac{1}{(z+k)^n}.$$

Assuming absolute convergence of this series, we can interchange the order of summation. For $|w| > 1$ we have

$$(8.8.30) \qquad \tfrac{1}{2} \sum_{n=2}^{\infty} (-1)^{n-1} \frac{n-1}{n(n+1)} \frac{1}{w^n} = (w + \tfrac{1}{2}) \log\left(1 + \frac{1}{w}\right) - 1.$$

Hence, if

(8.8.31)                     $|z + k| > 1, \quad k = 0, 1, 2, \cdots,$

we obtain

(8.8.32)          $\omega(z) = \sum_{k=0}^{\infty} \left\{ (z + k + \tfrac{1}{2}) \log \left( 1 + \frac{1}{z+k} \right) - 1 \right\}.$

The double series is due to J. P. M. Binet (1786–1856), the simple logarithmic series to Christof Gudermann (1798–1851), who was the teacher of Weierstrass.

It remains to examine the convergence of these series. In passing from (8.8.29) to (8.8.32) we had to assume condition (8.8.31) in order to be able to use the series (8.8.30). Actually this condition suffices for the convergence of all three series representations of $\omega(z)$. For (8.8.30) is an alternating series whose coefficients are steadily decreasing to zero. It follows that

$$(|w| + \tfrac{1}{2}) \log \left( 1 + \frac{1}{|w|} \right) - 1 < \frac{1}{12 |w|^2},$$

so that (8.8.31) implies that

(8.8.33)    $\sum_{k=0}^{\infty} \left\{ (|z + k| + \tfrac{1}{2}) \log \left[ 1 + \frac{1}{|z+k|} \right] - 1 \right\} < \tfrac{1}{12} \sum_{k=0}^{\infty} \frac{1}{|z+k|^2}.$

The series in the right member is obviously convergent for every $z$ which is not a negative integer or zero. It follows that (8.8.31) implies the convergence of the series (8.8.32). But if we first introduce absolute values in the double series (8.8.29) and then interchange the order of summation, we can still carry out the summation with respect to $n$ and obtain a majorant of the same type as (8.8.33) except that the factor $\frac{1}{12}$ is replaced by

$$\log \frac{r}{r-1} \quad \text{where} \quad r = \min_k |z + k|.$$

Hence, as long as $r > 1$ the series (8.8.29) and (8.8.26) are convergent.

We still have to examine $\omega(z)$ for large values of $z$. By (8.8.33) it is sufficient to consider

(8.8.34)                     $M(z) \equiv \sum_{k=0}^{\infty} |z + k|^{-2}.$

Suppose first that $\Re(z) \geq 0$. Then the terms of the series are monotone decreasing to zero and the series is comparable with the integral $\int_0^{\infty} |z + u|^{-2} du$. More precisely, by Cauchy's integral test for convergence we have

$$M(z) < |z|^{-2} + \int_0^{\infty} |z + u|^{-2} du.$$

The integral is elementary and its value may be given as

$$\frac{\theta}{y} \quad \text{or} \quad \frac{\theta}{\sin\theta}\cdot\frac{1}{r}, \quad z = re^{i\theta}, \quad -\pi < \theta < \pi,$$

where both fractions are assigned their common limit $1/x$ for $y = 0,\ \theta = 0$. It follows that

(8.8.35) $$M(z) < \tfrac{1}{2}\pi\,|\,z\,|^{-1} + |\,z\,|^{-2}, \quad \Re(z) \geqq 0.$$

In the left half-plane we may proceed as follows: Suppose in fact that $-m - 1 \leqq x < -m,\ 0 \leqq m,\ 0 < y$. A simple calculation shows that

$$M(z + m + 1) + M(-z - m) = \sum_{k=-\infty}^{\infty} |\,z + k\,|^{-2} > M(z).$$

Here $\Re(z + m + 1) \geqq 0$, $\Re(-z - m) > 0$, so that the corresponding $M$-functions may be estimated with the aid of (8.8.35), replacing $z$ by $z + m + 1$ and $-z - m$ respectively. Since

$$|\,z + m + 1\,| \geqq |\,y\,|, \quad |\,-z - m\,| > |\,y\,|,$$

we get

(8.8.36) $$M(z) < \pi\,|\,y\,|^{-1} + 2y^{-2}, \quad \Re(z) < 0.$$

Going back to (8.8.33) we obtain the final estimate

(8.8.37) $$|\,\omega(z)\,| < \begin{cases} \tfrac{1}{24}\pi\,|\,z\,|^{-1} + \tfrac{1}{12}\,|\,z\,|^{-2}, & \Re(z) \geqq 0,\ |\,z\,| > 1, \\ \tfrac{1}{12}\pi\,|\,y\,|^{-1} + \tfrac{1}{6}y^{-2}, & \Re(z) < 0,\ |\,\Im(z)\,| > 1. \end{cases}$$

Thus we see that $\omega(z) \to 0$ if $z$ tends to infinity in such a manner that its distance from the negative real axis becomes infinite. This completes our discussion of the asymptotic properties of $\omega(z)$ and of formula (8.8.24), but it does not exhaust the available information concerning $\omega(z)$, such as Stirling's asymptotic series. We shall not pursue this subject any further except for noting the important formula

(8.8.38) $$\lim_{z\to\infty} \frac{\Gamma(z + a)}{z^a\Gamma(z)} = 1.$$

Here $z$ tends to infinity in such a manner that its distance from the negative real axis becomes infinite, $z^a = \exp(a\log z)$ where the imaginary part of the logarithm lies between $-\pi$ and $\pi$, and $a$ is either fixed or may depend upon $z$ such that $a^2/z \to 0$ as $z \to \infty$. This relation follows readily from (8.8.24). A simple consequence is the formula

(8.8.39) $$(-1)^n \binom{-\alpha}{n} = \frac{1}{\Gamma(\alpha)}\,n^{\alpha-1}[1 + o\,(1)].$$

## EXERCISE 8.8

**1.** Prove (8.8.15).

**2.** If $a$, $b$, $c$, $d$ are complex numbers, none of which is a negative integer, and if $a + b = c + d$, prove that the infinite product

$$\prod_{n=1}^{\infty} \frac{(n + a)(n + b)}{(n + c)(n + d)}$$

is absolutely convergent, and that its value is expressible in terms of Gamma functions.

**3.** Prove that $\Gamma(\frac{1}{2} + iy)$ tends exponentially to zero as $y \to \pm\infty$.

**4.** What is the value of $\Gamma(\frac{1}{2})$?

**5.** What is the value of

$$\prod_{k=1}^{\infty} \left\{ \frac{2k}{2k + 1} \, e^{\frac{1}{2k}} \right\}?$$

**6.** Using formulas (8.8.20) and (6.2.11), and the results of Problems 4 and 5, prove the so-called *duplication formula* of the Gamma function

$$\pi^{\frac{1}{2}} \Gamma(2z) = 2^{2z-1} \Gamma(z) \Gamma(z + \tfrac{1}{2}).$$

**7.** Find a solution of the difference equation $f(z + 1) = R(z)f(z)$ in terms of Gamma functions, if $R(z)$ is a given rational function.

**8.** Show that $\Psi'(z)$, the logarithmic derivative of the Gamma function, satisfies the difference equation

$$g(z + 1) - g(z) = \frac{1}{z}.$$

**9.** Prove that $\qquad \Psi'(1 - z) - \Psi'(z) = \pi \cot \pi z.$

**10.** Deduce from Hankel's contour integral that·

$$\frac{1}{\Gamma(z)} = \frac{1}{2\pi i} \int_L e^t t^{-z} \, dt, \quad \Re(z) > 0,$$

where the integral is taken along the vertical line $\Re(t) = \alpha > 0$ from $\alpha - i\infty$ to $\alpha + i\infty$, and the integral is a Cauchy principal value for $0 < \Re(z) \leq 1$.

**11.** The series (8.8.32) may be differentiated term by term. Use this to obtain estimates of $\omega'(z)$ and show that

$$\Psi'(z) = \log z - \frac{1}{2z} + O\left(\frac{1}{z^2}\right)$$

for $|z| \to \infty$, $|\arg z| \leq \pi - \varepsilon$, $0 < \varepsilon$.

**12.** Prove (8.8.38).

## COLLATERAL READING

As a general reference see

WHITTAKER, E. T., and WATSON, G. N. *A Course of Modern Analysis*, Fourth Edition, Chaps. 5, 7, and 12. Cambridge University Press, New York, 1952.

There is much information to be found in

VIVANTI, G. *Teoria delle Funzioni Analitiche.* Ulrico Hoepli, Milan, 1901. German, much augmented translation by GUTZMER, A. *Theorie der eindeutigen analytischen Funktionen.* B. G. Teubner, Leipzig and Berlin, 1906. (This treatise contains a bibliography of close to 700 original papers.)

See also

HILLE, E. "Essai d'une Bibliographie de la Représentation Analytique d'une Fonction Monogène," *Acta Mathematica*, Vol. 52 (1928), pp. 1–80.

An excellent, well-documented presentation of the theory of the Gamma function is to be found in the following paper:

JENSEN, J. L. W. V. "An Elementary Exposition of the Theory of the Gamma Function," *Annals of Mathematics*, Series 2, Vol. 17 (1916), pp. 124–166. (This paper was translated from the original Danish article by T. H. GRON-WALL, who added much to the bibliography and to the proofs.)

# 9

# THE CALCULUS OF RESIDUES

**9.1. The residue theorem.** We return to Cauchy's theorem. The applications of this theorem to representation questions are by no means exhausted by the results listed in Sections 8.1 and 8.3, but we have now in mind questions of a different nature. In particular, we would like to amplify Theorem 7.3.3 for the case in which the single-valued function $f(z)$ has isolated singularities. The resulting residue theorem of Cauchy gives access to a vast range of applications of a highly diversified nature. A few of these will be listed in this chapter.

DEFINITION 9.1.1. *Let $f(z)$ be holomorphic in $0 < |z - a| < R$ and represented there by the series*

$$f(z) = \sum_{n=-\infty}^{\infty} a_n(z - a)^n;$$

*then $a_{-1}$ is called the residue of $f(z)$ at $z = a$.*

The term "residue" refers to the fact that

(9.1.1) $$a_{-1} = \frac{1}{2\pi i} \int_C f(t)\, dt,$$

where the integral is taken along the positively oriented circle $|t - a| = r < R$.

THEOREM 9.1.1. *Suppose that $f(z)$ is holomorphic inside and on a "scroc" $C$, save for a finite number of isolated singularities, none of which lie on $C$. Then*

(9.1.2) $$\int_C f(t)\, dt = 2\pi i\, \mathbf{S}[f;\ C_i],$$

*where $\mathbf{S}[f;\ C_i]$ is the sum of the residues of $f(z)$ in the interior of $C$.*

*Proof.* Let $f(z)$ have the isolated singular points $z_1, z_2, \cdots, z_n$, in $C_i$, and let $f(z)$ be holomorphic everywhere else in $C^*$. Let $c_k$ be a small circle $|t - z_k| = r_k$, where $r_k$ is less than the distance of $z_k$ from $C$ and such that the circles $c_k$ are exterior to each other. Theorem 7.3.3 gives

$$\int_C f(t)\, dt = \sum_{k=1}^{n} \int_{c_k} f(t)\, dt = 2\pi i \sum_{k=1}^{n} a_{-1,\,k},$$

where $a_{-1,\,k}$ is the residue of $f(z)$ at $z = z_k$. This is the desired result.

For the actual computations of residues, the following observations may be useful. Suppose that $f(z)$ has an isolated singularity at $z = a$ where we want

241

to compute the residue. If $f(z)$ is a rational function we can always use the method of partial fractions, and if all residues are required, and not merely the residue at $z = a$, this may be the least time-consuming device. Even if this method is used, some of the suggestions given below may shorten the work. In general, there are three cases to be considered.

I. $f(z)$ *has a simple pole at* $z = a$. Then the residue equals

$$(9.1.3) \qquad\qquad a_{-1} = \lim_{z \to a} (z - a)f(z).$$

In the case of a rational function, the procedure is immediate: we multiply the function by $(z - a)$. This cancels the vanishing factor in the denominator, and we can set $z = a$ in the product. As an example, the function

$$\frac{z^2 - 2z}{z^2 - 1} = \frac{z^2 - 2z}{z + 1} \frac{1}{z - 1} = \frac{z^2 - 2z}{z - 1} \frac{1}{z + 1}$$

has simple poles at $z = +1$ and $-1$. The residue at $z = +1$ equals $-\frac{1}{2}$, because this is the limit of the fraction $(z^2 - 2z)/(z + 1)$ as $z \to +1$. Similarly, the residue at $z = -1$ is found to be $-\frac{3}{2}$.

For transcendental functions we can often read off a local representation of $f(z)$ as the quotient of two holomorphic functions. Suppose that

$$f(z) = \frac{h(z)}{g(z)},$$

where $g(z)$ and $h(z)$ are holomorphic in some neighborhood of $z = a$ and $g(a) = 0$, $h(a) \neq 0$. Then

$$(z - a)f(z) = \frac{z - a}{g(z) - g(a)} h(z),$$

and the residue equals

$$(9.1.4) \qquad\qquad a_{-1} = \frac{h(a)}{g'(a)}.$$

An example is given by $\cot z$ at $z = 0$, where

$$z \cot z = \frac{z}{\sin z} \cos z \to 1 \quad \text{as} \quad z \to 0.$$

II. $f(z)$ *has a pole of order* $m$ *at* $z = a$. Here we set

$$p(z) = (z - a)^m f(z),$$

which is a holomorphic function in some neighborhood of $z = a$. We have

$$p(z) = p(a) + p'(a)(z - a) + \cdots + \frac{p^{(m-1)}(a)}{(m - 1)!} (z - a)^{m-1} + \cdots.$$

We obtain $f(z)$ by dividing this expansion by $(z-a)^m$. It follows that the required residue is now given by

$$(9.1.5) \qquad a_{-1} = \frac{p^{(m-1)}(a)}{(m-1)!}.$$

As an example, consider the function

$$\frac{z^3 + 5}{z(z-1)^3},$$

which has a triple pole at $z = 1$. Formula (9.1.5) applies and the residue is found to be

$$\frac{1}{2!} \frac{d^2}{dz^2} \frac{z^3 + 5}{z}\bigg|_{z=1} = 6.$$

This method is recommended in the case of rational functions. For transcendental functions the problem of evaluating the $(m-1)$th derivative at $z = a$ will normally lead to indeterminate forms requiring repeated use of l'Hospital's rule or of a similar device. Here the following procedure may be more practical: Suppose that

$$f(z) = \frac{h(z)}{g(z)},$$

where

$$h(z) = \sum_{n=0}^{\infty} h_n (z-a)^n, \quad g(z) = \sum_{n=0}^{\infty} g_n (z-a)^{n+m},$$

and $g_0 \neq 0$, $h_0 \neq 0$. Now

$$p(z) = \left[\sum_{n=0}^{\infty} h_n (z-a)^n\right] \bigg/ \left[\sum_{n=0}^{\infty} g_n (z-a)^n\right] \equiv \sum_{k=0}^{\infty} p_k (z-a)^k.$$

The problem is now reduced to that of finding $p_{m-1}$, which is the required residue. Since

$$\sum_{n=0}^{\infty} h_n (z-a)^n = \left[\sum_{j=0}^{\infty} g_j (z-a)^j\right] \left[\sum_{k=0}^{\infty} p_k (z-a)^k\right],$$

the Cauchy product rule gives the following system of equations for the computation of the unknown quantities $p_k$:

$$
\begin{aligned}
h_0 &= g_0 p_0, \\
h_1 &= g_1 p_0 + g_0 p_1,
\end{aligned}
$$

$$(9.1.6) \qquad \cdot \quad \cdot \quad \cdot \quad \cdot \quad \cdot \quad \cdot \quad \cdot$$

$$h_{m-1} = g_{m-1} p_0 + g_{m-2} p_1 + \cdots + g_0 p_{m-1},$$

$$\cdot \quad \cdot \quad \cdot \quad \cdot \quad \cdot \quad \cdot \quad \cdot \quad \cdot \quad \cdot \quad \cdot \quad \cdot$$

Since $g_0 \neq 0$, we can compute successively the unknowns $p_0$, $p_1$, $\cdots$. In the present case, however, we are interested in one single unknown, namely $p_{m-1}$, and, if we so desire, we can compute it directly using determinants. Since the coefficient matrix, formed by the $g$'s, is triangular, we see that the determinant of the first $m$ equations reduces to the product of the diagonal elements and, hence, equals $(g_0)^m$. It follows that the residue is given by

$$(9.1.7) \qquad a_{-1} = (g_0)^{-m} \begin{vmatrix} g_0 & 0 & 0 & \cdots & h_0 \\ g_1 & g_0 & 0 & \cdots & h_1 \\ g_2 & g_1 & g_0 & \cdots & h_2 \\ \cdot & \cdot & \cdot & \cdots & \cdot \\ g_{m-1} & g_{m-2} & g_{m-3} & \cdots & h_{m-1} \end{vmatrix} = p_{m-1}.$$

This formula is useful not merely for finding residues. It also gives the solution of the problem of expanding the quotient of two power series into a power series.

III. $f(z)$ *has an essential singularity at* $z = a$. In this case the methods developed above break down, and usually one has to resort to the Laurent series if it can be found. An example is given by the function

$$\exp\left\{\frac{a}{2}\left(z - \frac{1}{z}\right)\right\}.$$

According to Problem 4 of Exercise 8.3 the residue at $z = 0$ is $-J_1(a)$. That we succeed in finding the residue in the present case is due essentially to the fact that the Laurent series is the product of a power series in $z$ with a power series in $z^{-1}$, both of which are elementary.

After this long discussion of how to find residues we come to their applications. The latter are numerous. We start by giving applications of the residue theorem to the evaluation of definite integrals. This was a problem to which Cauchy devoted much attention and many memoirs. He developed an elaborate and highly diversified technique to cope with different types of integrals. A common feature of this technique is the deformation of a finite contour of integration (to which the residue theorem applies) by pushing part of the contour out to infinity or to singularities of the integrand. This naturally raises questions of convergence, and a part of the problem is to settle these questions. The following examples will give some idea of the power and scope of the method.

THEOREM 9.1.2.    *Suppose that* $f(z)$ *is a rational function,*

$$f(z) = \frac{P(z)}{Q(z)},$$

*where the degree of* $Q(z)$ *exceeds that of* $P(z)$ *by at least two, and suppose that no zero of* $Q(z)$ *is real. Then*

$$(9.1.8) \qquad \int_{-\infty}^{\infty} \frac{P(x)}{Q(x)}\, dx = 2\pi i\, \mathbf{S}\left[\frac{P}{Q}\, ;\, U\right],$$

*where* $U$ *is the upper half-plane.*

*Proof.* Let $R$ be so large that all the zeros of $Q(z)$ are interior to the circle $|z| = R$. Let $C$ be a "scroc" consisting of the semicircle $|z| = R$, $y > 0$, and the segment of the real axis from $-R$ to $+R$. The residue theorem shows that

$$\int_C \frac{P(z)}{Q(z)}\,dz = 2\pi i \, \mathbf{S}\left[\frac{P}{Q}\,;C_i\right] = 2\pi i \, \mathbf{S}\left[\frac{P}{Q}\,;U\right].$$

The assumption on the degrees of $P(z)$ and $Q(z)$ shows that we can find a constant $M$ such that

$$|f(Re^{i\theta})| \le MR^{-2}, \quad R > R_0.$$

Hence,

$$\left|\int_0^\pi f(Re^{i\theta})iRe^{i\theta}\,d\theta\right| \le \pi RMR^{-2} = \pi MR^{-1},$$

and this tends to zero with $R^{-1}$. On the other hand, as $R \to \infty$,

$$\int_{-R}^R f(x)\,dx \to \int_{-\infty}^\infty f(x)\,dx,$$

which exists by virtue of the assumption on the degrees of $P$ and $Q$. This proves (9.1.8).

An alternate formula is

(9.1.9)
$$\int_{-\infty}^\infty \frac{P(x)}{Q(x)}\,dx = -2\pi i \, \mathbf{S}\left[\frac{P}{Q}\,;L\right],$$

where $L$ denotes the lower half-plane. If $f(z)$ is real on the real axis, the poles of $f(z)$ are symmetric with respect to the axis, and the residues at conjugate imaginary poles are conjugate imaginary numbers. In this case the two formulas involve the same amount of work in evaluating the integral. If $f(x)$ is not real, however, one half-plane may be more advantageous than the other. In particular, we see that

(9.1.10)
$$\int_{-\infty}^\infty \frac{P(x)}{Q(x)}\,dx = 0$$

*if all the zeros of $Q(z)$ lie in the same half-plane, $U$ or $L$.*

Certain trigonometric integrals can also be handled by the calculus of residues. The following is an example:

THEOREM 9.1.3. *Suppose that $f(z)$ is a rational function of $e^{iz}$, and that*

(9.1.11)
$$\lim_{y \to +\infty} f(x + iy) = 0.$$

*Moreover, suppose that $f(z)$ has no poles on the real axis or on the upper half of the imaginary axis. Let $H$ denote the half-strip $0 < x < 2\pi$, $0 < y$. Then*

(9.1.12)
$$\int_0^{2\pi} f(x)\,dx = 2\pi i \, \mathbf{S}[f; H].$$

REMARK.    The assumption that there are no poles on the upper half of the imaginary axis is not essential, since the integral has the same value over any interval $(\alpha, \alpha + 2\pi)$. We merely replace $H$ by a semi-strip of width $2\pi$ on whose boundary there are no poles.

*Proof of Theorem 9.1.3.*    We have first to convince ourselves that $f(z)$ is a meromorphic function having only a finite number of poles in $H$. We can write $f(z)$ in the form

$$(9.1.13) \qquad f(z) = Ce^{piz} \prod_{j=1}^{m} [a_j - e^{iz}] \prod_{k=1}^{n} [b_k - e^{iz}]^{-1},$$

where $p$ is an integer and the $a_j$'s and the $b_k$'s need not be distinct complex numbers, though of course no $a_j$ can be equal to a $b_k$. The singular points are obviously the roots of the equations

$$e^{iz} = b_k, \quad k = 1, 2, \cdots, n.$$

If $0 < |b_k| < 1$, the corresponding equation has a single root in $H$; if $1 < |b_k|$, the roots are in the lower half-plane. The case $|b_k| = 1$ is excluded by assumption. Thus, there are at most $n$ singular points in $H$; they are all poles, and their total multiplicity does not exceed $n$. Further, formula (9.1.13) shows that $f(z)$ may be expanded in powers of $e^{iz}$ convergent for

$$y > \max_k \; [-\log |b_k|].$$

Moreover, by (9.1.11) this series cannot contain a constant term or any negative powers of $e^{iz}$. Thus, $f(z)$ tends exponentially to zero as $y \to +\infty$.

After these preparations we can attack the problem. There is a positive number $B$ such that all poles in $H$ have ordinate less than $B$. Let $H(B)$ be that part of $H$ in which $y < B$, and let $C$ be the boundary of $H(B)$ described in the positive sense. Then

$$\int_C f(z) \, dz = 2\pi i \, \mathbf{S}[f; H(B)] = 2\pi i \, \mathbf{S}[f; H].$$

Since $f(z)$ is periodic with period $2\pi$, we see that

$$i \int_0^B f(2\pi + iy) \, dy = i \int_0^B f(iy) \, dy.$$

Since the integrals along the vertical sides of $C$ are described in opposite sense, they have the sum zero no matter what value $B$ has. The integral along the upper horizontal boundary of $H[B]$ is less than a constant times $e^{-B}$ for large values of $B$. It follows that

$$\lim_{B \to \infty} \int_C f(z) \, dz = \int_0^{2\pi} f(x) \, dx,$$

and this proves (9.1.12).

If in (9.1.11) the limit relation holds for $-\infty$ instead of $+\infty$, the value of the integral is $-2\pi i\, \mathbf{S}[f;\, \bar{H}]$, where $\bar{H}$ is symmetric to $H$ with respect to the real axis. It is clear that a number of trigonometric integrals belong to this type.

Certain integrals of a mixed type, involving trigonometric functions and rational functions, can be handled by these methods. Thus, if $a > 0$,

$$(9.1.14) \qquad \int_{-\infty}^{\infty} \frac{e^{iax}}{1+x^2}\, dx = 2\pi i\, \mathbf{S}[f;\, U] = \pi e^{-a},$$

as we can prove by the argument used in deriving formula (9.1.2). Taking the real part, we get the well-known formula

$$\int_{0}^{\infty} \frac{\cos ax\, dx}{1+x^2} = \frac{\pi}{2} e^{-a}.$$

More of a challenge is offered by the integral

$$(9.1.15) \qquad J \equiv \int_{0}^{\infty} \frac{\sin x}{x}\, dx = \frac{\pi}{2}.$$

This is an improper integral, defined as the limit of $\displaystyle\int_{0}^{\omega}$ as $\omega \to \infty$. In order to obtain the result (9.1.15), we consider

$$(9.1.16) \qquad \int_{C} \frac{e^{iz}}{z}\, dz,$$

where $C = C(\varepsilon,\, R)$ is the contour shown in Figure 22. It consists of the semi-circles $|z| = R$, $y \geqq 0$, and $|z| = \varepsilon$, $y \geqq 0$, joined by the line segments

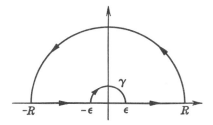

**Figure 22**

$(-R,\, -\varepsilon)$ and $(\varepsilon,\, R)$ of the real axis. This integral is obviously 0. The integral along the large semicircle tends to zero with $1/R$, for its absolute value is dominated by

$$\int_{0}^{\pi} e^{-R\sin\theta}\, d\theta = 2\int_{0}^{\pi/2} e^{-R\sin\theta}\, d\theta < 2\int_{0}^{\pi/2} e^{-2\theta R/\pi}\, d\theta = \frac{\pi}{R}(1 - e^{-R}).$$

Here we have used the inequality

$$\sin x \geq \frac{2}{\pi} x,$$

which is valid for $0 \leq x \leq \dfrac{\pi}{2}$ . The sum of the rectilinear integrals reduces to

$$2i \int_{\varepsilon}^{R} \frac{\sin x}{x} \, dx \rightarrow 2iJ$$

as $\varepsilon \rightarrow 0$, $R \rightarrow \infty$. Finally, the integral over the small semicircle, $\gamma$ say, equals

$$\int_{\gamma} \frac{dz}{z} + \int_{\gamma} (e^{iz} - 1) \frac{dz}{z} .$$

The first integral has the constant value $-\pi i$, the second one tends to zero with $\varepsilon$. Thus

$$2iJ - \pi i = 0,$$

and this is the desired result.

A totally different problem where, however, the same technique works, is that of evaluating[1]

$$(9.1.17) \qquad \mathbf{PV} \left\{ \frac{1}{\pi} \int_{-\infty}^{\infty} \frac{dt}{(t - \alpha - \beta i)t} \right\} = \frac{i \operatorname{sgn} \beta}{\alpha + i\beta},$$

$$\beta \neq 0, \quad \operatorname{sgn} \beta = \begin{cases} +1, \ \beta > 0, \\ -1, \ \beta < 0. \end{cases}$$

Using the same contour of integration as in (9.1.16) and setting $\alpha + \beta i = a$, we find for $R > |a|$

$$\int_{C} \frac{dz}{(z - a)z} = \frac{2\pi i}{a} \quad \text{or} \quad 0,$$

according as $\beta > 0$ or $\beta < 0$. On the other hand, letting $R \rightarrow \infty$, we see that the limit of the left-hand side is

$$\mathbf{PV} \left\{ \int_{-\infty}^{\infty} \frac{dt}{(t - a)t} \right\} + \frac{\pi i}{a} .$$

Equating these two expressions for the limit, we obtain (9.1.17). The integral (9.1.17) is important in the theory of conjugate functions, where

$$(9.1.18) \qquad \mathbf{PV} \left\{ \frac{1}{\pi} \int_{-\infty}^{\infty} \frac{f(t)dt}{t - x} \right\} \equiv \tilde{f}(x)$$

is referred to as the *conjugate* (or the *Hilbert transform*) of $f(t)$.

---

[1] The principal value is the limit of the integral over the range $|t| > \varepsilon$ as $\varepsilon \rightarrow 0$. In (9.1.18) the range is $|t - x| > \varepsilon$ instead. Cf. formula (7.4.4).

Finally, let us consider a couple of integrals involving multiple-valued functions. We start with

(9.1.19) $$\int_1^\infty (x^2 - 1)^{-\frac{1}{2}} \frac{dx}{x} = \frac{\pi}{2}.$$

Here we cut the $z$-plane along the real axis from $-\infty$ to $-1$ and from $+1$ to $+\infty$. We give $(z^2 - 1)^{\frac{1}{2}}$ its real positive value on the upper edge of the right cut. This makes the square root negative on the lower edge of the right cut, positive on the lower edge of the left cut, and negative on the upper edge. We then consider the integral

$$\int_C (z^2 - 1)^{-\frac{1}{2}} \frac{dz}{z},$$

where $C = C(\delta, \varepsilon, R)$ is the contour of Figure 23. It consists of two arcs from a large circle $|z| = R$, arcs from two small circles $|z \pm 1| = \varepsilon$, and four line segments, parallel to the real axis and at a distance $\delta$ from the latter. Since $(z^2 - 1)^{-\frac{1}{2}} = -i$ for $z = 0$ with our choice of the square root, we see that the

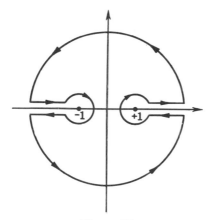

**Figure 23**

value of the integral is $2\pi$. We deform the contour by letting $\delta \to 0$, $\varepsilon \to 0$, $R \to \infty$, in this order. When $\delta \to 0$, we see that the rectilinear integrals now are taken along the real axis. Taking into account the signs of the square root on the different edges, and replacing $x$ by $-x$ in the integrals along the negative real axis, we find that the four integrals add up to

$$4\int_{1+\varepsilon}^R \to 4\int_1^\infty$$

as $\varepsilon \to 0$, $R \to \infty$. The integral along the big circle goes to 0 as $R^{-1}$, and the integrals along the small circles tend to 0 as $\varepsilon^{\frac{1}{2}}$. This gives (9.1.19).

Our last example is

$$(9.1.20) \qquad \int_0^\infty x^{a-1}(1+x)^{-1}\,dx = \frac{\pi}{\sin \pi a}, \quad 0 < a < 1,$$

where the power is given its principal value. Here we cut the plane along the positive real axis. Now $z^a$ is real positive on the upper edge of the cut but is multiplied by $\exp(2\pi ia)$ on the lower edge. The integral

$$\int_C z^{a-1}(1+z)^{-1}\,dz = -2\pi i e^{\pi ia},$$

when taken along the contour of Figure 24. It consists of arcs of two circles

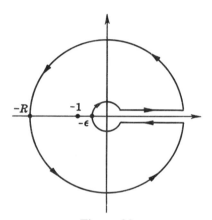

**Figure 24**

with center at $z = 0$, one of radius $\varepsilon$, the other of radius $R$, joined by two line segments parallel to the real axis and at a distance $\delta$ from the latter. The value of the integral of course equals $2\pi i$ times the residue at the pole $z = -1$. Here we let $\delta \to 0$, $\varepsilon \to 0$, $R \to \infty$. The integrals along the circles tend to zero as $\varepsilon \to 0$ and $R \to \infty$, since $0 < a < 1$. The sum of the integrals along the positive real axis is $[1 - \exp(2\pi ia)]$ times the left side of (9.1.20), so that this product equals $-2\pi i e^{\pi ia}$. The stated result is then immediate.

## EXERCISE 9.1

**1.** Evaluate the residues of the following functions at the points indicated:

    **a.** $\dfrac{z^3 + 3z}{z^2 - 1}, \quad z = 1.$         **b.** $\dfrac{z^2 - 1}{z^2 + 1}, \quad z = i.$

**c.**  $\cot z, \quad z = \pi.$          **e.**  $z^{-2}(\sin z)^{-1}, \quad z = 0.$

**d.**  $\dfrac{z^4 + 1}{z^2(z - 2)^3}, \quad z = 2.$          **f.**  $(z - \pi) \cot^2 z, \quad z = 0.$

**2.** Obtain the Maclaurin series of $\tan z$ through terms of order 5 with the aid of the identity $\cos z \tan z = \sin z$.

**3.** Obtain the Maclaurin series of $\sec z$ through terms of order 6, using $\sec z \cos z = 1$.

**4.** Expand the function in Problem 1a in powers of $z$ valid for $|z| < 1$. Use $(z^2 - 1)f(z) = z^3 + 3z$, and find all coefficients.

**5.** Evaluate by the calculus of residues

**a.**  $\displaystyle\int_{-\infty}^{\infty} \dfrac{dx}{x^2 - 2x + 5}.$          **c.**  $\displaystyle\int_{-\infty}^{\infty} \dfrac{x^2 - 1}{(x^2 + 1)^2} dx.$

**b.**  $\displaystyle\int_{-\infty}^{\infty} \dfrac{(x^2 + 2)\, dx}{(x^2 + 2x + 2)(x^2 - 2x + 2)}.$          **d.**  $\displaystyle\int_{0}^{\infty} \dfrac{x^2}{x^4 + 1} dx.$

**6.** If $f(z)$ is a rational function of $z^3$, vanishing at $\infty$ and having no poles on the positive real axis, then

$$\int_{0}^{\infty} f(x)\, dx$$

can be expressed in terms of residues. State an analogue of Theorem 9.1.2 for this case.

**7.** Given

$$\omega_n(t) = \overline{\omega_{-n}(t)} = \left(\frac{|n|}{\pi}\right)^{\frac{1}{2}} \prod_{k=1}^{n-1} \frac{t - ki}{t + ki} \frac{1}{t + ni}, \quad n = 1, 2, 3, \cdots.$$

Prove that $\{\omega_n(t)\}$ is an orthonormal system, in the sense that

$$\int_{-\infty}^{\infty} \omega_n(t)\overline{\omega_m(t)}\, dt = \delta_{mn}$$

where the Kronecker $\delta$'s equal 1 if $m = n$ and 0 otherwise.

**8.** Prove that

$$\mathrm{PV}\left\{\frac{1}{\pi}\int_{-\infty}^{\infty} \omega_n(t) \frac{dt}{t - x}\right\} = -i\, \mathrm{sgn}\, n \cdot \omega_n(x).$$

**9.** State and prove an analogue of Theorem 9.1.2 for integrals of the form

$$\int_{-\infty}^{\infty} e^{iax} R(x)\, dx,$$

where $a > 0$ and $R(x)$ is rational.

**10.** Since $2 \sin^2 x = \Re[1 - e^{2ix}]$, the method used in discussing the integral (9.1.15) can be used to evaluate

$$\int_0^\infty \left(\frac{\sin x}{x}\right)^2 dx.$$

Find its value.

**11.** Evaluate $\displaystyle\int_0^1 (1 - x^2)^{-\frac{1}{2}} dx$ by use of residues.

**12.** Evaluate $\mathbf{PV} \left\{ \dfrac{1}{\pi} \displaystyle\int_{-\infty}^\infty \dfrac{dt}{(t - a)^2(t - x)} \right\}$, $\quad \Im(a) \neq 0$.

**13.** Evaluate $\displaystyle\int_0^{2\pi} \dfrac{dx}{2 - \sin x}$.

**9.2. The principle of the argument.** The subject matter of this section centers around the logarithmic derivative of a meromorphic function.

THEOREM 9.2.1.    *Let $C$ be a "scroc," and let $f(z)$ be a function which is meromorphic in $C^*$, but which has neither zeros nor poles on $C$. Then*

(9.2.1)
$$\frac{1}{2\pi i} \int_C \frac{f'(z)}{f(z)} dz = Z_f - P_f,$$

*where $Z_f$ is the number of zeros of $f(z)$ in $C_i$, and $P_f$ is the number of poles of $f(z)$ there, zeros and poles being counted with their proper multiplicities.*

*Proof.*    The singularities of the integrand in $C_i$ are the zeros and poles of $f(z)$. If $z = a_j$ is one of these points, suppose that

$$f(z) = (z - a_j)^{m_j} f_j(z),$$

where $f_j(z)$ is holomorphic at $z = a_j$, $f_j(a_j) \neq 0$. $m_j$ is a positive integer if $z = a_j$ is a zero of $f(z)$ of multiplicity $m_j$, and $m_j$ is a negative integer if $z = a_j$ is a pole of multiplicity $-m_j$. Then we have

$$\frac{f'(z)}{f(z)} = \frac{m_j}{z - a_j} + \frac{f_j'(z)}{f_j(z)}.$$

Thus, the residue of the integrand at $z = a_j$ equals $m_j$. The residue theorem now gives

$$\sum_j m_j = Z_f - P_f$$

as the value of the integral.

The left-hand side of (9.2.1) suggests that it should be possible to restate the result just obtained in terms of properties of $\log f(z)$. This observation leads to another of the basic principles of complex function theory.

THEOREM 9.2.2.    [THE PRINCIPLE OF THE ARGUMENT.]  *Let $C$ be a "scroc,"
and let $f(z)$ be meromorphic in $C^*$, but having neither zeros nor poles on $C$.  When
$z$ describes $C$, the argument of $f(z)$ increases by a multiple of $2\pi$, namely*

$$(9.2.2) \qquad\qquad \triangle_C \arg f(z) = 2\pi(Z_f - P_f),$$

*where $Z_f$ and $P_f$ have the meaning stated above.*

*Proof.*    By assumption, $f(z) \neq 0, \infty$ on $C$.  At a point $t = a$ on $C$ we can
then select a determination of $\log f(a) = \log R(a) + i\Theta(a)$ by requiring, for
instance, that $0 \leq \Theta(a) < 2\pi$.  Since $f(z)$ is holomorphic on $C$, there exists a
circle $| z - a | < r(a)$ in which

$$\log f(z) = \log f(a) + \log \left\{ 1 + \frac{f(z) - f(a)}{f(a)} \right\}$$

$$(9.2.3)$$

$$= \log f(a) + \sum_{n=1}^{\infty} \frac{(-1)^{n-1}}{n} \left\{ \frac{f(z) - f(a)}{f(a)} \right\}^n$$

is holomorphic. It suffices that $f(z)$ is holomorphic, and that $| f(z) - f(a) | < | f(a) |$
in this circle.  At each point $z_0$ on $C$ we determine an $r(z_0)$ such that (1) $f(z)$ is
holomorphic in $| z - z_0 | < r(z_0)$, and (2) $| f(z) - f(z_0) | < | f(z_0) |$ in this circle.
Consider now the set of circles

$$| z - z_0 | < \tfrac{1}{2}r(z_0), \quad z_0 \in C.$$

They provide a covering of the bounded closed set $C$.  By the Heine-Borel
theorem there is a finite subcovering with circles at the points $a_1, a_2, \cdots, a_n$, say.
We may assume that the original point $a$ is included in this set of centers, that
$a_1 = a$, and that $a_j$ precedes $a_k$ on $C$ if $j < k$.  We now consider the circles

$$| z - a_k | < r(a_k), \quad k = 1, 2, \cdots, n.$$

These circles, of course, provide a more ample covering, and now we are sure
that the center $a_k$ of the $k$th circle lies in the interior of the $(k - 1)$th circle.
In particular, $a_n$ lies in the first circle.

In this first circle, $\log f(z)$ is uniquely determined and holomorphic, once
$\log f(a_1)$ is chosen.  Then $\log f(a_2)$ has a definite value, and $\log f(z)$ is represented
by (9.2.3) in the second circle, provided we replace $a$ by $a_2$.  In this manner we
can proceed from one circle to the next.  In each circle $\log f(z)$ is holomorphic
and is the analytic continuation of the function $\log f(z)$ defined in the preceding
circle.  Thus, we arrive at $z = a_1$, after completing the circuit, with a value for
$\log f(a_1)$ which may possibly differ from the original value.  The difference is an
integral multiple of $2\pi i$.

On the other hand, if we integrate along $C$,

$$\int_{a_1}^{z} \frac{f'(t)}{f(t)} dt = \log f(z) - \log f(a_1),$$

since both sides are locally holomorphic functions, having the same derivatives on $C$ and vanishing at $z = a_1$. It follows that

$$\int_C \frac{f'(t)}{f(t)} dt = i \triangle_C \arg f(z).$$

Comparing this with (9.2.1) we get the desired result.

In the applications of the principle of the argument, the following theorem due to Eugène Rouché (1832–1910) is often useful.

THEOREM 9.2.3.    *Under the assumptions of the preceding theorems, suppose that* $f(z)$ *can be written as the sum of two functions meromorphic in* $C^*$

$$f(z) = g(z) + h(z),$$

*such that* $g(z) \neq 0, \infty$ *on* $C$. *Further, suppose that*

(9.2.4)          $$| g(z) | > | h(z) |  \quad on \ C.$$

*Then the change in the argument of* $f(z)$ *when* $z$ *describes* $C$ *is the same as the change in the argument of* $g(z)$, *and the difference between the number of zeros and the number of poles is the same for both functions:*

(9.2.5)                    $$Z_f - P_f = Z_g - P_g.$$

*Proof.*    We have

$$f(z) = g(z)\left\{1 + \frac{h(z)}{g(z)}\right\}.$$

As $z$ describes $C$, the fraction $h(z)/g(z)$ always has an absolute value less than 1. Hence, if we write

$$w(z) = 1 + \frac{h(z)}{g(z)},$$

then on $C$ we have $| w(z) - 1 | < 1$. This implies that $| \arg w(z) | < \frac{1}{2}\pi$. Thus, $\arg w(z)$ has to return to its original value after $z$ has described $C$, and

(9.2.6)                    $$\triangle_C \arg f(z) = \triangle_C \arg g(z),$$

as asserted.  Formula (9.2.5) is a direct consequence of (9.2.2) and (9.2.6).

We shall give a couple of applications of Rouché's theorem.  The first is to give an alternate proof of Theorem 8.2.6, the fundamental theorem of algebra. We have

$$f(z) = z^n + \sum_{k=1}^{n} c_k z^{n-k},$$

and we set

$$g(z) = z^n, \quad h(z) = \sum_{k=1}^{n} c_k z^{n-k}.$$

Using the previously obtained estimate

$$| h(z) | \leqq \frac{c \, | \, z \, |^n}{| \, z \, | - 1}, \quad c = \max | \, c_k \, |,$$

we see that

$$| \, g(z) \, | > | \, h(z) \, | \quad \text{on} \quad | \, z \, | = R \quad \text{if} \quad R > 1 + c.$$

But $g(z)$ has $n$ (coincident) zeros inside the circle $| \, z \, | = R$. Rouché's theorem then asserts that $f(z)$ also has $n$ zeros inside this circle. This is the desired result.

The second application is to prove that the transcendental equation

(9.2.7)                    $\tan z = az, \quad a > 0,$

has two complex (purely imaginary) roots if $0 < a < 1$, together with infinitely many real roots, and only real roots if $1 \leqq a$. This equation arises in the propagation of heat in a sphere.[1] We set

$$f(z) = az - \tan z, \quad g(z) = az, \quad h(z) = -\tan z$$

and consider the square with vertices at

$$z = N\pi(\pm 1 \pm i),$$

where $N$ is an integer so large that

$$aN\pi > 1.005.$$

On the vertical sides of the square, $| \tan z | < 1$. On the horizontal sides we have

$$\max | \tan z | = \coth (N\pi) < 1.005 \quad \text{for} \quad N \geqq 1.$$

It follows that

$$| \, az \, | > | \tan z |$$

on the sides of the square, so that the numbers of zeros and poles in the square satisfy

$$Z_f = P_f + Z_g - P_g = 2N + 1.$$

---

[1] The transcendental equation (9.2.7) is one of many treated in the monumental *Théorie Analytique de la Chaleur* published by Fourier in 1822. Joseph Fourier (1768–1830), to whom we owe the first serious study of boundary value problems, and whose name is attached to Fourier series and Fourier transforms, led a rich, colorful life. He taught at the École Normale Supérieure and at the École Polytéchnique. In 1798–1799 he participated in Napoleon's expedition to Egypt, where he served as a commissioner in the French military government, chief of jurisdiction, and secretary of the Egyptian Institute in Cairo. He was the driving force of the scientific commission that made the first archeological description of Egypt. After his return to France he held the prefecture of the department of Isère for 14 years. While at Grenoble, Fourier fired the imagination of an eleven-year-old boy, Jean François Champollion (1790–1832), who twenty years later was to succeed in decoding the hieroglyphics. The mathematical institute of the University of Grenoble is named for Fourier.

Plotting in the Euclidean plane the graphs of

$$y = \tan x, \quad y = ax, \quad -N\pi \leq x \leq N\pi,$$

we see that the straight line has one and only one intersection with the tangent curve in each of the intervals

$$[-(k + \tfrac{1}{2})\pi, \, -k\pi], \; [k\pi, \, (k + \tfrac{1}{2})\pi], \quad k = 1, 2, \cdots, N - 1.$$

In addition, there are three intersections in $\left(-\dfrac{\pi}{2}, \dfrac{\pi}{2}\right)$ if $a \geq 1$, but only one if $0 < a < 1$. Thus, in the first case all roots are real. In the second case there are two complex roots. Since $f(-z) = -f(z)$, and $f(z)$ is real on the real axis, these two roots are of the form $\pm\beta i$. More precisely, $\beta$ is a root of

$$\tanh y = ay.$$

The proof of the following interesting extension of Theorem 9.2.1 is left to the student.

**THEOREM 9.2.4.** *Let $f(z)$ be meromorphic in $C^*$ with zeros at $a_1, a_2, \cdots, a_n$ and poles at $b_1, b_2, \cdots, b_m$, none of which lie on $C$. Let $g(z)$ be holomorphic in $C^*$. Then*

$$(9.2.8) \qquad \frac{1}{2\pi i} \int_C g(z) \frac{f'(z)}{f(z)} \, dz = \sum_{j=1}^{n} g(a_j) - \sum_{k=1}^{m} g(b_k),$$

*where each zero and pole occurs as often in the sum as is required by its multiplicity.*

The choice $g(z) = \log z$ is ordinarily not admissible in this theorem, but with suitable precautions the corresponding integral can be handled, and the result is basic in the modern theory of entire and meromorphic functions (see Volume II, Chapter 14). We shall prove Jensen's theorem.

**THEOREM 9.2.5.** *Let $f(z)$ be meromorphic in $|z| \leq R < \infty$, $f(z) \not\equiv 0$. Let the zeros of $f(z)$ in this disk be $a_1, a_2, \cdots, a_n$, and let the poles be $b_1, b_2, \cdots, b_m$, where each zero and pole is repeated as often as is indicated by its multiplicity. Further, let $f(0) \neq 0, \infty$. Then*

$$(9.2.9) \qquad \frac{1}{2\pi} \int_0^{2\pi} \log |f(Re^{i\theta})| \, d\theta = \log |f(0)| + \sum_{j=1}^{n} \log \frac{R}{|a_j|} - \sum_{k=1}^{m} \log \frac{R}{|b_k|}.$$

*Proof.* Without restricting the generality we may suppose that a narrow strip around the positive real axis is free from zeros and poles of $f(z)$. If this is not the case, we replace $z$ by $ze^{i\omega}$ for a suitable real number $\omega$. Such a rotation does not change any of the quantities entering in formula (9.2.9), but it may be chosen so as to free the positive real axis from zeros and poles. We can then form the integral

$$(9.2.10) \qquad J = \frac{1}{2\pi i} \int_C \log z \, \frac{f'(z)}{f(z)} \, dz,$$

taken over the contour of Figure 24. Here the logarithm is determined so that $\log(-1) = \pi i$. We apply Theorem 9.2.4 and obtain

$$(9.2.11) \qquad J = \sum_{j=1}^{n} \log a_j - \sum_{k=1}^{m} \log b_k.$$

In the present case $R$ is fixed, but we do let $\delta$ and $\varepsilon$ tend to zero. As $\varepsilon \to 0$, the integral along $\gamma$ tends to zero, since $\varepsilon \log \varepsilon \to 0$. As $\delta \to 0$, we get two integrals along the real axis from 0 to $R$. One of these is

$$\frac{1}{2\pi i} \int_0^R \log r \, \frac{f'(r)}{f(r)} \, dr,$$

the other is

$$-\frac{1}{2\pi i} \int_0^R (\log r + 2\pi i) \frac{f'(r)}{f(r)} \, dr,$$

and their sum is

$$-\int_0^R \frac{f'(r)}{f(r)} \, dr = \log f(0) - \log f(R).$$

There remains

$$\frac{1}{2\pi i} \int_\Gamma \log z \, \frac{f'(z)}{f(z)} \, dz$$

extended over $|z| = R$ in the positive sense, starting and ending at $z = R$. To this integral we apply integration by parts and obtain

$$\frac{1}{2\pi i} \triangle_\Gamma [\log z \log f(z)] - \frac{1}{2\pi i} \int_\Gamma \log f(z) \frac{dz}{z}.$$

The integrated part equals

$$\frac{1}{2\pi i} \{ (\log R + 2\pi i)[\log f(R) + (n - m)2\pi i] - \log R \log f(R) \}$$
$$= \log f(R) + (n - m)(\log R + 2\pi i).$$

Here we have used the fact that $f(z)$ has $n$ zeros and $m$ poles inside $\Gamma$; by Theorem 9.2.2 the increment of $\log f(z)$ is then $(n - m)2\pi i$. Thus we have also

$$(9.2.12) \quad J = \log f(0) + (n - m)(\log R + 2\pi i) - \frac{1}{2\pi} \int_0^{2\pi} \log f(Re^{i\theta}) \, d\theta,$$

since the two terms in $\log f(R)$ cancel. Equating the two expressions (9.2.11) and (9.2.12) for $J$, and rearranging, we obtain

$$(9.2.13) \quad \frac{1}{2\pi} \int_0^{2\pi} \log f(Re^{i\theta}) \, d\theta$$
$$= \log f(0) + \sum_1^n \log \frac{R}{a_j} - \sum_1^m \log \frac{R}{b_j} + (n - m)2\pi i.$$

Finally, taking real parts, we obtain formula (9.2.9).

## EXERCISE 9.2

**1.** If $f(z)$ is holomorphic in $C^*$, and if $b$ is a complex number such that $f(z) \neq b$ on $C$, what is the significance of the integral

$$\frac{1}{2\pi i} \int_C \frac{f'(z)}{f(z) - b} \, dz?$$

Extend this result to the case of a meromorphic function.

**2.** Prove that a polynomial in $z$ of degree $n$ takes on every value $b$ exactly $n$ times if multiple values are counted with their proper multiplicities. In what sense is this true for $b = \infty$? Show that at most $(n - 1)$ finite values $b$ can be multiple values.

**3.** Prove that a rational function takes every value the same number of times in the extended plane.

**4.** Prove that the equation

$$z \tan z = a, \quad a > 0,$$

has only real roots. Is this true also if $a < 0$?

**5.** Evaluate

$$\frac{1}{2\pi i} \int_C z \frac{f'(z)}{f(z)} \, dz$$

over a large circle if $f(z)$ is a polynomial.

**6.** Let $C: |z| = \frac{1}{2}$, and $|w| < \frac{3}{4}$, and consider the integral

$$f(w) = \frac{1}{\pi i} \int_C \frac{z(z + 1)}{z^2 + 2z - w} \, dz.$$

$f(w)$ is a holomorphic function (Why?) in the circle indicated (actually for $|w| < 1$). Find $f(0)$ and $f'(0)$.

**7.** Let $f(z)$ be holomorphic and bounded in the unit disk. Suppose that $f(a_n) = 0$, $n = 1, 2, 3, \cdots$, where $|a_n| < 1$. Use Theorem 9.2.5 to show that either

$$\sum_{n=0}^{\infty} [1 - |a_n|] < \infty$$

or $f(z) \equiv 0$. Is it necessary to assume $f(z)$ bounded?

**9.3. Summation and expansion theorems.** The calculus of residues enables us to express certain integrals as sums of residues of the integrand. Conversely, if $f(z)$ is a holomorphic function in a domain $D$, if $\{a_k \mid k = 1, 2, 3, \cdots, n\}$ are points in $D$, and if $\{\mu_k \mid k = 1, 2, 3, \cdots, n\}$ are given complex numbers, then

$$(9.3.1) \qquad \sum_{k=1}^{n} \mu_k f(a_k) = \int_C f(t) G(t) \, dt,$$

where $C$ is a "scroc" such that $\{a_k\} \subset C_i \subset C^* \subset D$, and $G(t)$ is any function meromorphic in $D$ with simple poles of residue $\mu_k$ at the points $a_k$, and such that these are the only poles of $G(t)$ in $C^*$. This formula, however, does not ordinarily tell us much about the sum in the left member of (9.3.1) that was not known in advance. Things become more interesting and also more difficult if we are dealing with infinite sums.

Series of the form

$$(9.3.2) \qquad \sum_{n=-\infty}^{\infty} f(n) \quad \text{or} \quad \sum_{n=-\infty}^{\infty} (-1)^n f(n),$$

where $f(t)$ is a meromorphic function having no poles at the integers, can be handled by the calculus of residues, provided that the behavior of $f(t)$ on a sequence of contours $C_N$ expanding to infinity is suitably restricted. Let $C_N$ be the sequence of squares with vertices at the points

$$(N + \tfrac{1}{2})[\pm 1 \pm i], \quad N = 1, 2, 3, \cdots.$$

The first series under (9.3.2) calls for a summatory function $G(t)$ having simple poles of residue 1 at the integers. Such functions are given by

$$(9.3.3) \qquad G_1(t) = \pi \cot \pi t \quad \text{and} \quad G_2(t) = 2\pi i [e^{2\pi i t} - 1]^{-1}.$$

To fix the ideas, let us take the first alternative. This leads to the integral

$$(9.3.4) \qquad J_N[f] \equiv \frac{1}{2i} \int_{C_N} \cot \pi t \, f(t) \, dt.$$

If $f(t)$ is a rational function, the integral exists for all large values of $N$. If $f(t)$ is a transcendental meromorphic function, however, there may be poles of $f(t)$ on $C_N$ for arbitrarily large values of $N$. Since the choice of $C_N$ is highly arbitrary, we shall assume that $C_N$ avoids the poles of $f(t)$. This can be accomplished, whenever necessary, by a slight deformation of the path in such a manner that the pole in question lies interior to $C_N$.

The value of $J_N[f]$, by the residue theorem, is the sum of two sets of residues, one corresponding to the poles of $\cot \pi t$, the other corresponding to the poles of $f(t)$. The first set gives

$$(9.3.5) \qquad \sum_{n=-N}^{N} f(n) \equiv S_N[f];$$

the second we write as

$$(9.3.6) \qquad \mathbf{S}_N\{\pi \cot \pi t \, [f(t)]\},$$

where the boldface brackets serve as a reminder that only the poles of $f(t)$ are to be considered. Thus

$$(9.3.7) \qquad J_N[f] = S_N[f] + \mathbf{S}_N\{\pi \cot \pi t \, [f(t)]\}.$$

This relation leads directly to

THEOREM 9.3.1.    *Let $f(t)$ be a meromorphic function having no poles at any of the integers.  Let*

$$(9.3.8) \qquad \lim_{N \to \infty} J_N[f] = 0,$$

*and let*

$$(9.3.9) \qquad \lim_{N \to \infty} \sum_{n=-N}^{N} f(n) \equiv S[f]$$

*exist as a finite number.  Then*

$$(9.3.10) \qquad S[f] = - \lim_{N \to \infty} \mathbf{S}_N \{ \pi \cot \pi t \, [f(t)] \}.$$

The quantity occurring in the left member of (9.3.8) was called the *integral residue* of $\pi \cot \pi t f(t)$ by Cauchy, who, as far back as 1827, realized its importance and derived the consequences listed here and below.

Suppose that $f(t)$ is a rational function, and that for large values of $t$

$$(9.3.11) \qquad f(t) = \frac{a_1}{t} + \frac{a_2}{t^2} + \cdots.$$

The conditions (9.3.8) and (9.3.9) are satisfied.  This is obvious if

$$a_1 = 0,$$

for then

$$| t^2 f(t) | \leq M$$

on $C_N$ for all large values of $N$, so that

$$| J_N[f] | \leq \frac{4(2N+1)}{(N + \tfrac{1}{2})^2} \, M \to 0 \quad \text{as} \quad N \to \infty.$$

Further, the series

$$\sum_{n=-\infty}^{\infty} f(n)$$

is absolutely convergent since the $n$th term is dominated by $Mn^{-2}$ for large $n$. If $a_1 \neq 0$, the situation is less obvious.  We note, however, that

$$\frac{1}{2i} \int_{C_N} \cot \pi t \, \frac{dt}{t} = 0$$

for every $N$.  In fact, for $| t | < 1$

$$(9.3.12) \qquad \frac{\pi}{t} \cot \pi t = \frac{1}{t^2} + \alpha_1 + \alpha_2 t^2 + \cdots,$$

since the origin is a double pole and the left side is an even function of $t$.  But this implies that the residue at $t = 0$ is 0. At $t = n$ the residue is $1/n$, and this is cancelled by the residue at $t = -n$. It follows that

$$J_N[f] = \frac{1}{2i} \int_{C_N} \cot \pi t \left[ f(t) - \frac{a_1}{t} \right] dt,$$

and here

$$\left| t^2 \left[ f(t) - \frac{a_1}{t} \right] \right| \le M, \quad |t| > R,$$

so that the previous argument applies. Further, for large $n$,

$$f(n) + f(-n) = \left\{ \frac{a_1}{n} + \frac{a_2}{n^2} + \cdots \right\} + \left\{ -\frac{a_1}{n} + \frac{a_2}{n^2} - \cdots \right\}$$

$$= 2 \frac{a_2}{n^2} + O\left( \frac{1}{n^3} \right),$$

so that (9.3.9) exists, even though the series is not absolutely convergent.

The following example illustrates Theorem 9.3.1 and is an important result: Let $z$ be a fixed complex number but not an integer or zero. The function

$$f(t) = \frac{1}{t - z}$$

obviously satisfies (9.3.11) and, hence, also the conditions of Theorem 9.3.1. It follows that

$$\frac{1}{2i} \int_{C_N} \frac{\cot \pi t}{t - z} dt = \sum_{n=-N}^{N} \frac{1}{n - z} + \pi \cot \pi z \to 0$$

as $N \to \infty$. Hence

(9.3.13)                          $$\pi \cot \pi z = \lim_{N \to \infty} \sum_{n=-N}^{N} \frac{1}{z - n}.$$

Here we can easily obtain an absolutely convergent series by adding $1/n$ to the $n$th term, $n \ne 0$. The added quantities cancel pairwise, so that

(9.3.14)                          $$\pi \cot \pi z = \frac{1}{z} + \sum_{n=-\infty}^{\infty}{}' \left\{ \frac{1}{z - n} + \frac{1}{n} \right\},$$

where the prime indicates that the value $n = 0$ is omitted.

Termwise differentiation of this series gives

(9.3.15)                          $$\frac{\pi^2}{\sin^2 \pi z} = \sum_{n=-\infty}^{\infty} \frac{1}{(z - n)^2},$$

as asserted in (8.5.2). Termwise integration also leads to important results. We conclude from (9.3.12) that

$$\pi \cot \pi t - \frac{1}{t}$$

has a removable singularity at $t = 0$. Hence, for $|z| \leq 1 - \delta$ and a rectilinear path of integration,

$$\int_0^z \left[ \pi \cot \pi t - \frac{1}{t} \right] dt = \sum_{n=-\infty}^{\infty}{}' \int_0^z \left\{ \frac{1}{t - n} + \frac{1}{n} \right\} dt,$$

or

(9.3.16)
$$\log \frac{\sin \pi z}{\pi z} = \sum_{-\infty}^{\infty}{}' \log \left[ \left( 1 - \frac{z}{n} \right) e^{\frac{z}{n}} \right],$$

where each logarithm has its principal value. Termwise integration is justified by Lemma 7.10.1, since we are integrating a uniformly convergent series. It is an easy matter to show that the resulting series is also uniformly convergent. In fact, Theorem 8.7.2 together with Problem 2, Exercise 6.2, shows that

$$\left| \log \left[ \left( 1 - \frac{z}{n} \right) e^{\frac{z}{n}} \right] \right| \leq -\log \left[ 1 - \left| \frac{z}{n} \right|^2 \right] < 2 \left| \frac{z}{n} \right|^2, \quad |z| \leq \tfrac{1}{2}n,$$

so that (9.3.16) is uniformly convergent. For $|z|$ sufficiently small, say for $|z| \leq \tfrac{1}{2}$, we can replace the sum of the logarithms by the (principal value of) the logarithm of the product, that is, after multiplication by $\pi z$,

(9.3.17)
$$\sin \pi z = \pi z \prod_{n=-\infty}^{\infty}{}' \left[ \left( 1 - \frac{z}{n} \right) e^{\frac{z}{n}} \right].$$

This relation between two entire functions must hold for all values of $z$ and not merely in the neighborhood of the origin. Compare formula (8.8.21), where this product occurs. Combining factors symmetric with respect to $n = 0$, we obtain the simple product

(9.3.18)
$$\sin \pi z = \pi z \prod_{n=1}^{\infty} \left( 1 - \frac{z^2}{n^2} \right),$$

due to Euler.

We turn now to the second class of series under (9.3.2). Here a suitable summatory function is

$$\frac{\pi}{\sin \pi z},$$

which has a simple pole with residue $(-1)^n$ at $z = n$ for every $n$. We set

(9.3.19)
$$A_N[f] = \frac{1}{2i} \int_{C_N} \frac{f(t)}{\sin \pi t} \, dt,$$

(9.3.20)
$$T_N[f] = \sum_{n=-N}^{N} (-1)^n f(n),$$

and we note the relation

(9.3.21)
$$A_N[f] = T_N[f] + \mathbf{S}_N \left\{ \frac{\pi[f(t)]}{\sin \pi t} \right\},$$

where the second term on the right is the sum of the residues of the integrand
at the poles of $f(t)$. We thus obtain

THEOREM 9.3.2.    *Let $f(t)$ be a meromorphic function having no poles at
the integers.  Let*

(9.3.22)                              $$\lim_{N \to \infty} A_N[f] = 0,$$

*and let*

(9.3.23)                              $$\lim_{N \to \infty} T_N[f] \equiv T[f]$$

*exist as a finite number.  Then*

(9.3.24)                $$T[f] = -\lim_{N \to \infty} \mathbf{S}_N \left\{ \frac{\pi[f(t)]}{\sin \pi t} \right\}.$$

This theorem also applies to rational functions $f(t)$ satisfying (9.3.11).
But in the present case we can go much farther, since

$$| \sin \pi(x + iy) |^{-1} < Ce^{-\pi|y|}, \quad | y | > 1.$$

We note the two following possibilities:  Let $a$ be a real number, $-\pi < a < \pi$,
and take

(9.3.25)            $$f_1(t) = R(t) \sin at, \quad f_2(t) = R(t) \cos at,$$

where $R(t)$ is a rational function satisfying (9.3.11) and such that $R(n)$ is finite
for all $n$.  A simple calculation shows that

(9.3.26)            $$| A_N[f] | < c_1 N^{-1} + c_2 \exp[-(\pi - | a | )N],$$

where the first term comes from the vertical sides of $C_N$ and the second one
comes from the horizontal sides.  $T_N[f_1]$ is the $N$th partial sum of the sine series

(9.3.27)            $$T[f_1] \sim \sum_{n=1}^{\infty} (-1)^n [R(n) - R(-n)] \sin na.$$

If the coefficient $a_1 = 0$ in (9.3.11), this series is absolutely convergent;  in any
event, it converges, since the series

$$\sum_{n=1}^{\infty} (-1)^n \frac{1}{n} \sin na = \sum_{n=1}^{\infty} \frac{1}{n} \sin n(a + \pi)$$

converges by the Corollary of Theorem 5.4.5.  On the other hand,

(9.3.28)        $$T[f_2] = R(0) + \sum_{n=1}^{\infty} (-1)^n [R(n) + R(-n)] \cos na$$

is always absolutely convergent.  We conclude from Theorem 9.3.2 that these two trigonometric series can be summed in finite form for $-\pi < a < \pi$. Taking $R(t) = (t - z)^{-1}$, $z$ not an integer, we get

(9.3.29) $$\sum_{n=1}^{\infty} (-1)^n \frac{n \sin na}{z^2 - n^2} = \frac{\pi}{2} \frac{\sin az}{\sin \pi z},$$

(9.3.30) $$\sum_{n=1}^{\infty} (-1)^n \frac{\cos na}{z^2 - n^2} = -\frac{1}{2z^2} + \frac{\pi}{2z} \frac{\cos az}{\sin \pi z}.$$

## EXERCISE 9.3

**1.** Find the sum of

$$\sum_{n=1}^{\infty} \frac{1}{n^2},$$

(**a**) by calculus of residues;   (**b**) by (9.3.15);   and (**c**) by (9.3.18).

**2.** Same question and same methods applied to

$$\sum_{n=1}^{\infty} \frac{1}{(2n - 1)^2}.$$

**3.** Find the sum of

$$\sum_{n=1}^{\infty} \frac{1}{n^4}.$$

**4.** Find the sum of

$$\sum_{n=1}^{\infty} (-1)^{n-1} \frac{1}{n^2}.$$

**5.** Find the sum of

$$\sum_{n=0}^{\infty} (n^2 + 1)^{-1}.$$

**6.** Take $f(t) = (t - z - \frac{1}{2})^{-1}$ in Theorem 9.3.1, and verify the expansion

$$\pi \tan \pi z = -\sum_{n=-\infty}^{\infty} \left[ \frac{1}{n + \frac{1}{2}} + \frac{1}{z - n - \frac{1}{2}} \right].$$

**7.** Use the preceding expansion to derive the infinite product of $\cos \pi z$.

**8.** Set $f(t) = (t - z)^{-1}$ in Theorem 9.3.2, and verify that

$$\frac{\pi}{\sin \pi z} = \frac{1}{z} + \sum_{-\infty}^{\infty}{}' (-1)^n \left[ \frac{1}{z - n} + \frac{1}{n} \right].$$

**9.** Verify that $\displaystyle\sum_{n=0}^{\infty} (-1)^n (2n + 1)^{-3} = \frac{\pi^3}{32}$.

**10.** Verify formulas (9.3.29) and (9.3.30).

**9.4. Inverse functions.** The calculus of residues also offers a simple and powerful approach to the theory of inverse functions, to the implicit function theorem, and to other types of inversion problems. The following theorem should be compared with Theorem 4.5.1, which deals with the same question. To simplify matters, we take $w_0 = 0$.

THEOREM 9.4.1.    *Suppose that $f(z)$ is holomorphic in $|z| < R$, that $f(0) = 0$, $f'(0) \neq 0$, and that $f(z) \neq 0$ for $0 < |z| < r \leq R$. Let $C$ be the circle $|z| = \rho$, $\rho < r$. Then*

$$(9.4.1) \qquad g(w) \equiv \frac{1}{2\pi i} \int_C \frac{tf'(t)}{f(t) - w} \, dt$$

*defines a holomorphic function of $w$, at least for $|w| < m = \min_\theta |f(\rho e^{i\theta})|$. For such values of $w$, $z = g(w)$ is the only solution of*

$$(9.4.2) \qquad f(z) = w$$

*which tends to zero with $w$.*

*Proof.*    For a fixed $w$ with $|w| < m$ and for $z$ on the circle $C$, we have

$$|f(z)| \geq m > |w|.$$

By Rouché's theorem, this implies that the two holomorphic functions

$$f(z) \quad \text{and} \quad f(z) - w$$

have the same number of zeros inside $C$. Since $f(z)$ has a single zero in this circle, namely at $z = 0$, we conclude that equation (9.4.2) has a single root, $g(w)$ say, inside $C$. By Theorem 9.2.4 this root is given by formula (9.4.1). To show that this root, $g(w)$, is a holomorphic function of $w$, at least for $|w| < m$, it suffices to note that

$$\frac{1}{f(t) - w} = \frac{1}{f(t)} + \frac{w}{[f(t)]^2} + \cdots + \frac{w^n}{[f(t)]^{n+1}} + \cdots.$$

This series converges uniformly with respect to $t$ and $w$ as long as $t \in C$ and $|w| \leq m(1 - \delta)$, $\delta > 0$, and it may be multiplied by the bounded function $tf'(t)$ and integrated term by term. The result is

$$(9.4.3) \qquad g(w) = \sum_{n=0}^{\infty} w^n \frac{1}{2\pi i} \int_C \frac{tf'(t)}{[f(t)]^{n+1}} \, dt.$$

We illustrate this theorem by the following example, which is of some independent interest: We take the cubic equation

$$(9.4.4) \qquad z^3 + 3z - w = 0,$$

and we plan to show that the root of this equation which tends to zero with $w$ is given by the hypergeometric function

$$(9.4.5) \qquad z = \tfrac{1}{3} w F(\tfrac{1}{3}, \tfrac{2}{3}, \tfrac{3}{2}, -\tfrac{1}{4} w^2) \quad \text{for} \quad |w| < 2.$$

For the notation, see Problem 5 of Exercise 5.4.

Equation (9.4.4) looks rather special, but the general cubic,

$$t^3 + a_1 t^2 + a_2 t + a_3 = 0, \quad \text{with} \quad 3a_2 \neq a_1{}^2,$$

may be reduced to this form by setting

$$3t = (3a_2 - a_1{}^2)^{\frac{1}{2}} z - a_1.$$

In the present case, we have $f(z) = z^3 + 3z$, and the coefficients $g_n$ in $g(w) = \Sigma\, g_n w^n$ are given by

(9.4.6)
$$g_n = \frac{3}{2\pi i} \int_C \frac{t^2 + 1}{t^n (3 + t^2)^{n+1}}\, dt,$$

where $C$ is the circle $|\,t\,| = \rho < \sqrt{3}$. If $n$ is even, $n = 2k$, the integrand is an even function of $t$, and the residue at $t = 0$ is necessarily zero; that is,

$$g_{2k} = 0, \quad k = 0, 1, 2, \cdots.$$

On the other hand, when $n$ is odd, $n = 2k + 1$, we find that

(9.4.7)
$$g_{2k+1} = 3^{-3k-1}\left\{ \binom{-2k-2}{k} + 3\binom{-2k-2}{k-1} \right\}$$

$$= (-1)^k 3^{-3k-1} \frac{(2k+2)(2k+3)\cdots(3k)}{k!} = (-1)^k 3^{-3k-1} \frac{(3k)!}{k!(2k+1)!}.$$

To reduce this to the hypergeometric form, we introduce the abbreviation

$$\alpha(\alpha + 1)(\alpha + 2) \cdots (\alpha + n - 1) = (\alpha, n).$$

With this notation,

$$k! = (1, k), \quad 3^{-3k}(3k)! = (\tfrac{1}{3}, k)(\tfrac{2}{3}, k)(1, k), \quad (2k + 1)! = 4^k (1, k)(\tfrac{3}{2}, k).$$

Formula (9.4.5) is an immediate consequence of these relations. The radius of convergence of the series is $R = 2$. The coefficients of the series have the sign of $(-1)^k$ so that all terms have the same argument on the imaginary axis. It follows from Theorem 5.7.1 that $w = \pm 2i$ are singular points of $g(w)$. That these points must be singularities of $g(w)$ could be foreseen, since they, and the point at infinity, are the only multiple values of $f(z)$. We have

$$f(\pm i) = \pm 2i, \quad f'(\pm i) = 0.$$

From the latter condition it follows that $g'(w) \to \infty$ as $w \to \pm 2i$.

The two other roots are also expressible by hypergeometric functions, and they are

(9.4.8)
$$\pm i\sqrt{3}\,F(-\tfrac{1}{6}, \tfrac{1}{6}, \tfrac{1}{2}, -\tfrac{1}{4}w^2) - \tfrac{1}{6}wF(\tfrac{1}{3}, \tfrac{2}{3}, \tfrac{3}{2}, -\tfrac{1}{4}w^2).$$

The same method applies to any trinomial equation

$$z^n + nz - w = 0$$

and leads to similar results.

Formula (9.4.1) defines a holomorphic function of $w$ under more general assumptions than those stated in the theorem. Suppose that

(9.4.9)          $$f(0) = f'(0) = \cdots = f^{(k-1)}(0) = 0, \quad f^{(k)}(0) \neq 0.$$

Then the argument used in the proof of the theorem shows that for small values of $\mid w \mid$, there are $k$ roots of the equation

$$f(z) = w$$

inside the circle $C$, and these roots coalesce for $w = 0$. If the roots are

$$z_1(w), z_2(w), \cdots, z_k(w),$$

then the integral (9.4.1) and the series (9.4.3) represent

(9.4.10)          $$g(w) = z_1(w) + z_2(w) + \cdots + z_k(w).$$

Thus the sum of the roots is a holomorphic function of $w$. Similarly we see that every sum of positive integral powers of the roots is a holomorphic function of $w$. Indeed, we have

(9.4.11)          $$G_m(w) \equiv \sum_{j=1}^{k} [z_j(w)]^m = \frac{1}{2\pi i} \int_C \frac{t^m f'(t)}{f(t) - w} \, dt.$$

It is proved in the theory of equations that any symmetric function of the roots of an algebraic equation is expressible in one, and only one, way as a polynomial in the power sums. In particular, this applies to the so-called *elementary symmetric functions*. We conclude that the elementary symmetric functions of $z_1(w), z_2(w), \cdots, z_k(w)$ are holomorphic functions of $w$ in some neighborhood of the origin. Thus, if we form the polynomial in $Z$

$$[Z - z_1(w)][Z - z_2(w)] \cdots [Z - z_k(w)] = Z^k + g_1(w)Z^{k-1} + \cdots + g_k(w),$$

then the coefficients $g_1(w), g_2(w), \cdots, g_k(w)$ are holomorphic functions of $w$ in a neighborhood of $w = 0$. Thus we have proved:

THEOREM 9.4.2.    *If $f(z)$ is holomorphic in $\mid z \mid < R$ and (9.4.9) holds, then for small values of $\mid w \mid$ the equation*

$$f(z) = w$$

*has $k$ roots $z_1(w), \cdots, z_k(w)$, which tend to zero with $w$. These roots also satisfy an algebraic equation of degree $k$*

(9.4.12)          $$Z^k + g_1(w)Z^{k-1} + \cdots + g_k(w) = 0,$$

*whose coefficients are holomorphic functions of $w$ tending to zero with $w$.*

Theorem 9.4.2 is due to Cauchy (1831). It is better known as Weierstrass's *Vorbereitungssatz*.

So far our discussion has not thrown any light on the analytic nature of the roots themselves.

THEOREM 9.4.3.    *Under the assumptions of Theorem 9.4.2 there exists a function $g(W)$, holomorphic for $|W|$ sufficiently small, such that for any fixed small value of $w$, the $k$ determinations of $g(w^{1/k})$ represent the roots*

$$z_1(w), z_2(w), \cdots, z_k(w)$$

*in some order. If $w$ describes a circuit about the origin, the roots are permuted cyclically.*

*Proof.*    We start by choosing an $a$ such that

$$k!a^k = f^{(k)}(0) \quad (\neq 0).$$

We then write

$$w = W^k, \quad f(z) = (az)^k \left[ 1 + \sum_{n=1}^{\infty} b_n z^n \right]$$

and observe that there exists a circular disk $|z| < r \leq R$ in which the function inside the brackets is different from zero. This function will then possess a $k$th root, holomorphic in the disk and uniquely determined by its value at the origin where we set the root equal to 1. We have then

$$\left[ 1 + \sum_{n=1}^{\infty} b_n z^n \right]^{1/k} = \sum_{m=0}^{\infty} \binom{1/k}{m} \left[ \sum_{n=1}^{\infty} b_n z^n \right]^m \equiv 1 + \sum_{p=1}^{\infty} c_p z^p.$$

It follows that there exists a $k$th root of $w$ such that

$$(9.4.13) \qquad F(z) \equiv az \left[ 1 + \sum_{p=1}^{\infty} c_p z^p \right] = W.$$

This equation can be solved for $z$ by the previous method, and we find that

$$(9.4.14) \qquad z = g(W) = \frac{1}{2\pi i} \int_C \frac{t F'(t)}{F(t) - W} \, dt$$

for $W$ in some neighborhood of the origin, and for a suitable choice of the contour $C$. Thus

$$F[g(W)] = W, \quad \{F[g(W)]\}^k = W^k = w,$$

or

$$(9.4.15) \qquad f[g(w^{1/k})] = w.$$

Now, the quantity $a$ of (9.4.13) is determined up to an arbitrary $k$th root of unity; if we replace $a$ by $\eta a$, $\eta^k = 1$, then $W$ is replaced by $\eta W$ and $w^{1/k}$ by $\eta w^{1/k}$, but the equation (9.4.15) remains valid. It follows that the $k$ roots $z_j(w)$ can be numbered in such a fashion that

$$(9.4.16) \quad z_j(w) = g[\omega^j w^{1/k}], \quad \omega = e^{2\pi i/k}, \quad 0 \leq \arg w^{1/k} < \frac{2\pi}{k}, \quad j = 1, 2, \cdots, k.$$

It is obvious from this representation that the roots are permuted cyclically when $w$ makes a circuit about the origin. This completes the proof.

The general implicit function theorem for analytic functions can also be handled by the calculus of residues. Here the problem is to find $w$ as a function of $z$, if it is given that there exists a relation

$$(9.4.17) \qquad\qquad F(z, w) = 0$$

between $z$ and $w$. Normally we can expect to get local solutions only. Suppose, for instance, that $(z_0, w_0)$ lies on the variety (9.4.17), and that we ask for a function $f(z)$, defined for values of $z$ close to $z_0$, such that $f(z_0) = w_0$ and

$$F(z, f(z)) \equiv 0 \quad \text{for} \quad |z - z_0| < r.$$

There is no restriction in assuming

$$z_0 = 0, \quad w_0 = 0.$$

With respect to $F(z, w)$ we shall suppose it to be a holomorphic function of $(z, w)$ in a neighborhood of $(0, 0)$. More precisely, we suppose that

$$(9.4.18) \qquad F(z, w) = \sum_{m=0}^{\infty} \sum_{n=0}^{\infty} a_{mn} z^m w^n, \quad a_{00} = 0, \quad a_{01} \neq 0,$$

and that the double series is absolutely convergent for

$$|z| \leq R_1, \quad |w| \leq R_2.$$

Here the restriction imposed on $a_{01}$ serves to ensure the existence of a unique solution $w = f(z)$, holomorphic in a neighborhood of $z = 0$ and tending to zero with $z$. We are now in a position to state and prove (a case of) the implicit function theorem.

THEOREM 9.4.4.   *If $F(z, w)$ is given by formula (9.4.18), then there exists a unique function $f(z)$, holomorphic in some neighborhood $|z| < \rho$ of $z = 0$, such that $f(0) = 0$ and*

$$F(z, f(z)) \equiv 0, \quad |z| < \rho.$$

*This function is represented by*

$$(9.4.19) \qquad\qquad f(z) = \frac{1}{2\pi i} \int_C w \, \frac{F_w(z, w)}{F(z, w)} \, dw,$$

*where $F_w(z, w)$ is the partial derivative of $F(z, w)$ with respect to $w$, and $C$ is a suitably chosen circle $|w| = r, r < R_2$.*

*Proof.*   We note that

$$F(0, w) = a_{01} w + a_{02} w^2 + \cdots + a_{0n} w^n + \cdots$$

has a simple zero at $w = 0$, and that there exists an $r_2$, $0 < r_2 \leqq R_2$ such that

$$F(0, w) \neq 0, \quad 0 < |w| < r_2.$$

As $z \to 0$,

$$F(z, w) \to F(0, w),$$

uniformly with respect to $w$ for $|w| \leqq R_2$. Hence, if we fix $r$, $0 < r < r_2$, there exists a number $r_1$, $0 < r_1 \leqq R_1$, such that

$$(9.4.20) \qquad |F(z, w) - F(0, w)| < |F(0, w)|, \quad |z| \leqq r_1, \quad |w| \leqq r.$$

By the theorem of Rouché, this implies that for each $z$ with $|z| \leqq r_1$ the functions of $w$

$$F(z, w) \quad \text{and} \quad F(0, w)$$

have the same number of zeros in $|w| < r$. Since this number is 1 in the case of $F(0, w)$, we conclude that $F(z, w)$ has a single zero, $f(z)$ say. By Theorem 9.2.4, this zero is given by (9.4.19), where we can take $C$ to be the circle $|w| = r$. The integral exists, since $F(0, w) \neq 0$ on $C$, and by (9.4.20) this implies that $F(z, w) \neq 0$ on $C$ for any choice of $z$ with $|z| \leqq r_1$. Formal differentiation with respect to $z$ gives

$$(9.4.21) \qquad f'(z) = \frac{1}{2\pi i} \int_C w \, \frac{F(z, w)F_{zw}(z, w) - F_z(z, w)F_w(z, w)}{[F(z, w)]^2} \, dw.$$

This integral exists since $F(z, w) \neq 0$ for the values of $z$ and $w$ under consideration, and a straightforward but tedious computation shows that the formal derivative is the limit of the difference quotient. Hence $f(z)$ is holomorphic, and the theorem is proved.

For such an important theorem there are naturally a number of proofs available. We observe that the method of successive approximations used in the proof of Theorem 4.5.1 also applies in the present more general situation. We owe to Cauchy still another method, based on the use of power series whose convergence is established with the aid of suitable majorants. Cauchy used that same method to prove existence theorems for differential equations. He called it *calcul des limites*; here the term *limites* has the meaning of "bounds" rather than that of limits in the strict sense of the word. We shall sketch a variant of this method since it is a powerful one having many applications.

It is convenient to rewrite the equation in the form

$$(9.4.22) \qquad w = b_{10}z + \sum_{j=0}^{\infty} \sum_{k=0}^{\infty}{}' b_{jk}z^j w^k \equiv G(z, w),$$

where the prime indicates that $j + k \geq 2$, and

$$b_{jk} = -\frac{a_{jk}}{a_{01}}.$$

We try to solve the equation

$$w = G(z, w)$$

by a power series for $w$ in $z$:

$$(9.4.23) \qquad w = \sum_{n=1}^{\infty} c_n z^n.$$

Substitution of this series in (9.4.22) gives the relation

$$\sum_{n=1}^{\infty} c_n z^n = b_{10} z + \sum_{j=1}^{\infty} \sum_{k=1}^{\infty} {}' b_{jk} z^j \left[ \sum_{n=1}^{\infty} c_n z^n \right]^k,$$

and this must be an identity in $z$ for small values of $|z|$. Here we expand the $k$th powers and collect terms. This is permitted if the multiple series involved are absolutely convergent. Assuming this to be the case, we are led to the following recurrence relations for the determination of the $c$'s:

$$(9.4.24) \qquad \begin{aligned} c_1 &= b_{10}, \\ c_2 &= b_{20} + b_{11} c_1 + b_{02} c_1{}^2, \\ c_3 &= b_{30} + b_{21} c_1 + b_{12} c_1{}^2 + b_{03} c_1{}^3 + b_{11} c_2 + 2 b_{02} c_1 c_2, \end{aligned}$$

$$\cdots \cdots \cdots \cdots \cdots \cdots \cdots \cdots \cdots \cdots$$

Thus, $c_n$ is a polynomial in $c_1, c_2, \cdots, c_{n-1}$, and it is a linear form in the coefficients $b_{jk}$ with $j + k \leq n$; moreover, all numerical coefficients are positive integers. These equations determine the coefficients $c_n$ uniquely. We note that if the $b_{jk}$'s are non-negative, so are the $c_n$'s. This fact will be the basis for the convergence proof. We start by making some remarks on majorants.

We say that a series

$$H(z, w) = h_{10} z + \sum_{j=0}^{\infty} \sum_{k=0}^{\infty} {}' h_{jk} z^j w^k$$

with non-negative coefficients is a *majorant* of $G(z, w)$, written

$$G(z, w) \ll H(z, w),$$

if for all $j, k$ we have

$$(9.4.25) \qquad |b_{jk}| \leq h_{jk}.$$

The problem of solving the equation

$$w = H(z, w)$$

by a power series

$$w = \sum_{n=1}^{\infty} \alpha_n z^n$$

leads to a system of equations of the form (9.4.24), where the $b_{jk}$ and the $c_n$ are replaced by $h_{jk}$ and $\alpha_n$ respectively. If these equations are solved, it is clear that the resulting sequence $\{\alpha_n\}$ consists of non-negative quantities, and, moreover, that

$$|c_n| \leq \alpha_n \quad \text{for all } n.$$

This is obviously true for $n = 1$ and is proved by induction for $n > 1$. Thus $G(z, w) \ll H(z, w)$ implies

$$\sum_{n=1}^{\infty} c_n z^n \ll \sum_{n=1}^{\infty} \alpha_n z^n.$$

Hence, the series on the left is absolutely convergent whenever the series on the right has this property. It remains to find a suitable majorant for $G(z, w)$.

We can obviously take

$$h_{jk} = |b_{jk}|,$$

and this gives in a sense the *least majorant* and, hence, also the best estimate of the radius of convergence of (9.4.23) obtainable by this method. We set

$$(9.4.26) \qquad B(z, w) \equiv |b_{10}| z + \sum_{j=0}^{\infty} \sum_{k=0}^{\infty}{}' |b_{jk}| z^j w^k,$$

and we let

$$w = \sum_{n=1}^{\infty} \beta_n z^n$$

be the corresponding solution of

$$w = B(z, w).$$

Then

$$G(z, w) \ll B(z, w), \quad \sum_{n=1}^{\infty} c_n z^n \ll \sum_{n=1}^{\infty} \beta_n z^n.$$

We shall now prove

THEOREM 9.4.5.    *Let $s$ and $t$ be any two positive numbers such that the series* (9.4.26) *converges for $z = s$, $w = t$. Then $R$, the radius of convergence of* (9.4.23), *satisfies the inequality*

$$(9.4.27) \qquad R \geq R_{st} = \min \left\{ s, \left[ \frac{t}{B(s, t)} \right]^2 s \right\}.$$

*Proof.*    We base the proof on the inequality

$$(9.4.28) \qquad B(\alpha^2 a, \alpha b) \leq \alpha^2 B(a, b),$$

valid for any $\alpha$ with $0 \leq \alpha \leq 1$ and for any positive numbers $a$, $b$ for which the right-hand side has a meaning. This inequality follows from

$$B(\alpha^2 a, \alpha b) = |b_{10}| a\alpha^2 + \sum_{j=0}^{\infty} \sum_{k=0}^{\infty}{}' |b_{jk}| a^j b^k \alpha^{2j+k}.$$

$\alpha^2$ can be factored out, and the remaining factor is $\leq B(a, b)$.

Suppose first that $t < B(s, t)$, and set

$$\alpha = \frac{t}{B(s, t)}.$$

Then

$$R_{st} = \alpha^2 s.$$

Suppose further that we know that for some $N \geq 1$

(9.4.29)
$$S_N \equiv \sum_{n=1}^{N} \beta_n (R_{st})^n < \alpha t.$$

This is obviously true for $N = 1$, since

$$\beta_1 R_{st} = |b_{10}| s\alpha^2 < B(s, t)\alpha^2 = \alpha t.$$

Now

$$S_{N+1} < B(R_{st}, S_N)$$

for, if we write the series involved, we see that all the terms in the $(N + 1)$th partial sum on the left cancel, by virtue of (9.4.24); after the common terms have been removed, there are still positive terms left on the right. Hence, by (9.4.28),

$$S_{N+1} < B(R_{st}, S_N) < B(R_{st}, \alpha t) = B(\alpha^2 s, \alpha t) < \alpha^2 B(s, t) = \alpha t.$$

Thus, (9.4.29) holds for every $N$, and the series (9.4.23) is absolutely convergent for all $z$ with $|z| \leq R_{st}$. This was proved under the assumption that $t < B(s, t)$. On the other hand, if $t \geq B(s, t)$, we set $\alpha = 1$, $R_{st} = s$ in the preceding argument. The conclusion is the same. This proves (9.4.27). From the absolute convergence of the series (9.4.23) for $|z| \leq R_{st}$ together with the estimate of its sum implied by (9.4.29), it follows that the result of substituting (9.4.23) into (9.4.29) is an absolutely and uniformly convergent multiple series for $|z| \leq R_{st}$. This series may then be rearranged as an ordinary power series by virtue of the Weierstrass double series theorem. Since the recurrence relations (9.4.24) hold, we have that (9.4.23) is the actual solution of the problem. This completes the proof.

　　The use of the least majorant (9.4.26) in the implicit function theorem and in corresponding existence theorems for systems of differential equations is due to Lindelöf and dates from 1896–1899.[1] The Cauchy majorant is given in Problem 4 of Exercise 9.4.

　　Formula (9.4.27) contains the two parameters $s$ and $t$. If $G(z, w)$ is a polynomial or an entire function in $z$ and $w$, then these parameters are completely

---

[1] Ernst Lindelöf (1870–1946) was the founder of the school of analysis in Finland. His many contributions are distinguished by a wealth of basically simple ideas of great carrying power, clearly and elegantly presented. His extension of the maximum principle (Phragmén-Lindelöf theorems) and his systematic study of majorants were basic for the development of function theory. His monograph *Le Calcul des Résidus et ses Applications à la Théorie des Fonctions* (Gauthier-Villars, Paris, 1905; reprinted by Chelsea Publishing Company, New York, 1947) is still the best in the field.

arbitrary, but even if the series (9.4.22) does not converge for all values of $z$ and $w$, we still have great freedom in the choice of $s$ and $t$. There arise questions of what is the optimal choice of $s$ and $t$ and of what is the exact value of $R$. We shall answer both questions under suitable restriction on $B(z, w)$.

THEOREM 9.4.6.   *Suppose that $B(z, w)$, defined by (9.4.26), is a polynomial or an entire function in $z$ and $w$ which either has degree $\geq 2$, or is transcendental, in $w$. Then the radius of holomorphy, $R$, of the solution of $w = B(z, w)$ which tends to zero with $z$, equals the positive root of the simultaneous equations*

$$(9.4.30) \qquad t = B(s, t), \quad B_t(s, t) = 1.$$

*The Maclaurin series of the solution is absolutely convergent for $|z| \leq R$.*

Proof.   We write

$$(9.4.31) \qquad b(z) \equiv \sum_{n=1}^{\infty} \beta_n z^n, \quad |z| < R.$$

Differentiation of

$$b(z) = B(z, b(z))$$

gives

$$b'(z) = \frac{B_z(z, b(z))}{1 - B_w(z, b(z))}.$$

It follows that for $0 < s < R$, $t = b(s)$, we have

$$B_t(s, t) < 1.$$

But $b(s) > \beta_1 s$, so that

$$B_t(s, t) > B_t(s, \beta_1 s),$$

and here the right member becomes infinite with $s$. We conclude that $R < \infty$. Further, $b(s)$ is an increasing function of $s$. Thus,

$$\lim_{s \to R} b(s) \equiv T$$

exists, and here $T$ must be finite since otherwise the increasing function $B_t(s, b(s))$ could not stay $\leq 1$. From this fact it also follows that the series (9.4.31) is absolutely convergent on $|z| = R$. It remains to prove that

$$B_t(R, T) = 1.$$

Suppose contrariwise that $B_t(R, T) < 1$. Since

$$\frac{\partial}{\partial t} \frac{t}{B(s, t)} = \frac{B(s, t) - t B_t(s, t)}{B^2(s, t)},$$

we can conclude that

$$\max_t \frac{t}{B(R, t)} > 1,$$

and this maximum would be reached at some point $t = T + \varepsilon$, $\varepsilon > 0$. Since $B(s, T + \varepsilon)$ is an increasing function of $s$, tending to $\infty$ with $s$, we can find an $s = R + \eta$, $\eta > 0$, such that

$$\frac{T + \varepsilon}{B(R + \eta, T + \varepsilon)} = 1.$$

Using $s = R + \eta$, $t = T + \varepsilon$ in (9.4.27), we could then conclude that the series (9.4.31) converges for $s = R + \eta$. This is impossible, however, and we have shown that $B_t(R, T) = 1$. This completes the proof of the theorem, and it shows also that $s = R$, $t = T$, the solutions of (9.4.30), give the optimal choice in (9.4.27).

## EXERCISE 9.4

**1.** If $g(w) = \dfrac{1}{\pi i} \displaystyle\int_C \dfrac{t(t + 1)}{t^2 + 2t - w}\, dt$   with   $C: |t| = 1,$

find the Maclaurin series of $g(w)$ and sum it in closed form.

**2.** The equation

$$z^3 + z^2 = w$$

has two roots $z_1(w)$ and $z_2(w)$, tending to zero with $w$. Represent these functions by integrals of the form (9.4.14) and find the first three terms in the expansions according to powers of $\sqrt{w}$.

**3.** Obtain the Maclaurin series of

$$\frac{1}{2\pi i} \int_C \frac{t^2(3t + 2)}{t^3 + t^2 - w}\, dt \quad \text{with} \quad C: |t| = \tfrac{1}{2},$$

through terms of the second order.

**4.** The Cauchy majorant of the function $G(z, w)$ of (9.4.22) is

$$C(z, w) = M \left\{ \left[ \left(1 - \frac{z}{s}\right)\left(1 - \frac{w}{t}\right) \right]^{-1} - 1 - \frac{w}{t} \right\},$$

if $|G(z, w)| \leq M$ for $|z| < s$, $|w| < t$. The solution of $w = G(z, w)$ is an algebraic function. Find this function explicitly and determine the radius of holomorphy of the branch that is holomorphic at $z = 0$.

**5.** Determine $R$ if $B(z, w)$ equals

        **a.** $\tfrac{1}{2}(z + w^2)$.        **b.** $\tfrac{1}{3}(z + w^3)$.

**6.** Solve the equation $w^3 + 3w - z = 0$ by a power series in $z$ and find the terms of degree $\leq 4$. Check your result with (9.4.5), where, however, $z$ and $w$ have to be interchanged.

## COLLATERAL READING

As a general reference see

LINDELÖF, E. *Le Calcul des Résidus et ses Applications à la Théorie des Fonctions*, Gauthier-Villars, Paris, 1905. Reprinted by Chelsea Publishing Company, New York, 1947.

Any French text on classical analysis contains numerous applications of the calculus of residues. See, for instance,

GOURSAT, E. *Cours d'Analyse*, Seventh Edition, Vol. II, Chap. XIV. Gauthier-Villars, Paris, 1949. (For an English translation of an earlier edition of Vol. II, Part I, see HEDRICK, E. R., and DUNKEL, O., *Functions of a Complex Variable*. Ginn and Company, Boston, 1916.)

There is much material on the calculation of principal values of integrals in

FRANKLIN, PHILIP. *Functions of Complex Variables*, Sections 76, 77. Prentice-Hall, Inc., Englewood Cliffs, New Jersey, 1958.

For the best majorant, see

LINDELÖF, E. "Démonstration Élémentaire de l'Existence des Fonctions Implicites," *Bulletin des Sciences Mathématiques*, Series 2, Vol. 23 (1899), pp. 68–75.

An existence proof for differential equations with analytic coefficients, based on the use of the best majorant, is to be found in

KAMKE, E. *Differentialgleichungen reeller Funktionen*, Sections 38, 75. Akademische Verlagsgesellschaft, Leipzig, 1930.

# Appendix A

# SOME PROPERTIES OF POINT SETS

The material in this Appendix is largely supplementary to the discussion of point sets in Chapters 1 and 2.

Given a set $S$ in the complex plane, the *diameter* of $S$ is defined by

$$\text{(A.1)} \qquad d[S] = \sup \left[ \, |z_1 - z_2| \, \big| \, z_1, z_2 \in S \right].$$

The set is *bounded* if $d[S] < \infty$. If $S$ is bounded and closed, there are two points $z_1$ and $z_2$ in $S$ such that $d[S] = |z_1 - z_2|$.

If $a$ is any point in the plane, the *distance* between $a$ and the set $S$ is

$$\text{(A.2)} \qquad d(a, S) = \inf \left[ \, |z - a| \, \big| \, z \in S \right].$$

This quantity is $\geq 0$; it is 0 if and only if $a \in \bar{S}$. If $S$ is closed, there exists a point $z_0 \in S$ such that $|z_0 - a| = d(a, S)$. We note that $d(a, S)$ is a continuous function of $a$ when $S$ is fixed.

Suppose now that $S_1$ and $S_2$ are two sets in the plane. We define the distance between $S_1$ and $S_2$ to be

$$\text{(A.3)} \qquad d(S_1, S_2) = \inf \left[ \, |z_1 - z_2| \, \big| \, z_1 \in S_1, z_2 \in S_2 \right].$$

If $S_1$ and $S_2$ are closed sets and at least one of them is bounded, then the distance is assumed, that is, there are points $z_1 \in S_1$ and $z_2 \in S_2$ such that $|z_1 - z_2| = d(S_1, S_2)$. If both sets are unbounded, such a pair of points need not exist.

A point $a \in S$ is an *isolated point* of $S$ if there is an $\varepsilon$-neighborhood of $a$ which does not contain any other point of $S$. The set $\{1/n \mid n = 1, 2, 3, \cdots\}$ consists of isolated points.

Next, we give an extension of Theorem 1.2.2 on nested intervals to nested sets.

**Theorem A.1.** *If $\{S_n\}$ is a sequence of non-void closed sets such that (1) $S_n \supset S_{n+1}$ for each $n$, (2) at least one set $S_n$ is bounded, and (3) $d[S_n] \to 0$ as $n \to \infty$, then there is one and only one point $z_0$ which belongs to all $S_n$.*

*Proof.* We may assume that $S_n$ is bounded for $n > m$. In each $S_n$ we select a point $z_n$. Then $z_k \in S_n$ for $k \geq n$. The bounded set $\{z_k \mid k \geq n\}$ has at least one limit point, $z_0$ say, and since $S_n$ is a bounded closed set for $n > m$, we see that $z_0 \in S_n$. This is evidently true for each $n$ so that $z_0 \in \cap S_n$. Moreover, it is the only point with this property, since, if $a \neq z_0$, we have ultimately $|z_0 - a| > d[S_n]$, and this prevents $a$ from belonging to $S_n$.

*Covering theorems* play an important role in analysis. The general setting is the following one: There is given an object $[S, C]$. Here $S = \{a\}$ is a set of points in the plane, and $C = \{C_a\}$ is a set of point sets, usually open, such that for each $a \in S$ there is a $C_a \in C$ with the property that $a \in C_a$. $C$ is known as a covering of $S$, since obviously

$$S \subset \bigcup C_a.$$

The problem is to decide whether a subset of $C$ already gives a covering of $S$, and, in particular, whether there exists a finite subset which has this property. It is easy to give examples of objects $[S, C]$ such that no proper subset of $C$ gives a covering of $S$. One such example is the set $S = \{n\}$ of positive integers with $C_n$ given as the disk $\mid z - n \mid < \frac{1}{3}$. Another instructive example is the set $S = \{1/n \mid n = 1, 2, 3, \cdots\}$, with $C_n$ given as

$$\left| z - \frac{1}{n} \right| < \frac{1}{3n(n+1)}.$$

It is easily seen that in both cases every single $C_n$ must be used to get a covering of $S$. We note that the first set is not bounded and that the second one is not closed. If in the second case we add the limit point $z = 0$ and a cover $C_0$, say $\mid z \mid < \varepsilon$, then we can obviously find a finite subcovering of the augmented set by using $C_0$ and $C_1, C_2, \cdots, C_n$ where $1/n < \varepsilon$. We shall prove the Heine-Borel theorem (Eduard Heine, 1821–1881).

**Theorem A.2.**    *Let $S$ be a bounded closed set, and let $\{C_a\}$ be a covering of $S$ by open sets such that $a \in C_a$. Then there exists a finite set of points $a_1, a_2, \cdots, a_n$ in $S$ such that*

**(A.4)**                    $$S \subset C_{a_1} \cup C_{a_2} \cup \cdots \cup C_{a_n}.$$

*Proof.*    Suppose, contrariwise, that there exists an object $[S, C]$ such that no finite subset of $C$ will suffice to cover $S$. Since $S$ is bounded by assumption, we may suppose that $S$ is contained in the square with vertices at $z = \pm a \pm ia$. The coordinate axes divide $S$ into four subsets, one or more of which may be void. At least one of these subsets will have the property that its closure cannot be covered by a finite subset of the corresponding sets $C_a$. We denote this closed set by $S_1$ and its covering by $C_1$. We have $S \supset S_1$, $C \supset C_1$, and $S_1$ is not covered by a finite subset of $C_1$. We now subject $S_1$ to the same treatment as $S$: we divide the corresponding square of side $a$ into four equal subsquares and select one of the resulting subsets whose closure is not covered by a finite subset of $C_1$. Let this closed set be $S_2$ and its covering $C_2$ so that $S_1 \supset S_2$, $C_1 \supset C_2$. We repeat this process indefinitely and obtain a nested sequence of closed sets $\{S_n\}$. The diameter of $S_n$ equals $a2^{1/2-n}$, and each $S_n$ has the property that its covering $C_n = \{C_a \mid a \in S_n\}$ fails to contain a finite subset which covers $S_n$. By Theorem A.1 there exists a unique point $z_0 \in \bigcap S_n$. To this point $z_0$, however, corresponds a set $C_{z_0} \in C$, and the open set $C_{z_0}$ contains an $\varepsilon$-neighborhood of $z = z_0$ for some $\varepsilon > 0$. But if $a2^{1/2-n} < \varepsilon$, the set $S_n$ lies entirely in this $\varepsilon$-neighborhood and is consequently completely covered by $C_{z_0}$. Thus, our assumption that the theorem is not valid for this particular object $[S, C]$ leads to a contradiction, and we conclude that the theorem must be true.

As an application of the Heine–Borel theorem we prove the following useful result.

**Theorem A.3.** *If $S$ is a bounded closed subset of a domain $D$, then the distance of $S$ from the complement of $D$ is positive.*

*Proof.* As usual, let $\mathbf{C}[D]$ denote the complement of $D$. Since every point of $D$ is an interior point, it follows that $d(a, \mathbf{C}[D]) = \delta(a) > 0$ if $a \in S$. If $C_a$ denotes the open disk $|z - a| < \frac{1}{2}\delta(a)$, the set $C = \{C_a\}$ is a covering of the closed bounded set $S$. There is consequently a finite subcovering. Let $\rho$ be the radius of the smallest disk of the latter. Then $d(S, \mathbf{C}[D]) \geq \rho$. For if $a$ is any point of $S$, then $a$ is interior to one of the disks $C_{a_k}$ of the finite subcovering. From

$$|a - a_k| < \tfrac{1}{2}\delta(a_k), \quad d(a_k, \mathbf{C}[D]) = \delta(a_k),$$

it then follows that $d(a, \mathbf{C}[D]) > \frac{1}{2}\delta(a_k) \geq \rho$.

## EXERCISE A

**1.** Prove that if $S$ is bounded and closed, then there exist points $z_1$ and $z_2$ in $S$ such that $|z_1 - z_2| = d[S]$.

**2.** Prove that if $S$ is closed, then there exists a $z_0 \in S$ such that $|a - z_0| = d(a, S)$.

**3.** Prove that if $S_1$ and $S_2$ are closed, and at least one of them is bounded, then there are points $z_1 \in S_1$ and $z_2 \in S_2$ such that $|z_1 - z_2| = d(S_1, S_2)$.

**4.** The sets $S_1 = [z \mid y = 0]$ and $S_2 = [z \mid y(x^2 + 1) = 1]$ are closed in the finite plane. What is $d(S_1, S_2)$? Is $d(S_1, S_2) = |z_1 - z_2|$ for certain points $z_k \in S_k$?

**5.** $S_1$ and $S_2$ are two closed sets such that $d(S_1, S_2) > 0$. If

$$F(z) \equiv \frac{d(z, S_1)}{d(z, S_1) + d(z, S_2)},$$

verify that $F(z)$ is continuous, and that $0 \leq F(z) \leq 1$, the value 0 being taken on for $z \in S_1$ and the value 1 for $z \in S_2$.

**6.** If $m$ and $n$ are positive integers, $|2m^2 - n^2| \geq 1$. Let $S$ be the set of rational numbers between 1 and 2. Construct a covering of $S$ by open intervals such that $\sqrt{2}$ does not belong to any of the intervals.

## COLLATERAL READING

As general references for the topology of the plane see

KERÉKJÁRTÓ, B. v. *Vorlesungen über Topologie.* Springer-Verlag, Berlin, 1923.

NEWMAN, M. H. A. *Elements of the Topology of Plane Sets of Points.* Cambridge University Press, New York, 1951.

# Appendix B

# SOME PROPERTIES OF POLYGONS

**B.1. The Jordan theorem.** In this Appendix we shall prove the Jordan curve theorem for the case of simple closed polygons (see Sections 2.3 and 2.4 for definitions). We shall also show the possibility of triangulating such polygons and shall discuss related results.

THEOREM B.1.1. *If $\Pi$ is a simple closed polygon, then the complement of $\Pi$ is the union of two mutually exclusive domains, $\Pi_e$ and $\Pi_i$, each having $\Pi$ as its complete boundary.*

The notation is so chosen that $\Pi_e$, known as the *exterior* of the polygon, is unbounded, and $\Pi_i$, the *interior* of $\Pi$, is bounded.

The proof will be given in several stages. We say that *a simple closed polygon $\Pi$ is of class J* if it has the separation property stated in the theorem. We start by showing that $J$ is not void.

LEMMA B.1.1. *Every convex polygon is of class $J$.*

*Proof.* This is an almost immediate consequence of Definition 2.3.3. Consider the polygon $\Pi = [P_1, P_2, \cdots, P_n, P_1]$. The straight line through $P_j$ and $P_{j+1}$ (where $P_{n+1} = P_1$) determines two open half-planes, $H_j$ and $H_j^\circ$, having this line as their common boundary. The polygon $\Pi$ is convex if, for each $j$, $\Pi$ is located in the closure of one of these half-planes; we choose the notation so that this half-plane is $H_j$. Then $\cap_j H_j$ is a domain, for it is open and connected, and it has $\Pi$ as its boundary. This is the domain $\Pi_i$ of Theorem B.1.1. Further, the complement of $\Pi \cup \Pi_i$ is

$$H_1^\circ \cup H_2^\circ \cup \cdots \cup H_n^\circ = \Pi_e.$$

This is a domain, for any three half-planes whose boundary lines are not all parallel have a connected union, since each such half-plane must overlap with at least one other half-plane. Now, in a convex polygon, at most two sides may be parallel to a given direction. We conclude that $\Pi_e$ is connected, and, as the union of open sets, it is also open. Hence, $\Pi_e$ is a domain, and its boundary is found to be $\Pi$. This completes the proof.

LEMMA B.1.2. *If $\Pi$ is of class $J$, if $P_1 \in \Pi_e$, $P_2 \in \Pi_i$, and if $\pi\colon z = z(t)$, $z(0) = P_1$, $z(1) = P_2$, is a polygonal line joining $P_1$ and $P_2$, then $\pi$ has at least one point in common with $\Pi$.*

*Proof.* Since $z(t)$ is continuous, we conclude that $z(t) \in \Pi_e$ for small values of $t$, and $z(t) \in \Pi_i$ for $t$ close to 1. Let $a$ be the supremum of the values $\alpha$ such that $z(t) \in \Pi_e$ for $0 \leq t \leq \alpha$. Then, clearly, $0 < a < 1$. Thus, it follows that $z(a) \in \overline{\Pi}_e \cap \overline{\Pi}_i$. On the other hand,

$$\partial\Pi_e = \overline{\Pi}_e \cap \overline{\mathbf{C}[\Pi_e]} = \overline{\Pi}_e \cap \overline{[\Pi \cup \Pi_i]} = \overline{\Pi}_e \cap \overline{\Pi}_i.$$

Hence, $z(a) \in \partial\Pi_e = \Pi$, and the lemma is proved.

LEMMA B.1.3.    *Let* $\Pi = [P_1, P_2, \cdots, P_n, P_1]$ *be of class* $J$. *Let* $d_{jk}$ *be the distance between the sides* $[P_j, P_{j+1}]$ *and* $[P_k, P_{k+1}]$ *where* $j \neq k-1$, $k$, $k+1$, *and let* $d = \min d_{jk}$. *Let* $\delta < d$ *and consider the circle with center* $P_j$ *and radius* $\delta$. *Then the sides* $[P_{j-1}, P_j]$ *and* $[P_j, P_{j+1}]$ *separate the interior of this circle into two sectors, one in* $\Pi_e$, *the other in* $\Pi_i$.

*Proof.*    The assumption on $\delta$ implies that the only parts of $\Pi$ which lie interior to the circle are a segment from each of the sides $[P_{j-1}, P_j]$ and $[P_j, P_{j+1}]$. These two sides, which meet at the center of the circle, clearly divide the interior of the circle into two open sectors, $S_1$ and $S_2$. Let $P \in S_1$. Since $P$ is not on $\Pi$, $P$ is either in $\Pi_e$ or in $\Pi_i$. Suppose $P \in \Pi_e$. Let $Q$ be another point of $S_1$. If the angle of $S_1$ at $P_j$ is $< \pi$, then $S_1$ is convex, $[P, Q] \in S_1$, and, since $\Pi_e$ is a domain, all points of $[P, Q]$ near enough to $P$ are in $\Pi_e$. If $Q$ were in $\Pi_i$, then by Lemma B.1.2 at least one point of $[P, Q]$ must lie on $\Pi$. This contradicts the choice of $\delta$; we conclude that $Q \in \Pi_e$ and, hence, that $S_1 \subset \Pi_e$. Next we consider $S_2$. Here the angle at $P_j$ is $> \pi$. Since $\Pi$ is the common boundary of $\Pi_e$ and $\Pi_i$, every point of $\Pi$ must be limit point of points in $\Pi_i$. In particular, this is true for that part of $\Pi$ which is interior to the circle. The corresponding points of $\Pi_i$ must be in $S_2$ since $S_1 \subset \Pi_e$. Suppose that $Q \in S_2$. We can then find a point $P \in S_2 \cap \Pi_i$ such that $[P, Q] \subset S_2$; the argument used above shows that $Q \in \Pi_i$. Hence, $S_2 \subset \Pi_i$. It is obvious how this argument has to be modified if the angle of $S_1$ at $P_j$ is $> \pi$ instead.

In view of this lemma, it makes sense to speak of the exterior angle of $\Pi$ at $P_j$. It is the angle at $P_j$ of the sector $S_1$ in the above notation, that is, of the sector which belongs to $\Pi_e$.

LEMMA B.1.4.    *Let* $\Pi$ *be a polygon of class* $J$. *Let* $z_1$ *and* $z_2$ *be two distinct points of* $\Pi$, *and let* $\delta$ *be a positive number. Then there exists a polygonal line* $\pi$, *joining* $z_1$ *and* $z_2$ *and contained in* $\Pi_e$ *except for the endpoints, such that no point of* $\pi$ *has a distance from* $\Pi$ *exceeding* $\delta$.

*Proof.*    The lemma is of interest only for small values of $\delta$. We assume that $3\delta < d = \min d_{jk}$, in the notation of Lemma B.1.3. We start by constructing a simple closed polygon $\Pi°$ having the same number of sides as $\Pi$ such that $\Pi° \subset \Pi_e$ and $d(\Pi, \Pi°) < \delta$. We get such a polygon by the following construction: By Lemma B.1.3, the exterior angle $\omega_j$ of $\Pi$ at $P_j$ is uniquely defined. On the bisector of $\omega_j$ we mark the point $Q_j$ such that $d(P_j, Q_j) = \delta$. We then take $\Pi° = [Q_1, Q_2, \cdots, Q_n, Q_1]$. See Figure 25. This is clearly a closed polygon; its vertices are in $\Pi_e$ and $d(\Pi, \Pi^\circ) < \delta$. If not every point of $[Q_j, Q_{j+1}]$ were in $\Pi_e$, then by Lemma B.1.2 there would be at least one point $P$ of $\Pi$ on $[Q_j, Q_{j+1}]$. Suppose that this point $P$ lies on $[P_k, P_{k+1}]$. Then we would have $d_{jk} < \delta$, since it is clear that $k \neq j - 1$, $j$, $j + 1$. But this inequality contradicts $3\delta < d = \min d_{\alpha\beta}$, and we may conclude that $\Pi^\circ \subset \Pi_e$, as asserted. It remains to show that $\Pi^\circ$ is simple. The contrary assumption would imply that two non-adjacent sides of $\Pi^\circ$, say $[Q_j, Q_{j+1}]$ and $[Q_k, Q_{k+1}]$, have a point in common. But each side of $\Pi^\circ$ is at a distance less than $\delta$ from the corresponding side of $\Pi$ so that $d_{jk} < 2\delta$ and again we have a contradiction. Hence $\Pi^\circ$ is simple.

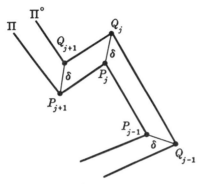

**Figure 25**

Without restricting the generality we may assume $d(z_1, z_2) \geq 3\delta$. To construct $\pi$, we note that there is a point $Z_1$ on $\Pi^\circ$ such that $d(z_1, Z_1) = d(z_1, \Pi^\circ) < \delta$ and a point $Z_2$ on $\Pi^\circ$ such that $d(z_2, Z_2) = d(z_2, \Pi^\circ)$ where $Z_1 \neq Z_2$. The two points $Z_1$ and $Z_2$ now determine complementary subarcs $\pi_1$ and $\pi_2$ of $\Pi^\circ$. Suppose that $\pi_1$ starts at $Z_1$ and ends at $Z_2$, while $\pi_2$ starts at $Z_2$ and ends at $Z_1$. We can then take $\pi$ equal to either of the following paths:

$$[z_1, Z_1] \cup \pi_1 \cup [Z_2, z_2] \quad \text{or} \quad [z_2, Z_2] \cup \pi_2 \cup [Z_1, z_1].$$

This completes the proof.

We come now to the observation that a polygon of class $J$ may be "spliced" by replacing a side by a polygonal line, satisfying certain conditions, in such a manner that the resulting polygon is still of class $J$. This is the sense of the next lemma.

**LEMMA B.1.5.** *Let $\Pi_1 = [P_1, P_2, \cdots, P_n, P_1]$ and $\Pi_2 = [P_1, P_2, Q_1, \cdots, Q_m, P_1]$ be two simple polygons of class $J$ with $\Pi_2$ in the interior of $\Pi_1$ except for the common side. Then the polygon $\Pi_3 = [P_1, Q_m, \cdots, Q_1, P_2, P_3, \cdots, P_n, P_1]$ is also of class $J$.*

*Proof.* We define the two sets:

$$S_1 = (\Pi_1)_e \cup (P_1, P_2) \cup (\Pi_2)_i, \quad S_2 = (\Pi_1)_i \cap (\Pi_2)_e,$$

where $(P_1, P_2)$ is the open line segment joining $P_1$ and $P_2$. $S_1$ is open, for if $P$ is a point in either the first or the third component of $S_1$, then there is a full neighborhood of $P$ in the same component, while if $P \in (P_1, P_2)$ the union of suitably chosen semicircular domains in $(\Pi_1)_e$ and $(\Pi_2)_i$ together with an open line segment on $(P_1, P_2)$ constitutes a full $\varepsilon$-neighborhood of $P$ in $S_1$. Further, $S_1$ is arcwise connected, for the components separately have this property, and, if $Z_1 \in (\Pi_1)_e$, $Z_2 \in (\Pi_2)_i$, and $P$ is a point of $(P_1, P_2)$, then we can join $Z_1$ with $P$ in $(\Pi_1)_e$ and $P$ with $Z_2$ in $(\Pi_2)_i$. The union of the two paths is a path in $S_1$. Hence $S_1$ is a domain.

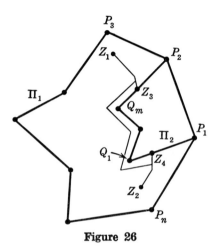

**Figure 26**

$S_2$, as the intersection of two open sets, is also open. If $Z_1$ and $Z_2$ are two points of $S_2$, they may be joined by a polygonal line $\pi_0$ in $(\Pi_1)_i$. This path may possibly enter $\Pi_2 \cup (\Pi_2)_i$, but it cannot stay there. Counting from $Z_1$, suppose that the first entry is at $Z_3$ on $\Pi_2$ and that the last exit is at $Z_4$ also on $\Pi_2$. $Z_3$ and $Z_4$ determine two subarcs of $\Pi_2$, one of which, $\pi_1$ say, lies entirely in $(\Pi_1)_i$. The closed set $\pi_1$ has a positive distance $d_0$ from $\Pi_1$. As in the proof of Lemma B.1.4, let $d = \min d_{jk}$, where the $d_{jk}$ refer to $\Pi_2$, and let $\delta$ be chosen subject to $0 < \delta < \min (d, d_0)$. Lemma B.1.4 now allows us to find a path $\pi$, joining $Z_3$ with $Z_4$ in $(\Pi_2)_e$, such that the distance of any of its points from $\pi_1$ does not exceed $\delta$. This will also ensure that $\pi$ lies in $(\Pi_1)_i$. We have now a path joining $Z_1$ and $Z_2$ which lies in $S_2$, except for the two points $Z_3$ and $Z_4$; this path is made up of $\pi$ together with the two arcs of $\pi_0$ from $Z_1$ to $Z_3$ and from $Z_4$ to $Z_2$. It is now clear that we can modify this path in the neighborhoods of $Z_3$ and $Z_4$ in such a manner that the resulting path stays away from $Z_3$ and $Z_4$. We can, for instance, omit the parts of the path which lie within circular $\delta$-neighborhoods of $Z_3$ and $Z_4$, and then join the loose ends by line segments in the intersections of the neighborhoods with $(\Pi_2)_e$. The resulting path $\pi_2$ joins $Z_1$ and $Z_2$ in $S_2$. Thus, $S_2$ is also a domain.

Every point of the plane not on $\Pi_1$ is either in $(\Pi_1)_e$ or in $(\Pi_1)_i$, every point not on $\Pi_2$ is either in $(\Pi_2)_e$ or in $(\Pi_2)_i$, and in each case the alternatives are mutually exclusive. It follows then from the definitions of $S_1$ and $S_2$ that every point not on $\Pi_3$ is either in $S_1$ or in $S_2$, and again the alternatives are mutually exclusive. Further, $\Pi_3$ is evidently a simple closed polygon and $\Pi_3$ is the common boundary of $S_1$ and $S_2$. It follows that $\Pi_3$ is also of class $J$, and $(\Pi_3)_e = S_1$, $(\Pi_3)_i = S_2$.

*Proof of Theorem* B.1.1.    After these preparations we are ready for the proof. We use an induction argument. Suppose that we know that every simple closed polygon with not more than $n$ sides belongs to the class $J$. We have such information for $n = 3$. Let $\Pi$ be a simple closed polygon having $(n + 1)$ sides. We may assume that $\Pi$ is not convex. This means that we can find two vertices, $P_j$ and $P_k$, of $\Pi$ where $j < k$, with the following properties: (i) the straight line through $P_j$ and $P_k$ separates the plane into two half-planes, the closure of one of which contains all of $\Pi$, and (ii) the open line segment $(P_j, P_k)$ has no point in common with $\Pi$. To obtain such a pair of vertices we examine the lines of support of $\Pi$. It is sufficient to consider lines of support parallel to the various line segments $[P_\alpha, P_\beta]$, $1 \leq \alpha, \beta \leq n + 1$, $\alpha \neq \beta$. If $P_\alpha = z_\alpha$, $P_\beta = z_\beta$, then a line parallel to $[P_\alpha, P_\beta]$ has the equation

$$L(\alpha, \beta; \rho): \quad \Im[(\bar{z}_\beta - \bar{z}_\alpha)(z - z_\alpha)] - \rho = 0.$$

There exists an interval $[\rho_1, \rho_2]$ such that $L(\alpha, \beta; \rho) \cap \Pi$ is void for $\rho < \rho_1$ or $\rho_2 < \rho$ but non-void for $\rho_1 \leq \rho \leq \rho_2$. The two lines $L(\alpha, \beta; \rho_1)$ and $L(\alpha, \beta; \rho_2)$ are the lines of support of $\Pi$ parallel to the given direction. We now examine the set $L(\alpha, \beta; \rho_k) \cap \Pi$ for $k = 1$ and $2$. If such a set is connected, that is, reduces to a single point or a closed line segment, then we discard the line of support in question. On the other hand, if the intersection has at least two components, then we can obviously find a pair of vertices of $\Pi$ on $L(\alpha, \beta; \rho_k)$ having the desired properties. Since the polygon is not convex, the set $L(\alpha, \beta; \rho_k) \cap \Pi$ cannot be connected for every choice of $\alpha, \beta, k$.

Having found vertices $P_j$ and $P_k$ with the required properties, we note that these vertices determine two subarcs, $\pi_1$ and $\pi_2$, of $\Pi$. Here

$$\pi_1 = [P_j, P_{j+1}, \cdots, P_k], \quad \pi_2 = [P_k, P_{k+1}, \cdots, P_{j-1}, P_j].$$

Each of these subarcs has at least 2 and at most $(n - 1)$ sides; they do not intersect and have only their endpoints in common. Now the polygons $\Pi_1$ and $\Pi_2$ obtained by adding the line segment $[P_j, P_k]$ to $-\pi_1$ and to $\pi_2$ have at most $n$ sides and, hence, belong to the class $J$ since they are simple closed polygons. Disregarding the common side, we see that either $\Pi_1$ is interior to $\Pi_2$ or vice versa. Suppose that the first alternative holds. We have then the situation described in Lemma B.1.5, and we conclude that $\Pi \equiv \Pi_3$ also belongs to the class $J$. This completes the induction proof.

**B.2. Triangulation.** Let $\Pi$ be a simple closed polygon of $n$ sides. A line segment joining two nonadjacent vertices of $\Pi$ is called a *diagonal* if it lies in the interior of $\Pi$ except for the endpoints. Such a diagonal divides the polygon into two subpolygons, $\Pi_1$ and $\Pi_2$, having a side in common, namely the diagonal in question, such that each of $\Pi_1$, $\Pi_2$ has fewer than $n$ sides. We may then attempt to subdivide $\Pi_1$ and $\Pi_2$ by diagonals of $\Pi$. We say that $\Pi$ *can be triangulated if there exists a system of* $(n-3)$ *diagonals of* $\Pi$ *which subdivide* $\Pi$ *into* $(n-2)$ *triangles.*

Any convex polygon $[P_1, P_2, \cdots, P_n, P_1]$ can be triangulated, for each of the line segments $[P_1, P_3], [P_1, P_4], \cdots, [P_1, P_{n-1}]$ is a diagonal, and

$$[P_1, P_j, P_{j+1}, P_1], \quad j = 2, 3, \cdots, n-1,$$

are the corresponding triangles. This property, however, is not restricted to convex polygons.

THEOREM B.2.1.    *Every simple closed polygon can be triangulated.*

*Proof.* We shall give an induction proof, and we start by observing that the theorem is true for quadrilaterals. For if $[P_1, P_2, P_3, P_4, P_1]$ is not convex, then one (and only one) of the interior angles is $> \pi$. If this is the angle at $P_1$, then $[P_1, P_3]$ is a diagonal splitting the quadrilateral into two triangles. Suppose it be known that every simple closed polygon with not more than $n$ sides can be triangulated, and let $\Pi$ be a non-convex simple closed polygon with $(n+1)$ sides. Let $P_1$ be a vertex of $\Pi$ where the interior angle is $> \pi$. Such a vertex must exist. From $P_1$ we draw the ray bisecting the interior angle. Let $Q$ be the first intersection of this ray with $\Pi$, not counting $P_1$. If $Q$ happens to be a vertex, then $[P_1, Q]$ is a diagonal subdividing $\Pi$ into two polygons, each having at most $n$ sides, and, consequently, capable of being triangulated. This gives a triangulation of $\Pi$.

Normally, such a stroke of luck does not happen. Suppose then that $Q$ is an interior point of the side $[P_j, P_{j+1}]$. We note first that either $j \neq 2$ or $j \neq n$, since otherwise $\Pi$ would be a triangle. Secondly, we observe that the sector whose sides are $[P_1, P_2]$ and $[P_1, P_{n+1}]$ produced and whose angle is the interior angle of $\Pi$ at $P_1$, must contain at least one vertex of $\Pi$ in its interior, if $n \geq 3$. We shall prove that there is a vertex in this sector which is "visible" from $P_1$. For this purpose we slide $Q$ along $[P_j, P_{j+1}]$, keeping an eye on the ray $(P_1, Q)$ during the process. If $j = 2$, we let $Q$ move toward $P_3$; if $j = n$, $Q$ moves toward $P_{n-1}$ instead; if neither case is present, the direction of the motion is arbitrary. There are two possibilities: (1) $Q$ can be moved to one of the endpoints of $[P_j, P_{j+1}]$ without $(P_1, Q)$ having any contact with $\Pi$. Then $[P_1, Q]$ with $Q = P_j$ or $P_{j+1}$ is a diagonal, and $\Pi$ is subdivided into polygons having at most $n$ sides and, hence, capable of triangulation. (2) Before reaching the endpoint of the side, $(P_1, Q)$ meets $\Pi$. Suppose this happens for the first time when $Q = Q_0$. Then $(P_1, Q_0)$ has one or more points in common with $\Pi$, possibly even one or more line segments. There are two alternatives, of which the first one is unfavorable: (2a) $(P_1, Q_0)$ coincides with one of the sides $[P_1, P_{n+1}]$ or $[P_1, P_2]$ produced. We then try again, now moving $Q$ in the opposite direction. We note that this alternative cannot arise when we have only one degree of freedom ($j = 2$ or $n$), and, if we can move in either direction, then it can arise for at most one of these. The contrary assumption would imply that the interior angle of $\Pi$ at $P_1$ is $< \pi$. (2b) There is a point $P_k$ of $(P_1, Q_0) \cap \Pi$ which is nearest to $P_1$. Then $P_k$ must be a vertex of $\Pi$, and $[P_1, P_k]$ is a diagonal of $\Pi$. Again we are through. Since this exhausts the possibilities, we see that we are always able to draw a first diagonal and thus get the triangulation process started. This completes the proof.

It remains to say something about the general polygon $\Pi = [P_1, P_2, \cdots, P_n, P_1]$. Here we assume $P_j \neq P_{j+1}$ for each $j$, but otherwise no restrictions are imposed on the "vertices" $P_j$. Such a polygon may intersect itself, and there may be line segments which are traversed several times in the same or the opposite sense. A triangle described twice, either in the same or in opposite sense, is a perfectly good hexagon. Likewise $[P_1, P_2, P_1, P_2, P_1]$ is an admissible quadrilateral. In spite of the great variety of possibilities, we can always decompose a closed polygon into simpler entities. We state the following result without proof:

THEOREM B.2.2.    *Given a closed oriented polygon $\Pi$. Then there exist a finite collection of simple closed polygons, $\Pi_1, \Pi_2, \cdots, \Pi_j$, and a finite set of double segments* $[A_1, B_1, A_1], \cdots, [A_k, B_k, A_k]$, *such that*

$$\Pi = \{\cup_1^j \Pi_\alpha\} \cup \{\cup_1^k [A_\beta, B_\beta, A_\beta]\}.$$

*Here several of the $\Pi_\alpha$'s may be identical and we may have $\Pi_\beta = -\Pi_\alpha$ for some choices of $\alpha$ and $\beta$. The double segments may be repeated, some of the $A_\alpha$'s may coincide, and the same is true for the $B_\beta$'s. The representation is to be understood in the following sense: Every oriented line segment that is a side of a $\Pi_\alpha$ or belongs to a double segment also occurs on the left with the same orientation as a segment of one of the sides of $\Pi$, and its endpoints are either vertices or points of self-intersection of $\Pi$. Conversely, every side $[P_j, P_{j+1}]$ of $\Pi$ is accounted for on the right by piecing together equally oriented sides of the $\Pi_\alpha$'s and segments $[A_\alpha, B_\beta]$ or $[B_\beta, A_\alpha]$.*

## COLLATERAL READING

A proof of the Jordan curve theorem for arbitrary Jordan curves which uses only the most elementary notions of topology can be found in:

ALEKSANDROV, P. S. *Combinatorial Topology*, trans. by HORACE KOMM, Vol. 1, Chap. II, pp. 39–64. Graylock Press, Rochester, New York, 1956.

# Appendix C

# ON THE THEORY OF INTEGRATION

**C.1. The Riemann integral.** The present section is intended for the student who needs a reminder of the theory of Riemann integration.

Let $f(t)$ be a real, bounded function defined on a finite interval $[a, b]$. Let

$$|f(t)| \leq M, \quad a \leq t \leq b,$$

and consider a partition $\pi \equiv \{t_k\}$ of $[a, b]$

$$\pi: \quad a = t_0 < t_1 < t_2 < \cdots < t_n = b.$$

In each interval $[t_{k-1}, t_k]$ we choose a point $s_k$, and we denote the set $\{s_k\}$ by $\sigma$. We then form the Riemann sum

(C.1.1)
$$S_{\pi, \sigma}[f] \equiv \sum_{k=1}^{n} f(s_k)(t_k - t_{k-1})$$

and set

(C.1.2)
$$\| \pi \| = \max_{k} (t_k - t_{k-1}).$$

This quantity is called the *norm* of $\pi$. We have then

$$-M(b - a) \leq S_{\pi, \sigma}[f] \leq M(b - a)$$

for any choice of $\pi$ and $\sigma$. Thus the set of Riemann sums $S_{\pi, \sigma}[f]$ is a bounded point set, and its diameter is at most $2M(b - a)$. Let $0 < \delta < b - a$, and let $F(\delta)$ be the closure of the set $\{S_{\pi, \sigma}[f]\}$ where $\| \pi \| \leq \delta$. Thus, $F(\delta)$ is a subset of the interval $[-M(b - a), M(b - a)]$. Then clearly $\delta < \varepsilon$ implies $F(\delta) \subset F(\varepsilon)$. The interesting case is that in which the diameter of $F(\delta)$ tends to zero with $\delta$, so that there is a single point, $J$ say, common to all the sets $F(\delta)$.

DEFINITION C.1.1.    *$f(t)$ is integrable in the sense of Riemann if and only if the intersection of the sets $F(\delta)$, $\delta > 0$, reduces to a single point $J$. If this is the case, we set*

(C.1.3)
$$\int_a^b f(t)\, dt = J.$$

The so-called fundamental theorem of the calculus can be formulated as follows:

THEOREM C.1.1.    *A continuous function $f(t)$ is integrable in the sense of Riemann over any finite interval $[a, b]$.*

288

This theorem is a special case of Theorem C.3.1 below. The student is advised to go over the proof of the latter theorem, setting $g(t) = t$. He will then have a proof of Theorem C.1.1.

The extension to multiple integrals is routine analysis, up to a point. Consider, for instance, the two-dimensional case, and suppose that $f(x, y)$ is bounded and defined in a closed bounded set $S$. We may enclose $S$ in a rectangle

$$R: \quad a \leq x \leq b, \quad c \leq y \leq d,$$

and define $f(x, y) = 0$ for $(x, y) \in R \ominus S$. A partition $\pi$ is now a subdivision of $R$ into smaller rectangles by means of a grid

$$a = x_0 < x_1 < x_2 < \cdots < x_m = b,$$
$$c = y_0 < y_1 < y_2 < \cdots < y_n = d.$$

We define its norm to be

$$\| \pi \| = \max_{j, k} [x_j - x_{j-1}, y_k - y_{k-1}].$$

In each of the subrectangles we choose a point $(\xi_j, \eta_k)$. The corresponding Riemann sum is then

(C.1.4) $$S_{\pi, \sigma}[f] = \sum_{j=1}^{m} \sum_{k=1}^{n} f(\xi_j, \eta_k)(x_j - x_{j-1})(y_k - y_{k-1}).$$

Definition C.1.1 extends right away to the present case and so does Theorem C.1.1, provided $f(x, y)$ is continuous in $R$. The situation is different if $f(x, y)$ is continuous merely in $S$. If the boundary of $S$ is sufficiently regular, however, say made up of a finite number of smooth arcs, it may be shown that $f(x, y)$ is integrable in the sense of Riemann over $S$ if it is continuous there.

**C.2. Functions of bounded variation.** Let $f(t)$ be a real-valued or complex-valued function defined in the finite interval $[a, b]$. Consider any partition $\pi$ of this interval by points $\{t_k\}$

$$a = t_0 < t_1 < t_2 < \cdots < t_n = b,$$

and form the corresponding sum

(C.2.1) $$S_\pi[f] = \sum_{k=1}^{n} |f(t_k) - f(t_{k-1})|.$$

DEFINITION C.2.1. *$f(t)$ is said to be of bounded variation in $[a, b]$ if the set of all sums $S_\pi[f]$ is bounded. The quantity*

(C.2.2) $$\sup |S_\pi[f]| \equiv V_a^b[f]$$

*is called the total variation of $f(t)$ in $[a, b]$. The class of all functions of bounded variation in $[a, b]$ is denoted by $BV[a, b]$.*

A sufficient condition that $f(t) \in BV[a, b]$ is that $f(t)$ be bounded and monotone in $[a, b]$. In this case $V_a^b[f] = |f(a) - f(b)|$. As we shall see below, any function of bounded variation is a linear combination of bounded monotone functions.

THEOREM C.2.1.    $BV[a, b]$ *is an algebra in the sense of Definition* 4.7.4.

*Proof.*    The algebraic operations are defined in the obvious manner, so that

$$(f + g)(t) = f(t) + g(t), \quad (\alpha f)(t) = \alpha f(t), \quad (fg)(t) = f(t)g(t).$$

It is clear that the sum and the scalar product are in $BV[a, b]$ and that

(C.2.3) $$V_a^b[f + g] \le V_a^b[f] + V_a^b[g], \quad V_a^b[\alpha f] = |\alpha| V_a^b[f].$$

Products are slightly more complicated. Let us first observe that an element of $BV[a, b]$ is necessarily bounded in $[a, b]$; in fact

(C.2.4) $$\sup |f| \le |f(a)| + V_a^b[f].$$

We have then

$$\sum_{k=1}^{n} |f(t_k)g(t_k) - f(t_{k-1})g(t_{k-1})| \le \sum_{k=1}^{n} |f(t_k)| |g(t_k) - g(t_{k-1})|$$
$$+ \sum_{k=1}^{n} |g(t_{k-1})| |f(t_k) - f(t_{k-1})|$$

so that

(C.2.5) $$V_a^b[fg] \le \sup |f| \cdot V_a^b[g] + \sup |g| \cdot V_a^b[f].$$

If $f(t) \in BV[a, b]$ and if $[\alpha, \beta] \subset [a, b]$, then $f(t) \in BV[\alpha, \beta]$ and

(C.2.6) $$V_\alpha^\beta[f] \le V_a^b[f],$$

as is seen from the definition of the total variation. This definition also yields the important relations

(C.2.7) $$|f(\beta) - f(\alpha)| \le V_\alpha^\beta[f],$$

(C.2.8) $$V_\alpha^\gamma[f] = V_\alpha^\beta[f] + V_\beta^\gamma[f], \quad a \le \alpha < \beta < \gamma \le b.$$

The variation

$$V_a^x[f], \quad a \le x \le b,$$

is a non-negative, never decreasing function of $x$. In fact

$$V_a^\beta[f] - V_a^\alpha[f] = V_\alpha^\beta[f] \ge |f(\beta) - f(\alpha)|$$

by (C.2.7) and (C.2.8). Here $a \le \alpha < \beta \le b$.

THEOREM C.2.2.    If $f(t) \in BV[a, b]$, *then there exist four non-negative, never decreasing functions* $f_k(t)$ *such that*

(C.2.9) $$f(t) = \sum_{k=1}^{4} i^{k-1} f_k(t).$$

*Proof.*    If $f(t) \in BV[a, b]$ is complex-valued, then its real and imaginary parts both belong to $BV[a, b]$, and

$$V_a{}^x\{\Re[f]\} \leq V_a{}^x[f], \quad V_a{}^x\{\Im[f]\} \leq V_a{}^x[f].$$

It is, consequently, enough to prove that any real-valued function of bounded variation is the difference of two non-negative, never decreasing functions $g(t)$ and $h(t)$. It is clear that if one such representation of $f(t)$ exists, then we can find infinitely many. Without restricting the generality we may assume that $f(a) \geq 0$. This can always be achieved by adding a constant. We set

(C.2.10)          $g(x) = f(a) + V_a{}^x[f], \quad h(x) = f(a) + V_a{}^x[f] - f(x).$

It is clear that $g(x)$ has the desired properties.

Now $h(a) = 0$, and for $a \leq \alpha < \beta \leq b$,

$$h(\beta) - h(\alpha) = V_a{}^\beta[f] - f(\beta) - V_a{}^\alpha[f] + f(\alpha) = V_\alpha{}^\beta[f] - [f(\beta) - f(\alpha)] \geq 0$$

by (C.2.7). Since obviously $f(x) = g(x) - h(x)$, the theorem is proved.

THEOREM C.2.3.    *A function of bounded variation has finite left- and right-hand limits, $f(t - 0)$ and $f(t + 0)$, for every t, $a < t < b$, and the points where at least one of the following inequalities holds:*

$$f(t - 0) \neq f(t), \quad f(t) \neq f(t + 0), \quad f(t - 0) \neq f(t + 0),$$

*form a countable set.*

*Proof.*    By the preceding theorem it suffices to prove this for a never decreasing function. If $h > 0$,

$$f(t - h) \leq f(t) \leq f(t + h).$$

It follows that

(C.2.11)          $\lim_{h \to 0+} f(t - h) \equiv f(t - 0), \quad \lim_{h \to 0+} f(t + h) \equiv f(t + 0)$

must exist, and

$$f(t - 0) \leq f(t) \leq f(t + 0).$$

We call

(C.2.12)     $S_-[t, f] = |f(t) - f(t - 0)|, \quad S_+[t, f] = |f(t + 0) - f(t)|$

the *left saltus* and the *right saltus* of $f$ at $t$. A point $t = \alpha$ is a point of discontinuity of $f(t)$ if and only if

(C.2.13)                    $S_-[\alpha, f] + S_+[\alpha, f] \neq 0.$

By considering a partition of $[a, b]$ using $t = \alpha$ as one of the partition points, one sees that for every $\alpha$

$$S_-[\alpha, f] + S_+[\alpha, f] \leq V_a{}^b[f],$$

and this may be strengthened to

(C.2.14)                    $\sum\{S_-[\alpha, f] + S_+[\alpha, f]\} \leq V_a{}^b[f],$

where now the summation extends over any finite or infinite set of points $\alpha$ in the interval $[a, b]$. From this inequality we conclude that (C.2.13) can hold at most in a countable set of $\alpha$-values.

We have assumed $[a, b]$ to be a finite interval, but extensions to infinite intervals are immediate. We say that $f(t) \in BV[a, \infty]$ if $f(t) \in BV[a, \omega]$ for every finite $\omega$ and if

$$\lim_{\omega \to \infty} V_a^\omega[f] \equiv V_a^\infty[f]$$

is finite. All the results stated above hold also for infinite intervals.

### EXERCISE C.2

**1.** If $f(t) \in BV[a, b]$ and $\inf |f(t)| > 0$, show that $[f(t)]^{-1} \in BV[a, b]$ and

$$V_a^b[f^{-1}] \leq [\inf |f|]^{-2} V_a^b[f].$$

**2.** A norm may be introduced in the algebra $BV[a, b]$ by setting

$$\| f \| = |f(a)| + 2 V_a^b[f].$$

Verify that this is a norm, using (C.2.4) and (C.2.5).

**3.** Is the resulting normed algebra a complete metric space?

**4.** Fill in the missing details in the proof of Theorem C.2.3.

**5.** A function $f(t)$ is defined in $[0, 1]$ by $f(t) = 0$ if $t$ is irrational and $f(t) = q^{-3}$ if $t$ is rational, $t = p/q$ with $p$ and $q$ relatively prime. Show that $f(t)$ is discontinuous at the rational points, continuous elsewhere, and that $f(t) \in BV[0, 1]$.

**6.** Let $a = \alpha_1 < \alpha_2 < \cdots < \alpha_n = b$, let $f(t)$ have a constant value $\beta_k$ in each of the intervals $(\alpha_{k-1}, \alpha_k)$, and let $f(\alpha_k) = \gamma_k$. Such a function is known as a *step function*. Compute the total variation and the sum of the saltus.

**C.3. The Riemann-Stieltjes integral.** An important generalization of the Riemann integral was introduced by Stieltjes[1] in 1894.

Consider two functions, one continuous, the other of bounded variation, $f(t) \in C[a, b]$ and $g(t) \in BV[a, b]$. Both may be complex-valued. We recall the definition of the modulus of continuity

$$\mu(\delta; f) = \max \{|f(x_1) - f(x_2)| \mid x_1, x_2 \in [a, b], |x_1 - x_2| \leq \delta\},$$

---

[1] Thomas Jan Stieltjes (1856–1894), a Dutch astronomer who became a mathematician and made a career in France, where he was a professor at Toulouse. His early work deals with number theory, asymptotic series, and special functions. His last work and crowning masterpiece, a memoir on continued fractions, contained epoch-making ideas, not really understood until long after his death. The integral and the moment problem named after him were presented in this paper.

which is a positive, monotone increasing function, tending to zero with $\delta$, unless $f(t)$ is a constant so that $\mu(\delta; f) \equiv 0$. Consider a partition $\pi = \{t_k\}$ of $[a, b]$

$$\pi: \quad a = t_0 < t_1 < t_2 < \cdots < t_n = b,$$

and set $\sigma = \{s_k\}$ where $s_k$ is a point in $[t_{k-1}, t_k]$, $k = 1, 2, \cdots, n$. We then define

(C.3.1)
$$S_{\pi,\sigma}[f, g] = \sum_{k=1}^{n} f(s_k)[g(t_k) - g(t_{k-1})].$$

Finally we set

(C.3.2)
$$\| \pi \| = \max_k (t_k - t_{k-1})$$

and refer to $\| \pi \|$ as the norm of $\pi$.

THEOREM C.3.1.    *There exists a quantity $J$ such that*

(C.3.3)     $$| J - S_{\pi,\sigma}[f, g] | \leq 2\mu(\delta, f) V_a^b[g] \quad when \quad \| \pi \| \leq \delta.$$

DEFINITION C.3.1.    *$J$ is called the Stieltjes integral of $f(t)$ with respect to $g(t)$ over $[a, b]$ and is denoted by*

(C.3.4)
$$\int_a^b f(t) \, dg(t).$$

*Proof.*    Consider first two sums corresponding to the same partition $\pi$ of norm $\leq \delta$ with two different sets $\sigma$, $\sigma_1$ and $\sigma_2$ say. Since $| s_{k,1} - s_{k,2} | \leq \delta$ for every $k$, we have

$$| S_{\pi,\sigma_1}[f, g] - S_{\pi,\sigma_2}[f, g] | \leq \sum_{k=1}^{n} | f(s_{k,1}) - f(s_{k,2}) | \, | g(t_k) - g(t_{k-1}) |$$
$$\leq \mu(\delta, f) V_a^b[g]$$

by the definition of the total variation of $g$. Suppose next that $\pi_1$ and $\pi_2$ are two partitions such that $\pi_2$ is a subdivision of $\pi_1$ (often called a "refinement" of $\pi_1$), and consider two sums corresponding to these partitions, say $S_{\pi_1, \sigma_1}[f, g]$ and $S_{\pi_2, \sigma_2}[f, g]$. Let

$$T_1 = f(s)[g(\beta) - g(\alpha)]$$

be a term in the first sum and suppose that the interval $[\alpha, \beta]$ is subdivided by points $\{t_{\alpha, j}\}$ in passing from $\pi_1$ to $\pi_2$. We have then

$$\sum_j [g(t_{\alpha, j}) - g(t_{\alpha, j-1})] = g(\beta) - g(\alpha).$$

The term $T_1$ of the first sum is replaced by a group of terms

$$T_2 = \sum_j f(s_{\alpha, j})[g(t_{\alpha, j}) - g(t_{\alpha, j-1})]$$

in the second sum. Since $| s_{\alpha, j} - s | \leq \delta$, we have

$$| T_1 - T_2 | = \left| \sum_j [f(s) - f(s_{\alpha, j})][g(t_{\alpha, j}) - g(t_{\alpha, j-1})] \right| \leq \mu(\delta, f) V_\alpha^\beta[g],$$

and by (C.2.8)

$$| S_{\pi_1, \sigma_1}[f, g] - S_{\pi_2, \sigma_2}[f, g] | \leq \mu(\delta, f) V_a^b[g].$$

Let us now take two arbitrary sums corresponding to partitions $\pi_1$ and $\pi_2$ and sets $\sigma_1$ and $\sigma_2$. The union of the sets of division points of $\pi_1$ and $\pi_2$ defines a partition, $\pi_3$, which is a refinement of both $\pi_1$ and $\pi_2$. For $\pi_3$ we choose arbitrarily a set $\sigma_3$, having one point in each subinterval of $\pi_3$. We have then

$$| S_{\pi_1, \sigma_1}[f, g] - S_{\pi_2, \sigma_2}[f, g] | \leq | S_{\pi_1, \sigma_1}[f, g] - S_{\pi_3, \sigma_3}[f, g] |$$

(C.3.5)

$$+ | S_{\pi_2, \sigma_2}[f, g] - S_{\pi_3, \sigma_3}[f, g] | \leq 2\mu(\delta, f) V_a{}^b[g].$$

Let $F(\delta)$ be the closure of the set of all sums $S_{\pi, \sigma}[f, g]$ with $\| \pi \| \leq \delta$. By (C.3.5) the diameter of this set does not exceed $2\mu(\delta, f) V_a{}^b[g]$ and thus tends to zero with $\delta$. Further, $F(\delta) \subset F(\varepsilon)$ if $\delta < \varepsilon$. By the basic property of nested closed sets there is one and only one point $J$ common to all these sets, and this is the desired integral. Since $J \in F(\delta)$, its distance from any element of $F(\delta)$ cannot exceed the diameter of $F(\delta)$, and this is the assertion of (C.3.4). This completes the proof.

We refer to $f(t)$ as the *integrand*, $g(t)$ as the *integrator*. This type of integration is a *bilinear operation*, for *the integral is linear both in the integrand and in the integrator*:

(C.3.6) $$\int_a^b [\alpha f_1(t) + \beta f_2(t)] \, dg(t) = \alpha \int_a^b f_1(t) \, dg(t) + \beta \int_a^b f_2(t) \, dg(t),$$

(C.3.7) $$\int_a^b f(t) \, d[\alpha g_1(t) + \beta g_2(t)] = \alpha \int_a^b f(t) \, dg_1(t) + \beta \int_a^b f(t) \, dg_2(t).$$

*The integral is also additive with respect to the interval of integration:*

(C.3.8) $$\int_a^b f(t) \, dg(t) = \int_a^c f(t) \, dg(t) + \int_c^b f(t) \, dg(t), \quad a < c < b.$$

These properties are immediate consequences of the definition of the integral as a limit of sums. The definition also leads to important estimates for the integral. Thus the inequality

(C.3.9) $$\left| \int_a^b f(t) \, dg(t) \right| \leq \max | f(t) | \cdot V_a{}^b[g]$$

holds, since it holds for the sums. Going back to (C.2.7), we see that

$$| f(s)[g(\beta) - g(\alpha)] | \leq | f(s) | \, V_\alpha{}^\beta[g] = | f(s) | \, \{ V_a{}^\beta[g] - V_a{}^\alpha[g] \}.$$

We are thus led to the inequality

(C.3.10) $$\left| \int_a^b f(t) \, dg(t) \right| \leq \int_a^b | f(t) | \, dV_a{}^t[g].$$

The right member is also a Stieltjes integral with the integrand $|f(t)|$ and the integrator $V_a{}^t[g]$, the total variation of $g(\cdot)$ over the interval $[a, t]$. This inequality is quite often written

(C.3.11)
$$\left| \int_a^b f(t)\, dg(t) \right| \leq \int_a^b |f(t)|\ |dg(t)|$$

instead, where $|dg(t)|$ is an abbreviation for $dV_a{}^t[g]$.

If $g(t)$ is continuous and has a continuous derivative, we have

(C.3.12)
$$\int_a^b f(t)\, dg(t) = \int_a^b f(t) g'(t)\, dt;$$

the right member is an ordinary Riemann integral. The right member exists under more general assumptions, but the equality does not necessarily hold.

Finally we have a formula for integration by parts involving an interchange of the roles of the integrand and the integrator, namely,

$$S_{\pi,\,\sigma}[f, g] = \sum_{k=1}^n f(s_k)[g(t_k) - g(t_{k-1})]$$

$$= f(s_n)g(b) - f(s_1)g(a) - \sum_{k=1}^{n-1} g(t_k)[f(s_{k+1}) - f(s_k)].$$

Here the first two terms in the last member differ arbitrarily little from the value $f(b)g(b) - f(a)g(a)$ if $\|\pi\|$ is small. Consequently

(C.3.13)
$$\int_a^b f(t)\, dg(t) = f(b)g(b) - f(a)g(a) - \int_a^b g(t)\, df(t).$$

If we know merely that $f(t) \in C[a, b]$ and $g(t) \in BV[a, b]$, then this formula serves to define the integral on the right, but if $f(t)$ and $g(t)$ both belong to the class $C[a, b] \cap BV[a, b]$, then (C.3.13) is highly useful in evaluating Stieltjes integrals.

## EXERCISE C.3

**1.** If $g(t)$ is a step function (see Problem 6 of Exercise C.2) with jumps at $t = \alpha_k$, $a < \alpha_1 < \alpha_2 < \cdots < \alpha_n < b$, then

$$\int_a^b f(t)\, dg(t) = \sum_{k=1}^n f(\alpha_k)[g(\alpha_k + 0) - g(\alpha_k - 0)].$$

**2.** Verify (C.3.12).

**3.** If $g(t) \in C[a, b] \cap BV[a, b]$, evaluate

$$\int_a^b [g(t)]^k\, dg(t), \quad k \text{ integer}, \ k > 0.$$

## COLLATERAL READING

As a general reference see

WIDDER, D. V. *Advanced Calculus*, Chap. 5. Prentice-Hall, Inc., Englewood Cliffs, New Jersey, 1947.

For the two-dimensional Riemann integral see

BUCK, R. C. *Advanced Calculus*, Chap. 3. McGraw-Hill Book Company, Inc., New York, 1956.

More details concerning the Stieltjes integral and functions of bounded variation are to be found in

WIDDER, D. V. *The Laplace Transform*, Chap. I. Princeton University Press, Princeton, New Jersey, 1941.

# BIBLIOGRAPHY

AHLFORS, L. V. *Complex Analysis. An Introduction to the Theory of Analytic Functions of One Complex Variable.* McGraw-Hill Book Company, Inc., New York, 1953.

BEHNKE, H., and SOMMER, F. *Theorie der analytischen Funktionen einer komplexen Veränderlichen.* Springer-Verlag, Berlin, 1955.

BIEBERBACH, L. *Lehrbuch der Funktionentheorie,* Vols. I and II. B. G. Teubner, Leipzig and Berlin, 1930–1931. Reprinted by Chelsea Publishing Company, New York, 1945.

CARATHÉODORY, C. *Funktionentheorie,* Vols. I and II. Verlag Birkhäuser, Basel, 1950. Translated by STEINHARDT, F., as *Theory of Functions.* Chelsea Publishing Company, New York, 1954.

CHURCHILL, R. V. *Introduction to Complex Variables and Applications.* McGraw-Hill Book Company, Inc., New York, 1948.

COPSON, E. T. *An Introduction to the Theory of Functions of a Complex Variable.* Clarendon Press, Oxford, 1935.

FRANKLIN, P. *Functions of Complex Variables.* Prentice-Hall, Inc., Englewood Cliffs, New Jersey, 1958.

GOURSAT, E. *Cours d'Analyse,* Seventh Edition, Vol. II. Gauthier-Villars, Paris, 1949. Second French edition of Vol. II, Part I, translated by HEDRICK, E. R., and DUNKEL, O., as *Functions of a Complex Variable.* Ginn and Company, Boston, 1916.

HURWITZ, A., and COURANT, R. *Vorlesungen über allgemeine Funktionentheorie und elliptische Funktionen,* Second Edition. Verlag von Julius Springer, Berlin, 1925.

KNOPP, K. *Funktionentheorie,* Second Edition (2 vols. plus 2 vols. of problems). Sammlung Göschen, Berlin, 1918–1928. Translated by BAGEMIHL, F., as *Theory of Functions.* Dover Publications, Inc., New York, 1945.

LINDELÖF, E. *Le Calcul des Résidus et ses Applications à la Théorie des Fonctions.* Gauthier-Villars, Paris, 1905. Reprinted by Chelsea Publishing Company, New York, 1947.

NEHARI, Z. *Conformal Mapping.* McGraw-Hill Book Company, Inc., New York, 1952.

SAKS, S., and ZYGMUND, A. *Analytic Functions.* Monografie Matematyczne, Vol. 28, Warsaw and Wroclaw, 1952.

SPRINGER, G. *Introduction to Riemann Surfaces.* Addison-Wesley Publishing Company, Inc., Reading, Massachusetts, 1957.

THRON, W. J. *Introduction to the Theory of Functions of a Complex Variable.* John Wiley & Sons, Inc., New York, 1953.

297

TITCHMARSH, E. C. *The Theory of Functions*, Second Edition. Oxford University Press, London, 1939.

VALIRON, G. *Théorie des Fonctions*. Masson et Cie, Paris, 1948.

WATSON, G. N. *Complex Integration and Cauchy's Theorem*. Cambridge Tracts in Mathematics and Mathematical Physics, No. 15. Cambridge University Press, London, 1914.

WHITTAKER, E. T., and WATSON, G. N. *A Course of Modern Analysis*, Fourth Edition. Cambridge University Press, London, 1952.

# Index

# INDEX